technical analysis 2016-2018

Giorgio Piola

Formula 1

Half a century of technology
AND MORE

Monte Carlo 2018 was my 50th Grand Prix in the principality. 50 years have passed since my first appearance in the paddock in 1969 at that race I saw live for the first time thanks to generosity of Gianni Cancellieri who made a kid's dream come true, giving him the opportunity to take up a new profession that over the years has evolved continuously.

From the first sketches at my school desk, inspired by those of a great Italian draughts-

man, Bruno Nestola, a mentor and a source of useful advice, through to the super-detailed cutaways drawn on acetate slides for which more than 30 works were required; then came a drastic simplification of my style with the arrival in the Eighties of the first relatively unsophisticated fax machines and then the fateful encounter with the computer in 1998.

The advent of colour and Photoshop marked another turning point in this work: from the black and white volume reserved exclusively for the specialist press we arrived at a book published by Giorgio Nada Editore, one that was more demanding and in two languages and above all des-

tined for the general public A constant feature of the work has been the use of freehand drawings used as the basis for subsequent colouration. Another important stage came with the more recent creation of animated sequences, which began in 2008.

Over the years, with increasing restrictions on access to the various cars and above all the irresistible advent of communication via the Internet, the work of technical analysis has become ever more difficult and focussed on the immediacy of the publication of scoops rather than the painstaking investigation and analysis that characterised the earlier editions of this book. The latest turning point came with the exclusive collaboration with the web site Motorsport.com, with both drawings and animations.

After all this and in the year in which I celebrate my golden wedding anniversary with F1 and with 800 Grands Prix under my belt, I have come to the decision to conclude this experience with *Analisi Tecnica* with a long-awaited volume that encapsulates the last three seasons.

Clearly, the structure of the book has changed and the "Chassis History" chapter has been dropped as over the years there has been increasingly little to write about given the abolition of the T-car and the standardization of the number of chassis built each season by the various teams. We have instead retained the three tables with data regarding all the cars that provide a valid comparison of the various years.

These are among the first drawings I did at school. The first, Chris Amon's 1968 Ferrari, and the second, a Lotus 49 in the style of the drawings published by in Auto Italiana Bruno Nestola who was so generous with his advice when I was contributing to *Grand Prix* which he edited.

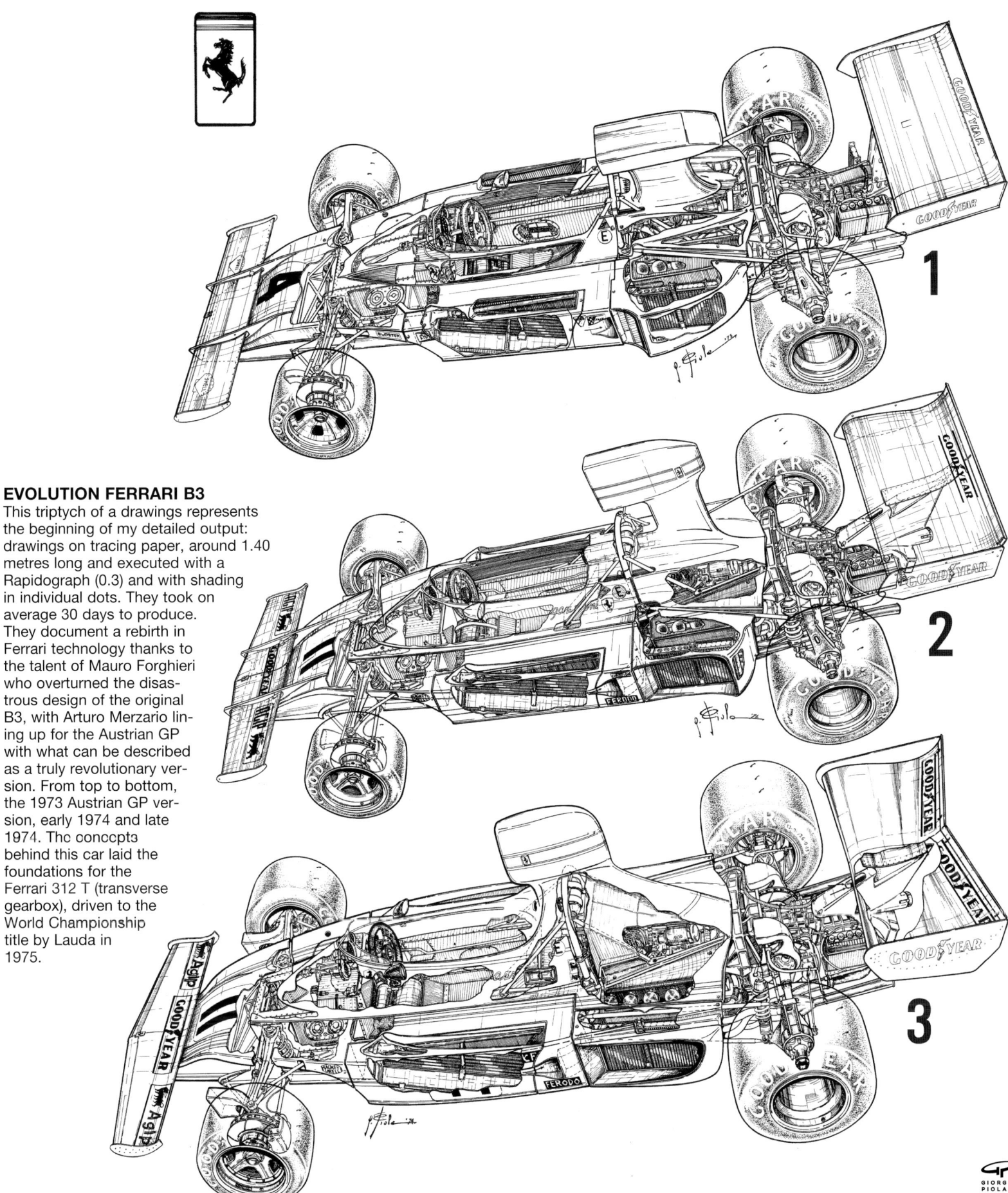

EVOLUTION FERRARI B3

This triptych of a drawings represents the beginning of my detailed output: drawings on tracing paper, around 1.40 metres long and executed with a Rapidograph (0.3) and with shading in individual dots. They took on average 30 days to produce. They document a rebirth in Ferrari technology thanks to the talent of Mauro Forghieri who overturned the disastrous design of the original B3, with Arturo Merzario lining up for the Austrian GP with what can be described as a truly revolutionary version. From top to bottom, the 1973 Austrian GP version, early 1974 and late 1974. The concepts behind this car laid the foundations for the Ferrari 312 T (transverse gearbox), driven to the World Championship title by Lauda in 1975.

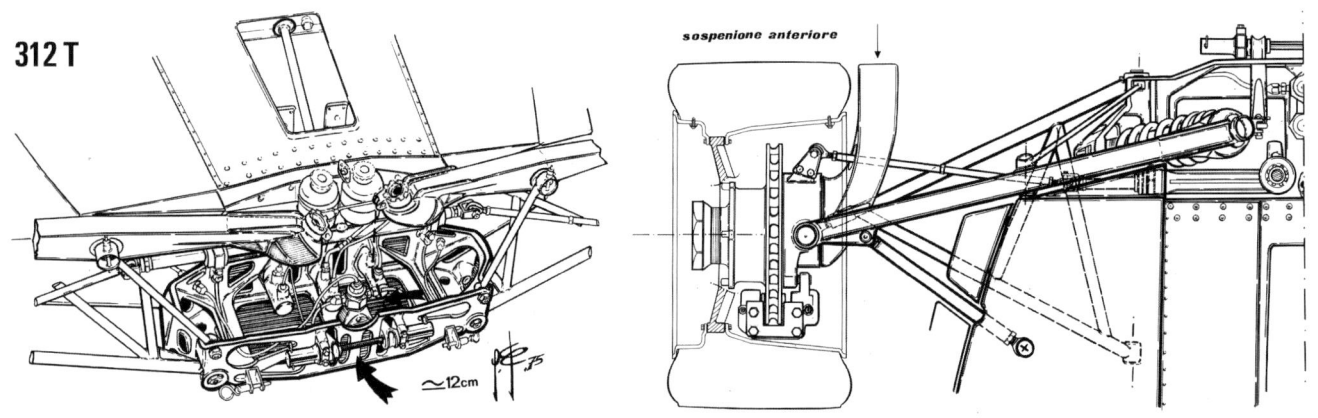

312 T

sospensione anteriore

~12cm

FERRARI 312T FRONT SUSPENSION

The complete freedom to photograph details of the car allowed very detailed drawings to be produced in the 1970s and 1980s, like this one of the Ferrari 312T front suspension and the cast magnesium upright..

TYRRELL P34 6-WHEELER

An important turning point for me was the opportunity to produce technical drawings for a kind of sponsored press pack for Tyrrell, thanks to a bit of good luck: that of having sat next to Ken Tyrrell on a flight to Brazil, as well having an excellent relationship with Derek Gardner, the team's designer.
A curiosity: this drawing was part of a complete series that originally had no less than 32 reference numbers that on the sheets of tracing paper could not be cancelled. The result was that no periodical of the time published it. Now, with Photoshop, the cancellation of the numbers was child's play.

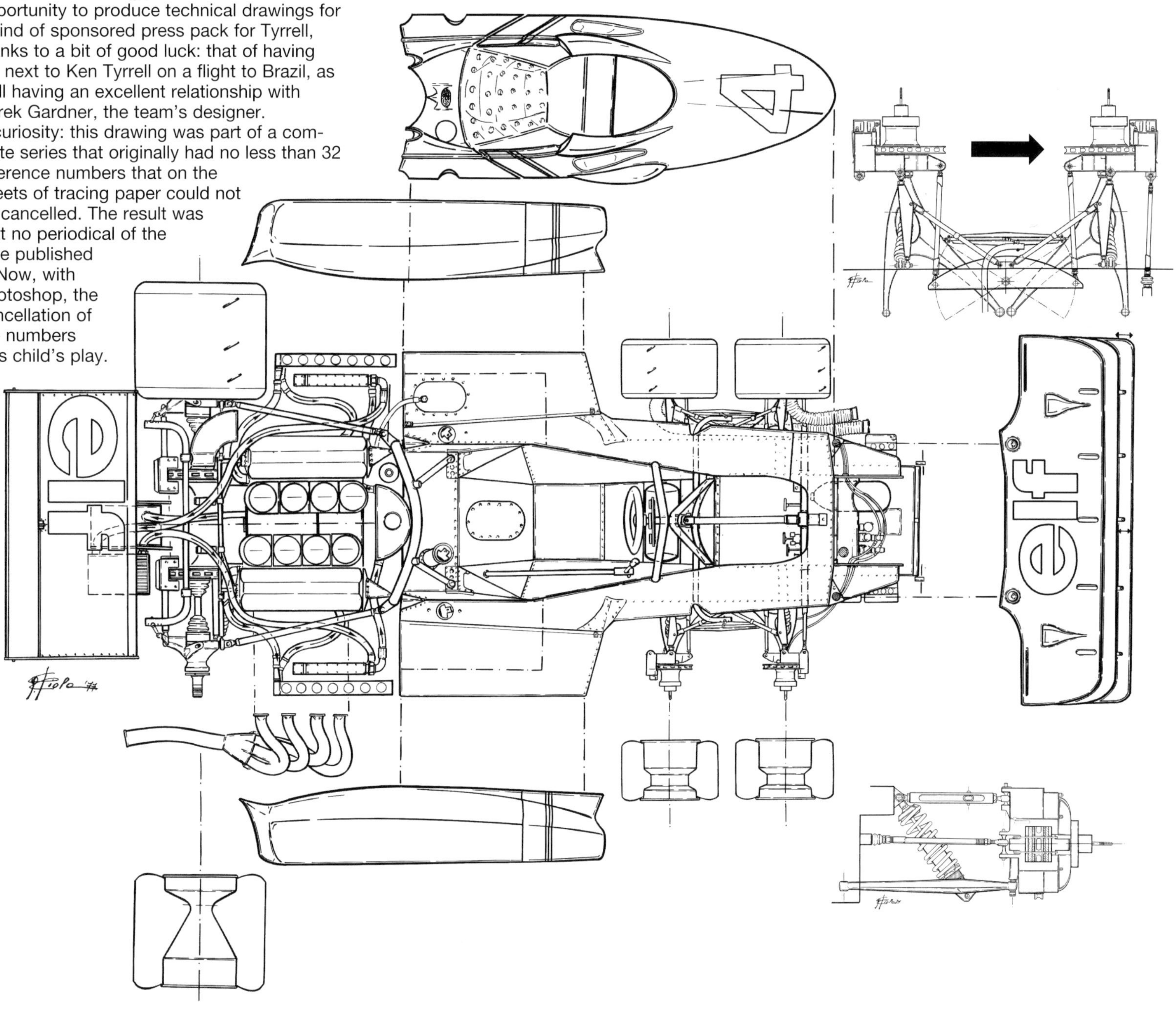

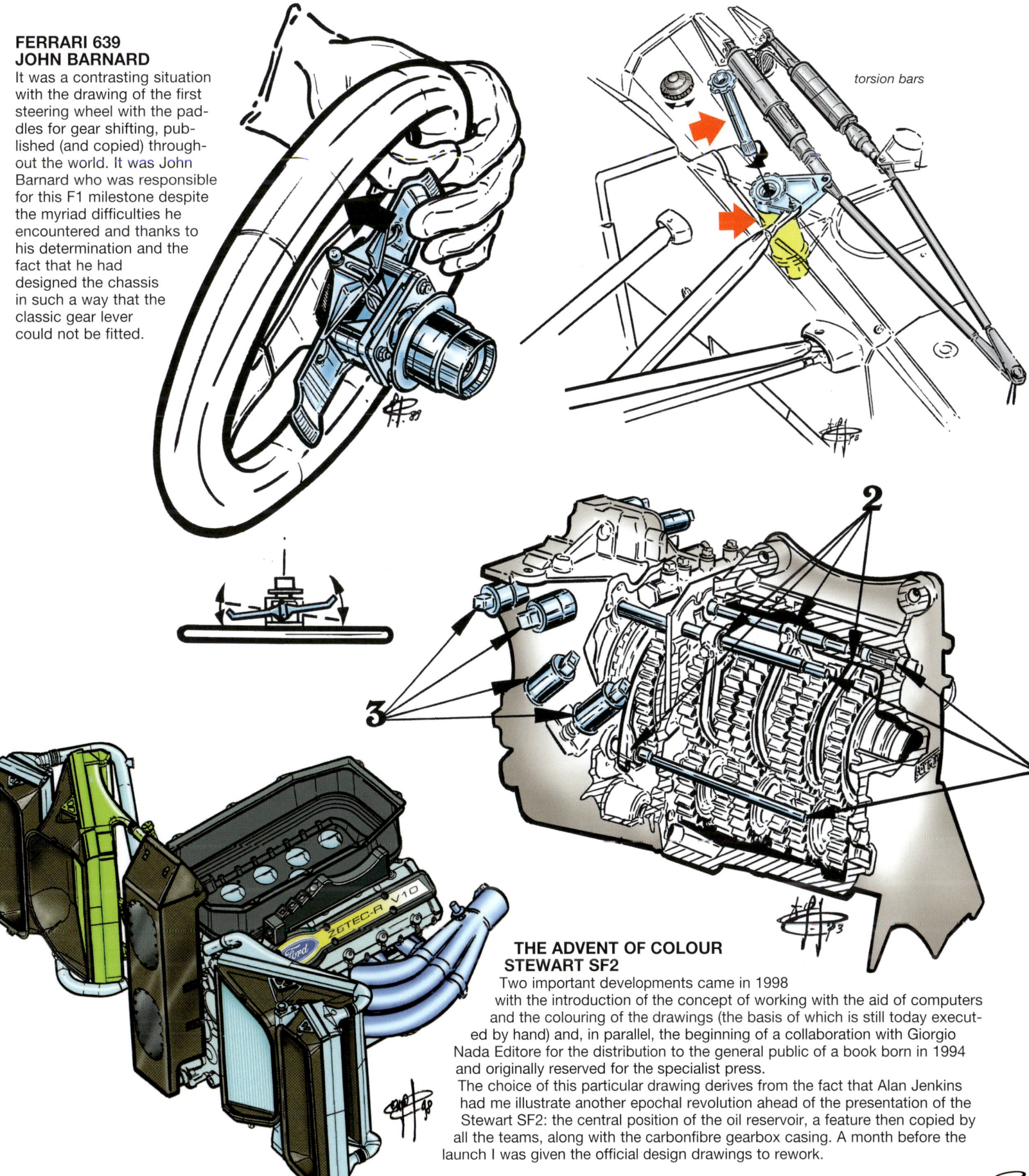

FERRARI 639
JOHN BARNARD

It was a contrasting situation with the drawing of the first steering wheel with the paddles for gear shifting, published (and copied) throughout the world. It was John Barnard who was responsible for this F1 milestone despite the myriad difficulties he encountered and thanks to his determination and the fact that he had designed the chassis in such a way that the classic gear lever could not be fitted.

torsion bars

2

3

1

THE ADVENT OF COLOUR
STEWART SF2

Two important developments came in 1998 with the introduction of the concept of working with the aid of computers and the colouring of the drawings (the basis of which is still today executed by hand) and, in parallel, the beginning of a collaboration with Giorgio Nada Editore for the distribution to the general public of a book born in 1994 and originally reserved for the specialist press.
The choice of this particular drawing derives from the fact that Alan Jenkins had me illustrate another epochal revolution ahead of the presentation of the Stewart SF2: the central position of the oil reservoir, a feature then copied by all the teams, along with the carbonfibre gearbox casing. A month before the launch I was given the official design drawings to rework.

2016 SEASON

The 2016 Formula 1 season saw the return of an American teams, Haas, after an absence of no less than 39 years.

In the mid-Seventies, from 1974 to 1977 to be precise, two US teams sporadically participated in the World Championship: Parnelli, which competed in just 16 Grands Prix in three seasons, and Penske, with 39 races through to the end of the 1977 championship. In the 1980s the American-owned but British-based Force Haas team (no relation) contested two seasons.

The merit goes to Gene Haas who following success in Nascar decided back in July 2014 to make the courageous move to Formula 1, taking a very wise and professional approach.

The team's debut was deliberately delayed until 2016, even though Haas had taken over the Manor facility at Banbury in England as the team's European operational base from the December of 2014.

Having entered on tip-toe, Haas achieved its objective by securing a place (8th, immediately behind Toro Rosso and McLaren, but ahead of a blue-blooded team such as Renault) among the Formula 1 teams and obtaining two excellent results in the two opening races in 2016.

As one team arrived in F1, another left: Manor, formerly Marussia, ironically at the end of what had been the best season for the small outfit. McLaren's third season with Honda power proved to be another disaster, the British team sharing some of the blame with its insistence on the "size zero" philosophy without taking into account that the Honda Power Unit had been completely reconfigured in a search for missing power. Ferrari was given a veritable cold shower after having secured three wins the previous season and igniting hopes of returning to the top in F1. Instead,2016 brought not a single victory and the team slipped a place in the Constructors' Championship standings behind Red Bull. What was positive, however, was that the SF16 T saw a return to push-rod front suspension and despite its limitations did serve as the basis for the 2017 car.

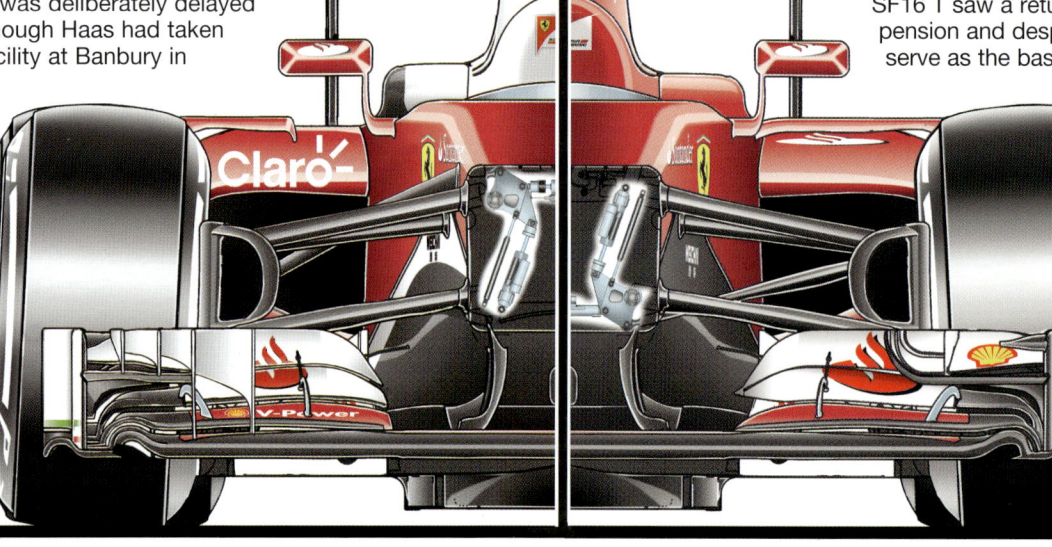

FERRARI SF16-H
While failing to win a single race, the SF16 T served as the starting point for the successive SF70 H: it marked a return to push-rod front suspension (left) after Ferrari had adopted the pull-rod layout in 2012.

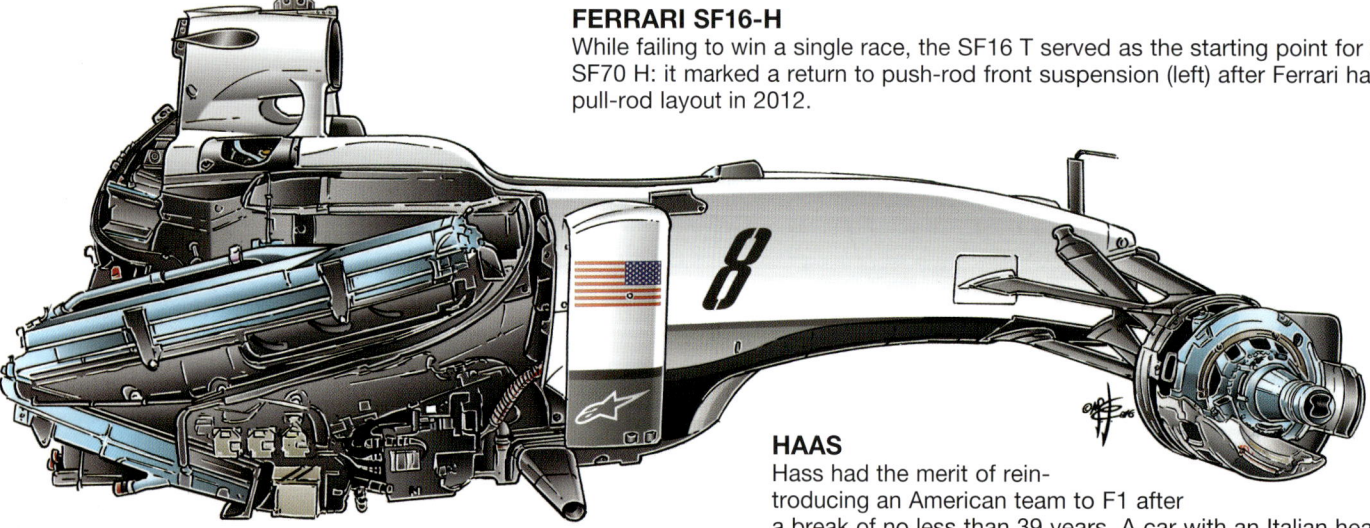

HAAS
Hass had the merit of reintroducing an American team to F1 after a break of no less than 39 years. A car with an Italian heart: Ferrari engine and running gear, Dallara chassis.

2017 SEASON

Having decided to group together the last three seasons has allowed us to temper the crushing disappointment of Ferrari's 2016 season, with the Maranello-based firm returning to a leading role in both 2017 and 2018. Not only did Ferrari compete on a level footing with the undisputed number one, Mercedes, but for two consecutive seasons the SF70H and the SF71H were the most interesting cars in the field in terms of new and avant-garde technical features. In 2017, the year that coincided with the greatest technical revolution of the last two decades in F1, Ferrari introduced innovations that set trends for 2018, regaining the technical leadership it had lost back in 2008. The merit was due to an envelope-stretching approach to the regulations with a pinch of genius from Rory Byrne who had returned to the group led by Mattia Binotto, Simone Resta and above all the aerodynamicists David Sanchez

Ferrari SF16-H

FERRARI 2017

The 2017 season saw Ferrari running the most revolutionary car thanks to an interpretation of the regulations that took the cockpit protection structures outside rather than inside the sidepods (SF16-H on the left), concealed within an aerodynamic appendage (traced in yellow).

Ferrari SF70H

and Enrico Cardile. The technical revolution introduced by the FIA in 2017 significantly modified the appearance of the cars, which became more "aggressive" thanks to wider tyres and greater downforce produced in part due to more liberal aerodynamic regulations and which led to extremely spectacular races. We had expected new features from Adrian Newey but this time it was Ferrari's turn to surprise, although Mercedes did make a mark with the wheelbase dimension increased by no less than 300 mm after having once again foregone using the rake configuration adopted instead by Force India, once again finishing in an excellent 4th place in the Constructors' Championship.

The negative sequence at the end of the season weighed heavily in Ferrari's duel with Mercedes (Singapore, Sepang and Suzuka), with errors and a lack of reliability, all too clearly evident on the start lines in the last two races.

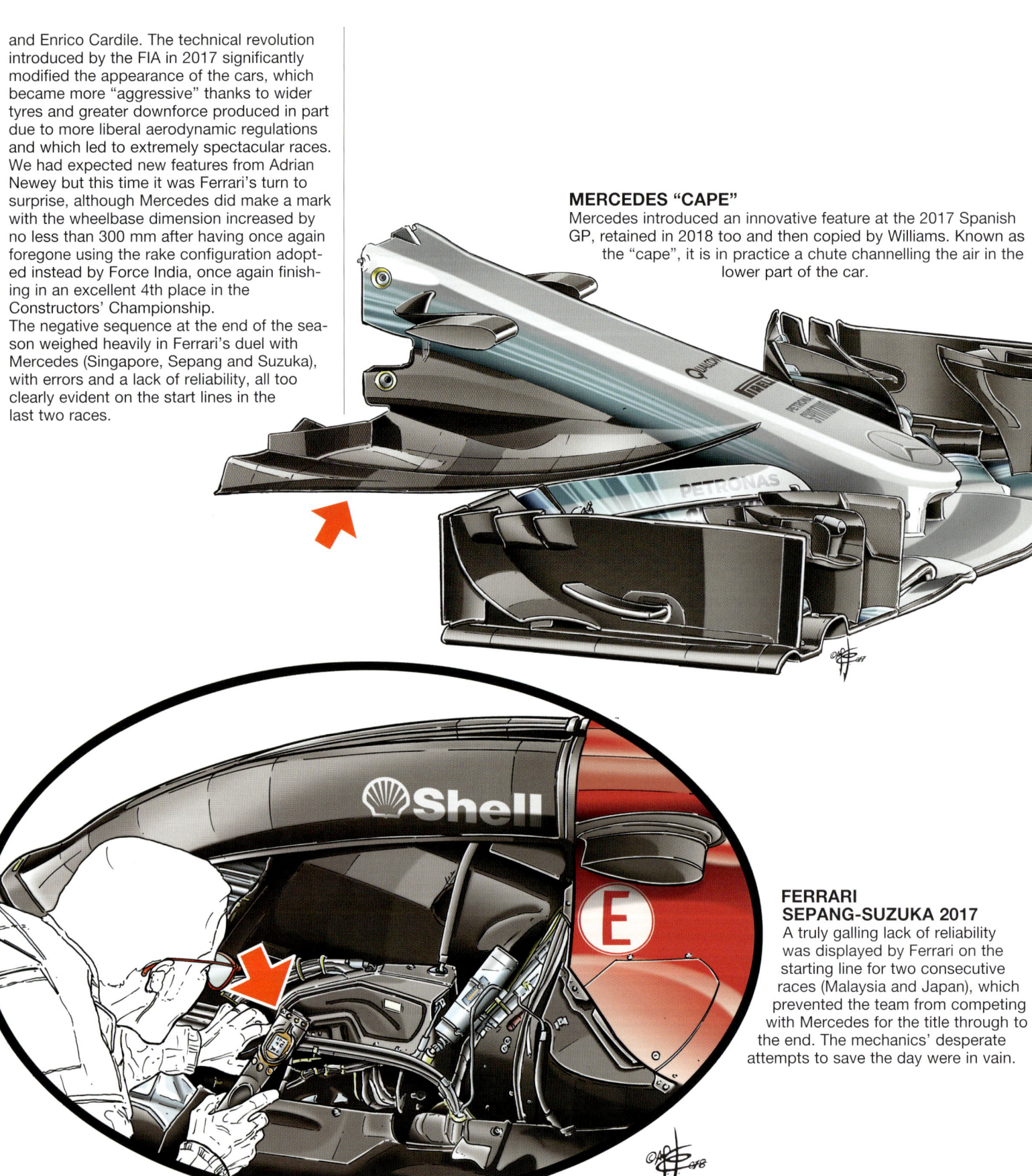

MERCEDES "CAPE"
Mercedes introduced an innovative feature at the 2017 Spanish GP, retained in 2018 too and then copied by Williams. Known as the "cape", it is in practice a chute channelling the air in the lower part of the car.

FERRARI
SEPANG-SUZUKA 2017
A truly galling lack of reliability was displayed by Ferrari on the starting line for two consecutive races (Malaysia and Japan), which prevented the team from competing with Mercedes for the title through to the end. The mechanics' desperate attempts to save the day were in vain.

2018 SEASON

The 2018 season can rightly be considered as the most interesting of the last 20 years in both sporting and technical terms, despite Mercedes' latest clean sweep (the fifth consecutive, equalling Ferrari between 1999 and 2004). The merit goes in part to the talent of the drivers in the three best teams (along with the "toothless" Alonso in the shameful McLaren), comparable to the era of Senna, Prost, Mansell and Piquet. That merit is shared, however, by the considerable technical progress made at the highest levels, with a substantial balance of power between Mercedes and Ferrari along with an improved Red Bull despite the announcement of the divorce from Renault in 2019. Renault and Haas both made notable progress the latter in its third F1 season, at the expense of Force

India which after maintaining a fantastic fourth place in both of the previous two seasons slumped to seventh due to a chronic lack of funds for actuating the developments planned by Andy Green's staff. Above all, there was the return of the Alfa Romeo marque (if only on the cylinder heads of the Ferrari engine) with Sauber, the Maranello power unit ensuring that the Swiss team was always very competitive. Naturally, along with the engine, the gearbox was also Ferrari, as was the braking system, although there were new and interesting features that have been analysed in the New Features 2018 chapter. The season also saw the re-ignition of disputes that interrupted the armistice that had lasted throughout 2017, a season in which the FIA had intervened to block developments that broke the regulations without offi-

cial communications but rather by sending clarifying letters to all the teams.
The technical analysis of the last two Ferrari seasons has two almost diametrically opposed readings. From an extremely positive technical point of view, it should be noted that in 2018 just three cars (Renault, Force India and McLaren) did not copy the configuration with the separation of the leading edge of the sidepods from the protection structures introduced by Ferrari in 2017. On the other hand, there is a sense of regret over the failure to bring back to Maranello even one of the titles on offer over the two season, despite having a car, especially in 2018, that was the equal of its Mercedes rival and actually more consistent in terms of performance on the various World Championship circuits.

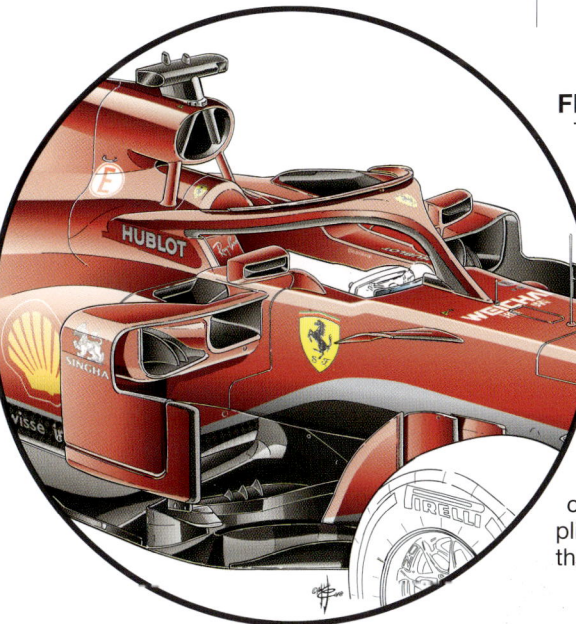

FERRARI SF71H
The SF71 H was the most competitive Ferrari of recent years and boasted another new feature, the rear-view mirrors equipped with a vent improving the efficiency of the second air intake at the top of the sidepods.

THE RETURN OF ALFA ROMEO
The Alfa Romeo marque made its return to F1 in 2018, appearing on the cylinder heads of the Ferrari engines supplied to Sauber as well as on the bodywork of the Swiss car.

THE END OF AN ERA
This book is actually the last in the series dedicated to the technical analysis of the Formula 1 seasons. It is clearly no coincidence that its launch in Monte Carlo coincides with my 50-year involvement in F1 (from the Monaco GP in 1969), but it also celebrates 25 years of collaboration with Philip Morris and 21 years with Giorgio Nada Editore to whom I am sincerely grateful, in particular to Leonardo Acerbi the editorial coordinator of this book. I would also like to thank Aimone Bolliger who has patiently worked on the layout, Paolo Rondelli who introduced me to the secrets of Photoshop way back in 1998, the various assistants who have followed one another over the years through to the current Gisella Nicosia, Alessia Bardino, Paolo Rondinelli and Giulia Giusto, the 3D experts, Camillo Morande and Generoso Annuziata for the animations and lastly

Francesco Pizzolante, archive. Naturally, I would also like to mention the colleagues who have made important contributions with texts and statistics: Franco Nugnes, Michael Schmidt, Kazuhito Kaway and Mark Hughes. It has been a great privilege to have met and respected over the course of my long career the true protagonists of F1: Colin Chapman, Patrick Head, Gordon Murray, Mauro Forghieri, John Barnard, Ross Brawn, Rory Byrne, Adrian Newey, Paddy Lowe… naturally the list could go on an on and could be embellished with priceless memories and experiences that went well beyond the strictly technical. Last but certainly not least, my thanks go to Gianni Cancellieri: it was he who received and published a drawing I did at school, transforming a dream into reality and shortly afterwards sending me to follow the Monaco Grand Prix in 1969.

A question of ENGINES

In the third year of the Power Units, the FIA permitted the constructors to fit two small supplementary exhausts to be placed no more than 100 mm from the principal one. The two served to increase the sound of the engines when the turbo pop-off valve opened. With the F1 calendar being extended to 21 races in 2016, the FIA authorised the use of five Power Units before grid penalties were imposed.

Each constructor could modify their Power Unit with 32 development tokens, just like the previous year, while the number of tokens had been due to fall to 25. The differences in performance between the various units convinced the FIA to allow the engineers to work with greater freedom, respecting the highly complicated tables that gave a very different weighting to each token depending on where it was applied.

Ferrari used 23 before the start of the 2016 season, while Mercedes, which had the most competitive unit, used just 19 and could count on greater development during the course of the season. Honda immediately spent 18 tokens, while surprisingly Renault used just 7, not having resolved the reliability issues at Viry Chatillon.

In fact, Red Bull, by no means satisfied with the performance of the Renault engine, had done all it

could to leave the French constructor, contesting the contract after ferocious arguments, but in the end Helmut Marko and Christian Horner had to return to Canossa because Mercedes and Ferrari had refused to supply their Power Units to the Milton Keynes team which they saw as a potential rival.

Toro Rosso instead succeeded in escaping, rejecting the Renault engines and turning to Ferrari once the FIA had conceded a waiver to the regulations. The STR11 was fitted with the Prancing Horse's 2015 Power Unit and the 059/4 six-cylinder was not subjected to further development during the course of the season.

Maranello provided the previous year's engines given that it was not in a position to supply a fourth team given that, along with the Ferrari works cars, also equipped the Sauber and Hass teams, while Mercedes equipped the Silver

Arrows, along with Force India, Williams and Manor.

Renault therefore focused on rebuilding its relationship with Red Bull, which had renamed its power units Tag Heuer, and on the return of the Enstone team to the French marque after having bought out Lotus, while Honda remained an exclusive supplier to McLaren.

The Ferrari 061 Power Unit was the object major work in the search for performance, with as mentioned, 23 tokens being spent. The designer Lorenzo Sassi adopted variable geometry intake trumpet a year later than the other manufacturers because he had kept the compressor intercooler located in the V between the cylinder banks.

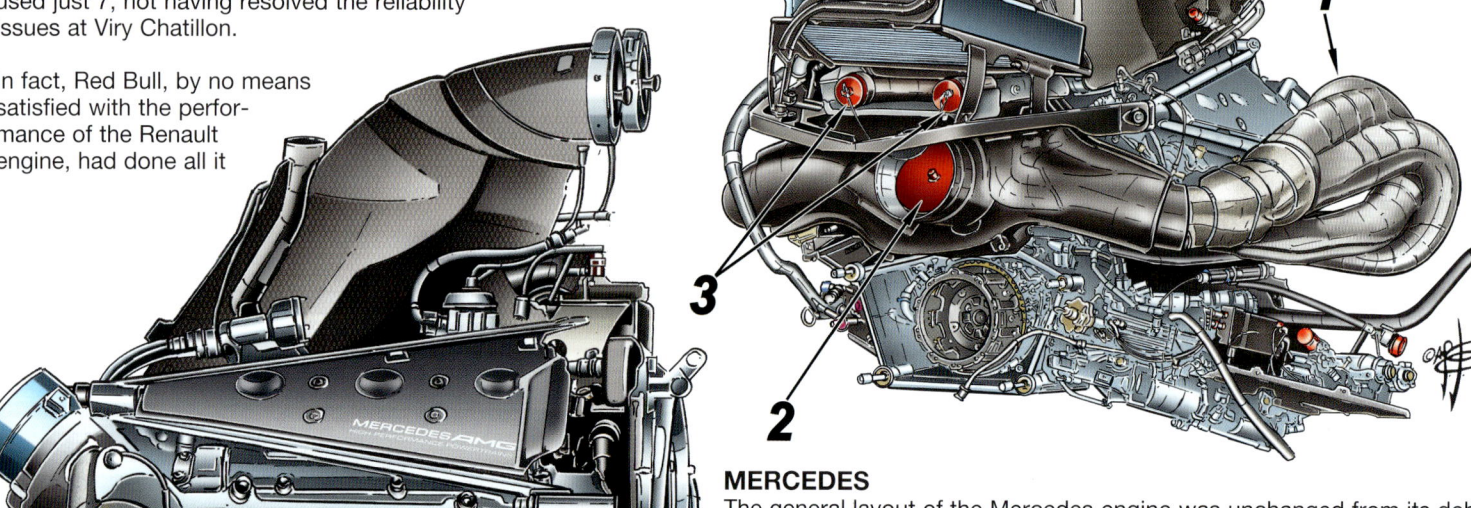

MERCEDES

The general layout of the Mercedes engine was unchanged from its debut in 2014. The following season the exhausts were no longer contained in a kind of lung (indicated by the arrow). In the third year of the Power Unit the FIA permitted the constructors to fit two small additional exhausts (3) to be set no more than 100 mm from the principal exhaust (2). The two terminal served to augment the sound of the engine when the turbo pop-off valve opened. The cooling of the Power Unit assembly was very efficient with a heat exchanger located almost horizontally (4) and cooled by a section of the intake manifold.

Considering that the packaging of the Power Unit had been revolutionised, the intercooler was shifted and split into two elements, the first located between chassis and engine and the second on the left-hand side.

The injection system was also revised and for the first time the fuel pump was made from composite materials. In the face of an increase in power, the cooling system also had to be revised, with the radiators mounted in V-formation.

Given that each unit had to last five Grands Prix, the block was machined from billet, following the advice of Wolf Zimmermann, the specialist who had arrived from Mercedes and who drew on the collaboration of AvL of Graz.

During the season, the Prancing Horse collected 11 podium finishes with five second places, but no wins, despite having spent 3 tokens in Russia, working on the combustion chambers and camshafts and used a new turbo in Canada that cost 2 tokens. In Austria, there was a modification to the cylinder head costing 1 token associated with the debut of a new Shell fuel, while the team spent its last 3 tokens at Spa on a further evolution of the combustion chamber. Sebastian Vettel used a sixth engine in Singapore given that the torsion bar problem experienced in qualifying had forced him to start from the back of the grid in any case, but the 061 had proved to be very reliable despite engine rotations, especially at the start of the season have suggested that the five units would not be sufficient.

On the strength of its superiority, Mercedes spent 2 tokens in Sochi to improve the fuel system and a further 5 at Spa where it made the final seasonal update, counting on not having to use the final 6 tokens. In qualifying and in certain stages of the race the PU106C Hybrid power unit used the "magic button" that provided the two drivers, Nico Rosberg and Lewis Hamilton, with an overboost that created a huge gap with respect to their rivals, with Toto Wolff able to intervene to put the two World Championship contenders the on the same level when technical differences were created.

The situation at Renault was more difficult and it modified the MGU-H in Sochi and then spent 3 tokens at Monaco and therefore the French engineers had to focus on reliability rather than incrementing performance. This policy bore fruit given that the Enstone and Red Bull drivers avoided penalties in 2016.

Those who suffered most from grid penalties were the drivers using Honda power, obliged to start from the back of the grid and cope with continual power unit failures. Jenson Button totalled 35 grid penalties, while Fernando Alonso accumulated no less than 609 in Belgium and 45 in Malaysia. McLaren had shrewdly begun homologating two power units on the same race weekend from 2015, thus paying the penalty only on one occasion and having a new unit ready for the successive race.

This was a practice that Mercedes had already exploited in Belgium, when the Brackley-based team homologated no less than three power units (fourth, fifth and sixth engines, along with the sixth MGU-K and the eighth MGU-H and turbo-compressor unit) ensuring Hamilton a supply of components that was then useful for completing the season without further surprises.

The engine failure the British driver had suffered when leading in Malaysia did not affect the rotation of the power units but ignited the world championship hopes of his teammate Rosberg who won his first and only title before retiring from F1.

The three-pointed star comprehensively dominated the 2016 season, winning no less than 19 races out of 21, leaving the remaining two in Spain and Malaysia to Tag Heuer (Renault). Ferrari endured a poor season with the SF16-H, with the car falling behind in chassis and aerodynamic terms; both the Prancing Horse and Honda were therefore left with nothing to celebrate.

2017

2017 was an important year as it saw the F1 debut of the wider tyres and more extreme aerodynamics. In terms of the engines, the most important novelty was that of the 5 kg increase in the fuel tank capacity to avoid drivers being forced to lift off on the straights before the braking points to save fuel on the most critical circuits. Fuel consumption therefore rose from 100 to 105 kg per race, while the fuel flow rate was unchanged at 100 kg/h at 10500 rpm.

Moreover, in accordance with the teams, the FIA abolished development of the Power Units through tokens, a method considered to be too restrictive for those such as Renault and Honda who were trying to catch up with Mercedes and Ferrari. Given that the calendar featured 20 races again, it was decided that just four Power Units could be used which meant that the one homologated for the Australian GP could be followed by three completely different engines.

Penalties would only be applied following the use of the fifth internal combustion engine, MGU-H, MGU-K, battery pack, turbo or CPU. In order to cap costs it was decided to reduce costs for customer teams by 1 million dollars with respect to the 2016 prices and limits were imposed on research into lightening components, with restrictions on materials and dimensions. The sporting regulations also specified that only the last unit replaced and taken to a race could be used without penalties in the following races, thus eliminating the aberrations of previous years.

HONDA

Honda revolutionised its Power Unit by adopting architecture closer to that of the Mercedes:
1) carbonfibre plenum chamber channelling cool air from the airbox to the compressor (3), fitted at the front of the Japanese 6-cylinder, while the turbine (5) remained at the back in its original position. The two elements were connected by a shaft that frequently suffered from reliability issues.
(2) The position of the motor-generator high on the right-hand bank was unusual, while from the wastegate valve (4) emerge the two smaller exhausts permitted by the 2017 regulations.

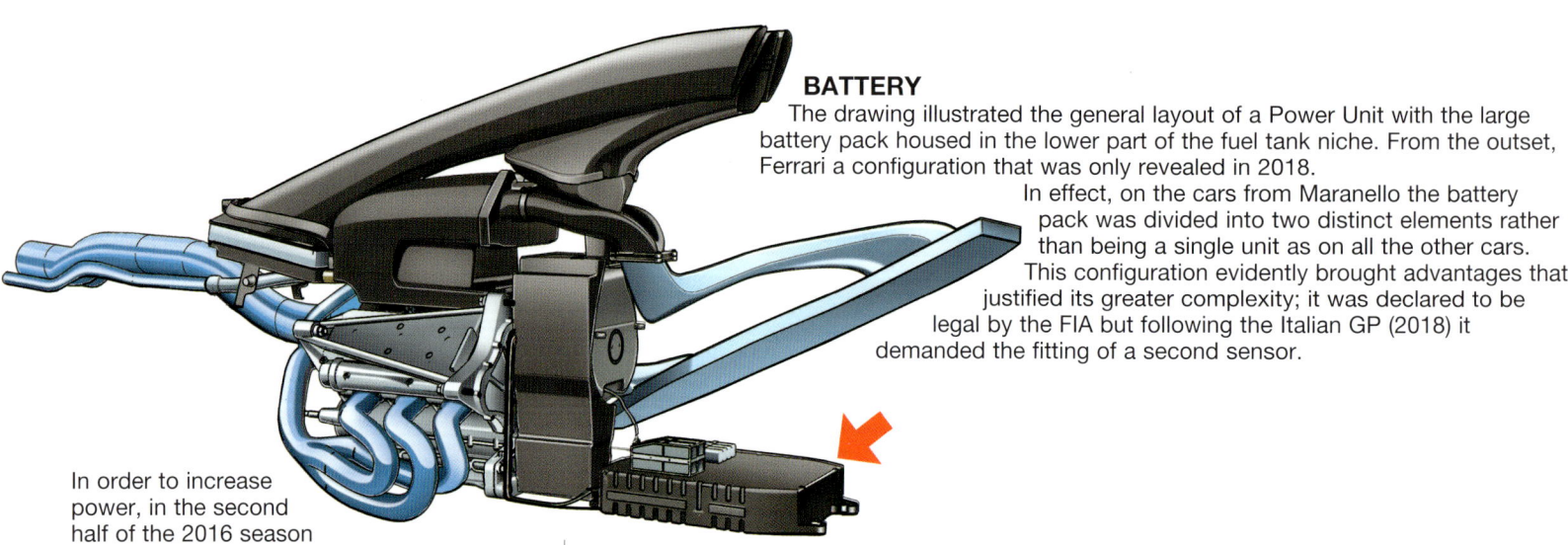

BATTERY

The drawing illustrated the general layout of a Power Unit with the large battery pack housed in the lower part of the fuel tank niche. From the outset, Ferrari a configuration that was only revealed in 2018.

In effect, on the cars from Maranello the battery pack was divided into two distinct elements rather than being a single unit as on all the other cars. This configuration evidently brought advantages that justified its greater complexity; it was declared to be legal by the FIA but following the Italian GP (2018) it demanded the fitting of a second sensor.

In order to increase power, in the second half of the 2016 season Mercedes has used a pre-combustion chamber that facilitated flame propagation, adopting a spark plug equipped with a cap in which fuel was injected at a pressure of 500 bar; by the end of the season the system was permitting a power output of around 980 hp.

In developing the 062/1 engine, Ferrari adopted something similar with the TJI (Turbulent Jet Ignition) fuel system derived from the Mahle system to which it added a new Magneti Marelli "dual anchor" injector that was later also adopted by Honda and which permitted through multi-injections to increase performance and reduce fuel consumption.

The Racing Department gave up on the 2016 engine, going back to the 059/4 genealogy. The Gestione Sportiva engineers focused on increasing turbo boost pressure, achieving peaks of 5 bar compared with the maximum of 3.5 bar of the previous configuration. The ERS's ability to recharge was therefore also enhanced.

Despite the Prancing Horse's best efforts, Mercedes had proved to be unbeatable in qualifying, while the Ferrari SF70H seemed to be able to rival the W08 EQ Power in terms of race pace. Red Bull had sent a letter with a request for clarification to the FIA: how was it possible to "burn" 5 kg of lubricant during a single qualifying session?

Considering that the F1 regulations did not permit aromatics or anti-knock additives, the Milton Keynes engineers suspected that Mercedes was exploiting the lubricant to "enrich" the fuel in the combustion chamber. How? By burning oil that oozed from the segments and "boosted" the petrol for the few qualifying laps that served to conquer pole position in Q2 and Q3 and in those few moments of the race in which the drivers were trying to overtake. There was talk of an increase in power of more than 70 hp with respect to the 2016 Power Units. Clearly, the Petronas chemists had been particularly

clever in perfectly "binding" two elements (oil and petrol) produced at Villastellone, in the Turin hinterland.

Ferrari took an even cleverer approach to the issue, skilfully exploiting every regulatory loophole and the Rossa in fact featured an oil tank for the engine lubricant and a second tank for the oil that was heated in the crankcase before being drawn into the 6-cylinder's plenum chamber and then the compressor, as the recirculation system had to be a closed circuit to prevent the dispersal of liquids around the circuit.

The FIA informed Newey's engineers that it was forbidden to "burn" oil to enrich the fuel, although it was understood that all racing engines consume a quantity of lubricant during their use.

On the occasion of the Azerbaijan GP, the FIA therefore specified that the fuel could not be enriched with substances added to the oil because only the components of the homologated fuel could be introduced to the combustion chamber. Given that checks were by no means simple, this was merely a statement.

At this point, ahead of the Hungarian GP, the FIA's chief engineer for F1 Marcin Budkowski sent a new technical directive that specified that from Monza onwards maximum oil consumption would be capped at 0.9 litres per 100 km, anticipating the 2018 regulation that would lower the figure to 0.6 litre for the same distance.

For two races everything remained the same, however, and Mercedes decided to homologated it fourth and final engine of the season in Belgium, knowing that the Monza provision could not be retroactive and that Lewis Hamilton and Valtteri Bottas would therefore have been able to complete the season without falling foul of the restrictions on oil consumption. This had the effect of arousing fierce disputes in the paddock due with accusations of overt favouritism.

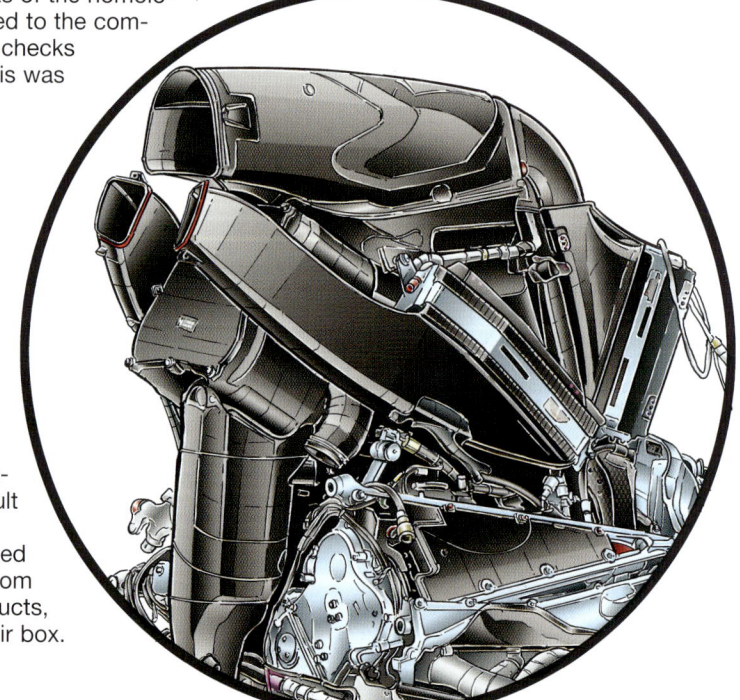

RENAULT

In order to solve the temperature problems with the Renault engine installation, twin ERS radiators were used on the Red Bulls, air being drawn both from the sides, via two separate ducts, and in part from the engine air box.

GIORGIO PIOLA

Mercedes was quick to point out that the W08 EQ Power+ engine already met the 2018 regulations and that there was nothing to worry about. Ferrari instead found itself in trouble as at Monza it had failed to come up with the super-engine expected for its home race and which should have been a powerful harbinger of things to come for the following year. The chief designer Sassi lost his job and the Prancing Horse seemed to losing its way, not just at its home track where it was subjected to the supremacy of the Silver Arrows.

The Asian trip that followed Monza was marked by a dramatic sequence of problems. At Singapore there was the crash at the start involving Sebastian Vettel and Kimi Raikkonen with Max Verstappen, which consequently favoured Hamilton's victory; then in Malaysia, the German was unable even to compete in qualifying as between free practice and the timed sessions, his team had had to replace his engine due to the failure of an air duct between the compressor and the engine. Vettel, whose car had been fitted with engine 5, was obliged to start from the back of the grid, while Raikkonen was unable to start due to a failure similar to that of his teammate. Worse was to come at Suzuka, when the German driver was left standing on the grid with a spark plug failure! These failures were all effects of the Prancing Horse's engineers' vane attempts to match the increased output of the Mercedes engine by pushing their Power Units beyond the limits of reliability, with all too evident results. Hamilton's accident in qualifying at Interlagos had led to the Brackley men deciding to introduce engine 5 in the Silver Arrow, a precursor to the 2018 power unit. The World Champion fought back through field from last to 4th, demonstrating that he could be competitive even with an engine designed to run with the 0.6 kg per 100 km oil consumption limit.

Mercedes as an engine builder had 12 wins in 2017, 9 with Hamilton and the others with Bottas, against Ferrari's 5, all with Vettel, and those of Renault, 2 for Verstappen and one for Ricciardo. Honda was extremely disappointing in its third F1 season and McLaren decided to interrupt its exclusive supply contract with the Japanese firm, foregoing the free engines and support to the tune of 100 million dollars a year in order to sign with Renault for 2018. The Japanese in the meantime had sealed an agreement with Toro Rosso, which was to change Power Unit for the third consecutive season.

Stoffel Vandoorne was the driver inflicted with the heaviest penalty points total with the Honda Power Unit: 10 engines, 12 turbos, 12 MGU-Hs, 9 MGU-Ks and 7 batteries and CPUs. The Japanese firm suffered from chronic unreliability, caused in part by the excessively extreme McLaren rear end that did not allow the RA617 H six-cylinder to breathe. In winter testing a problem with the oil tank had already emerged (later resolved) that was symptomatic of the issues with the design. It was no coincidence that the consultant Gilles Simon left Honda because he was not being listened to, the Frenchman preferring to return to the head of the FIA engineers.

Renault endured a gruelling start to the season due to the lightened MGU-H, which could not cope with the stresses and the engineers at Viry Chatillon therefore had to fall back on the 2016 version that was 5 kg heavier, a solution that required an additional cooling system that accounted for at least another kilo. It was due to these modifications that the hardly superlative RB13 also proved to overweight.

While Ricciardo's win in Russia had been fortunate, those of Verstappen in Malaysia and Mexico had shown an improvement in the French internal combustion engine, while the hybrid part continued to the cross the teams had to bear as when it worked well it permitted gratifying results, but all too often its created headaches for those who might otherwise have been in a position to challenge Mercedes and Ferrari. Red Bull's Australian driver had to make recourse to no less than 8 MGU-Hs and 9 were required by each of the two Toro Rosso STR12s to complete the season…

Franco Nugnes

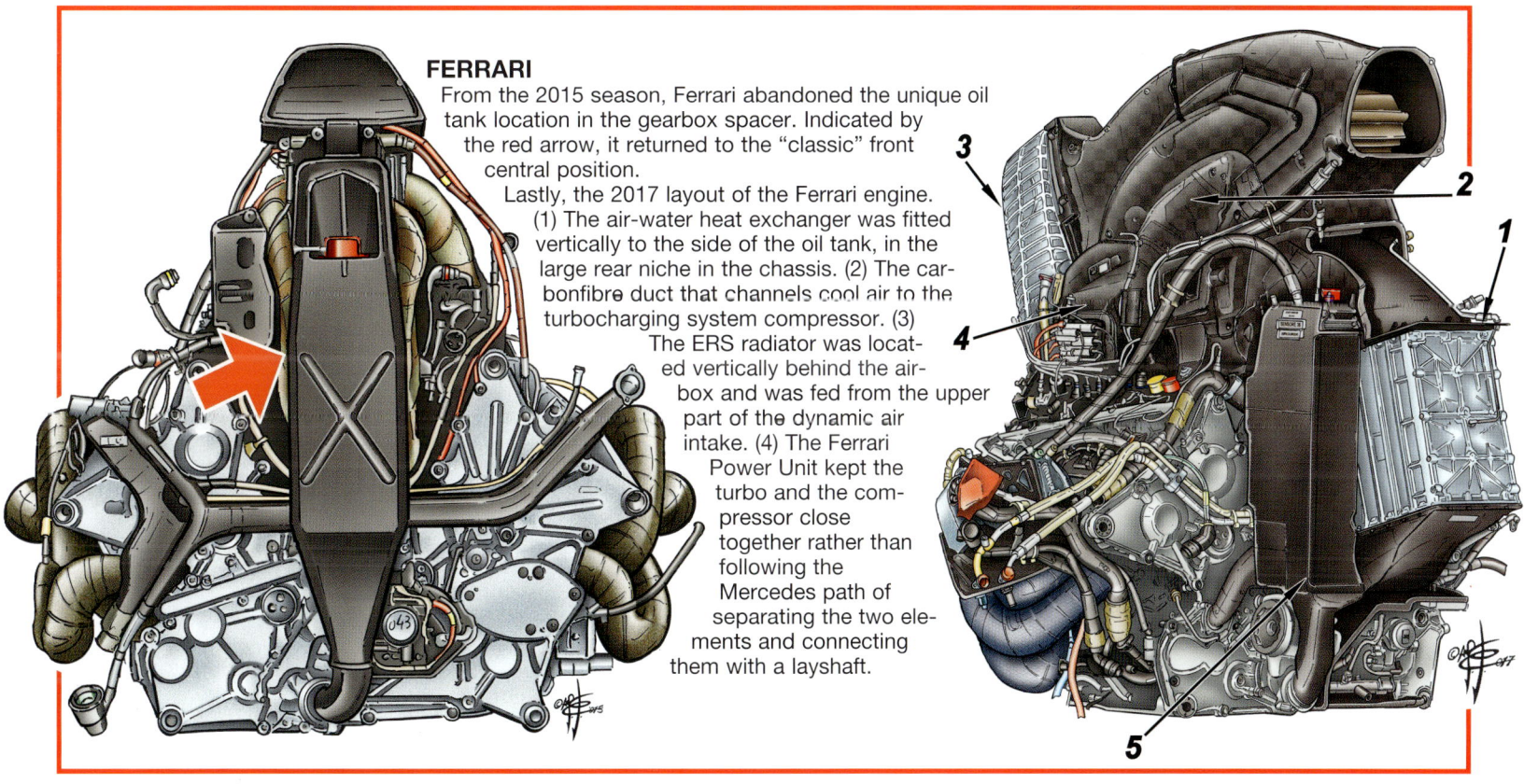

FERRARI

From the 2015 season, Ferrari abandoned the unique oil tank location in the gearbox spacer. Indicated by the red arrow, it returned to the "classic" front central position.

Lastly, the 2017 layout of the Ferrari engine. (1) The air-water heat exchanger was fitted vertically to the side of the oil tank, in the large rear niche in the chassis. (2) The carbonfibre duct that channels cool air to the turbocharging system compressor. (3) The ERS radiator was located vertically behind the airbox and was fed from the upper part of the dynamic air intake. (4) The Ferrari Power Unit kept the turbo and the compressor close together rather than following the Mercedes path of separating the two elements and connecting them with a layshaft.

Mercedes W07

Red Bull RB12

Ferrari SF16-H

Williams FW38

Force India VJM09

McLaren MP4-31

Toro Rosso STR11

Renault R.S.16

Sauber C35

Haas VF-16

Manor MRT05

GIORGIO PIOLA

Car TABLE 2016

		6-44 MERCEDES	3-26-33 RED BULL	5-7 FERRARI	11-27 FORCE INDIA	19-77 WILLIAMS	
		W07	**RB12**	**SF16-H**	**WJM09**	**FW38**	
CAR	Designers	Paddy Lowe Andy Cowell Aldo Costa	Adrian Newey Rob Marshall Dan Fallows	Mattia Binotto Simone Resta Lorenzo Sassi	Adrew Green Akio Haga	Pat Symonds Jakob Andreasen	
	Race engineers	Andrew Showling Tony Ross (6) Peter Bonington (44)	Paul Monagham Simon Rennie (3) Giampiero Lambiase (26-33)	Riccardo Adami (5) Dave Greenwood (7)	Bradley Joice (27) Tim Wright (11)	Dave Robson (19) Jonathan Eddolls (77)	
	Chief mechanic	Mattew Deane	Chris Gent Lee Stevenson	Francesco Ugozzoni	Andy McLaren Will Wickery	Mark Pattinson	
CHASSIS	Front track	1470 mm	1440 mm*	1470 mm	1480 mm	1480 mm	
	Rear track	1405 mm*	1410 mm*	1405 mm*	1410 mm	1420 mm	
	Front suspension	Push-rod 2+1 dampers and torsion bars	Push-rod 2+1 dampers and torsion bars	Push-rod 2+1 dampers and torsion bars	Push-rod 2+1 dampers and torsion bars	Push-rod 2+1 dampers and torsion bars	
	Rear suspension	Pull-rod 2+1 dampers and torsion bars	Pull-rod 2+1 dampers and torsion bars	Pull-rod 2+1 dampers and torsion bars	Pull-rod 2+1 dampers and torsion bars	Pull-rod 2+1 dampers and torsion bars	
	Dampers	Sachs	Multimatic	Sachs	Sachs	Williams	
	Brakes calipers	Brembo	Brembo	Brembo	A+P	A+P	
	Brakes discs	Brembo Carbon Industrie	Brembo	Brembo CCR Carbon Industrie	Hitco	Carbon Industrie	
	Wheels	BBS	O.Z.	BBS	BBS	O.Z.	
	Radiators	Secan	Marston	Secan	Secan	IMI Marston	
	Oil tank	middle position inside fuel tank	middle position inside fuel tank	middle position inside fuel tank	middle position inside fuel tank	middle position inside fuel tank	
GEARBOX		Longitudinal carbon	Longitudinal carbon	Longitudinal carbon	Longitudinal carbon	Longitudinal titanium	
	Gear selection	Semiautomatic 8 gears	Semiautomatic 8 gears	Semiautomatic 8 gears	Semiautomatic 8 gears	Semiautomatic 8 gears	
	Clutch	Sachs	A+P	Sachs	A+P	A+P	
	Pedals	2	2	2	2	2	
ENGINE		Mercedes PU106C	RBR - TAG Heuer RB12 2016	Ferrari 059/5	Mercedes PU106C	Mercedes PU106C	
	Total capacity	1600 cmc	1600 cmc	1600 cmc	1600 cmc	1600 cmc	
	N° cylinders and V	6 - V90°	6 - V90°	6 - V90°	6 - V90°	6 - V90°	
	Electronics	Mercedes	Magneti Marelli	Magneti Marelli	Mercedes	Mercedes	
	Fuel	Petronas	Total	Shell	Petronas	Total	
	Oil	Petronas	Total	Shell	Petronas	Total	
	Dashboard	Mercedes	Red Bull	Magneti Marelli	P.I.	Williams	

[1] non official value   *extimated value

GIORGIO PIOLA®

14-22-47 McLAREN	26-33-55 TORO ROSSO	8-21 HAAS	20-30 RENAULT	9-12 SAUBER	31-88-94 MANOR
MP4-31	**STR11**	**VF-16**	**R.S.16**	**C35**	**MR05**
Timo Goss Matt Morris Peter Prodomou	James Key Phil Charles	Rod Taylor Ben Agathangelou	Nick Chester Martin Tolliday	Eric Gandelin	John McQuillam Luca Furbatto
-	Marco Matassa Xevi Pijolar (33-26) Pierre Hamelin (55)	Ajo Komatsu Gary Cannon (8) Giuliano Salvi (21)	Julien Simon Chutemps (30) Chris Richards (20)	Graig Gardiner (9) Jorn Becker (21)	Juan Pablo Ramirez Josh Peckett (94) Juan Pablo Ramirez (31-88)
-	Domiziano Facchinetti	Stuart Crump	Robert Cherry	Reto Camenzind	Pete Vale
1470 mm*	1440 mm	1460 mm	1450 mm	1460 mm	1470 mm
1405 mm*	1410 mm	1400 mm	1420 mm	1400 mm	1405 mm*
Push-rod 2+1 dampers and torsion bars	Push-rod 2+1 dampers and torsion bars	Push-rod 2+1 dampers and torsion bars	Push-rod 2+1 dampers and torsion bars	Push-rod 2+1 dampers and torsion bars	Push-rod 2+1 dampers and torsion bars
Pull-rod 2+1 dampers and torsion bars	Pull-rod 2+1 dampers and torsion bars	Pull-rod 2+1 dampers and torsion bars	Pull-rod 2+1 dampers and torsion bars	Pull-rod 2+1 dampers and torsion bars	Pull-rod 2+1 dampers and torsion bars
McLaren	Koni	Sachs	Penske	Sachs	Sachs
Akebono	Brembo	Brembo	A+P	Brembo	A+P
Carbon Industrie Brembo	Brembo	Brembo	Hitco	Brembo	Carbon
Enkey	O.Z.	O.Z.	AVUS	O.Z.	APP tech
Calsonic - IMI	Marston	Calsonic	Marston	Calsonic	Secan
middle position inside fuel tank	middle position inside fuel tank	middle position inside fuel tank	middle position inside fuel tank	middle position inside fuel tank	middle position inside fuel tank
Longitudinal carbon	Longitudinal carbon	Longitudinal carbon	Longitudinal titanium	Longitudinal carbon	Longitudinal carbon
Semiautomatic 8 gears	Semiautomatic 8 gears	Semiautomatic 8 gears	Semiautomatic 8 gears	Semiautomatic 8 gears	Semiautomatic 8 gears
A+P	A+P	A+P	A+P	A+P	A+P
2	2	2	2	2	2
Honda 615H	Ferrari 060 2015	Ferrari 061	Renault RS16	Ferrari 060 201	Mercedes PU106C
1600 cmc	1600 cmc	1600 cmc	1600 cmc	1600 cmc	1600 cmc
6 - V90°	6 - V90°	6 - V90°	6 - V90°	6 - V90°	6 - V90°
McLaren el.sys.	Magneti Marelli	Magneti Marelli	Magneti Marelli	Magneti Marelli	Magneti Marelli
Mobil	Total[1]	Shell[1]	Total	Shell[1]	Shell
Mobil	Total[1]	Shell[1]	Total	Shell[1]	Shell
McLaren	Toro Rosso	Ferrari	Renault F1	Magneti Marelli	Magneti Marelli

New DEVELOPMENTS 2016

In the last season before the epochal revolution enacted in 2017, the teams were very busy introducing new features. Mercedes, which dominated the season, was a leader in this sense too.

During the winter break, the German team made a further leap forwards with even greater progress being made than in the 2015 pre-season and from the very debut of the W07 they laid claim to the 2016 championship title. The car was in fact crammed with novelties, some of them never seen before. As was the case with the interpretation of the rules relating to the chas-

sis to create a particularly efficient front suspension layout.
In a season of crushing disappointment for Ferrari, positive notes came from Toro Rosso which had the merit of introducing a rear-wing end plate feature that was the most widely admired in the field, even being copied by the all-conquering Mercedes.
McLaren was very active in producing a series of features designed by Peter Prodomou, with front and rear wings boasting a wealth of new solutions.

MERCEDES

In interpreting the regulations, the designer frequently have to find where concessions may be made, reading between the lines to see what is not actually written. In the 2014 season, the FIA had drastically lowered the height of the chassis at the point of the front A-A section from 625 to 525 mm, leaving unaltered the regulation that allowed holes to be opened in the structure to permit access to the suspension. Almost all the teams improved this area, even going as far as to create a kind of cover. Mercedes in particular took advantage of two loopholes in the technical regulations. The first had been exploited by Manor in 2015 to adapt the 2014 chassis given that the regulations did not require the tubs to be fabricated as a single structure. This was always the case of the safety rollbar and the lateral protection structures applied at a later date. On the 2015 Manor an extension had been added to the front part of the chassis (in the circle).
The second loophole concerned the lack of dimensional restrictions for any apertures, so as to create an area in which to work on the suspension "in the open", using the vanity panel to meet the dimensions that were specified in the regulations.
The feature was questionable in terms of the spirit of the regulations but unassailable from a practical point of view because it respected the safety norms as the frontal crash test had been passed, albeit without the voluminous vanity panel.

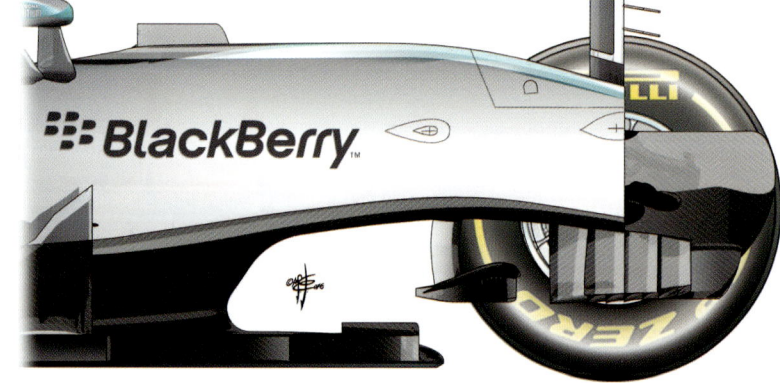

MERCEDES FRONT SUSPENSION

The Mercedes front suspension was at the centre of attention from its debut in experimental form at the 2015 Brazilian GP.

Despite the banning of FRIC during the course of the 2014 season, Mercedes went on to use hydraulic components to control the dynamic ride height of the W07, not only to improve driveability but also to ensure a more consistent aerodynamic set-up. Along with

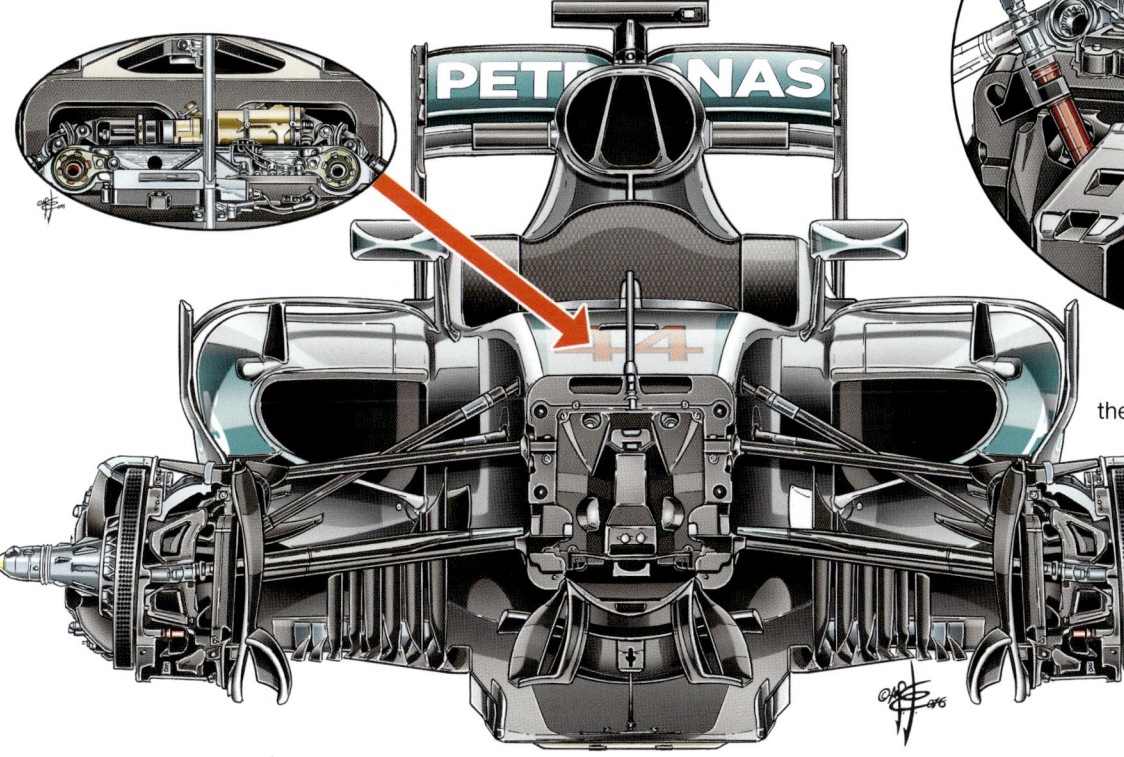

the third hydraulic damper, the anti-roll bar was also controlled electronically, a novelty with respect to the other cars.

The generously dimensioned transverse third hydraulic element, the object of intense scrutiny by the other teams, was jealously covered with a kind of rudimentary fairing.

FERRARI

The comparison between the Ferrari and Red Bull chassis clearly reveals how the Mercedes design was more extreme. Ferrari also adopted a kind of cover/vanity panel, but much smaller in size and equipped with ample apertures for accessing the suspension elements, still installed within the chassis.

RED BULL

A middle ground was represented by the Red Bull solution with a larger vanity panel than the Ferrari's, with the chassis structure satisfying both the letter and the spirit of the regulations. The third transverse element was exposed and permitted easy adjustment, but the shape of the aperture in the chassis respected the B-B and A-A section dimensions in full.

FERRARI PUSH-ROD/PULL-ROD FRONT SUSPENSION

Clearly, the push-rod layout for the front suspension was not in itself innovative, but it was for Ferrari, which after using a pull-rod configuration at the front end for four years returned for a push-rod design for the SF16-H like all the other cars. The narrow base, tuning fork-shaped lower wishbone was retained, while all the other suspension elements were radically modified. Note also the protections either side of the driver's head, 20 mm higher as per the regulations.

SF16-H   SF15-T

MERCEDES BARGEBOARD

The W07 proved to be crammed with new feature from its debut, some of them extreme and previously unseen. The bargeboards vanes ahead of the sidepods presented no less than six separate elements, with horizontal "teeth" in the lower part (highlighted in the oval), destined to manage the flow of air towards the lower area and the diffuser. Strangely, this innovative and effective feature was not taken up by other teams over the course of the season.

MERCEDES TORO ROSSO S-DUCT

Curiously, both Mercedes and Toro Rosso adopted the same stratagem to get around the 150 mm from the front axle restriction so as to gain advantages from the S-Duct mouth.

The enters on the Mercedes thanks to a kind of shark mouth that takes the air somewhat further forward than the Sauber and Red Bull and later the Force India and McLaren. The Toro Rosso design was more sophisticated, with intakes on both sides thanks to two NACA ducts, while a further two cooled the chassis interior.

MERCEDES END PLATE

Sochi saw the introduction of a new Mercedes feature, albeit in static form: this decomposition of the end section of the front wing end plates with two vertical slots to accelerate the air flow to the outside of the wheels. Its definitive track debut came in the following race in Barcelona.

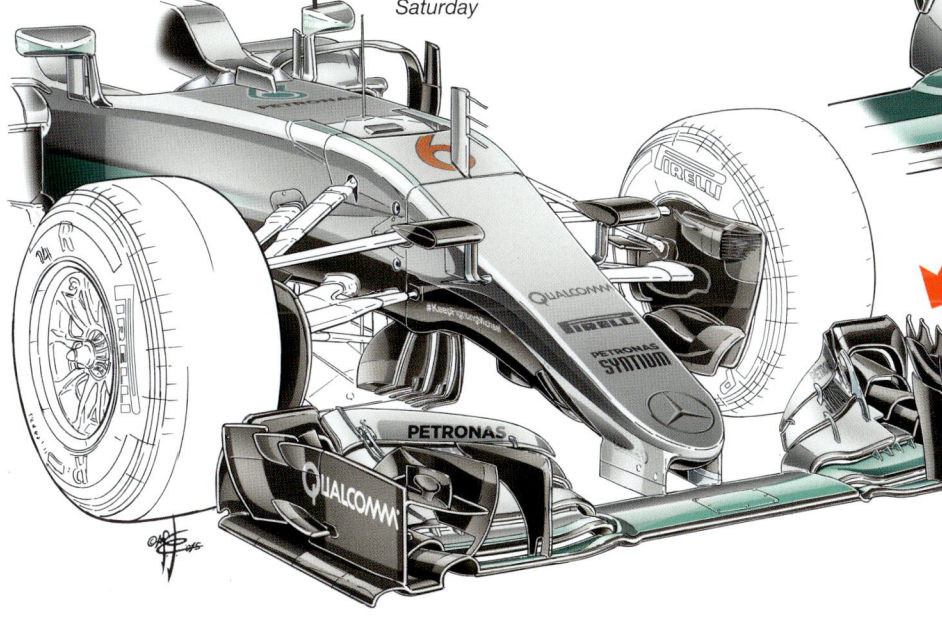

Saturday

TORO ROSSO END PLATES

From pre-season testing, Toro Rosso ran a rear wing for the STR11 equipped with new slots in the upper part of the end plates. This was the most "popular" technical feature of the year and was copied by almost all the other teams. A series of conspicuous horizontal clots that, alongside, appear open at the front while they are actually attached to one another by a slim mount. These slots modified the flow around the rear wing so as to reduce the toxic vortexes that increase drag.

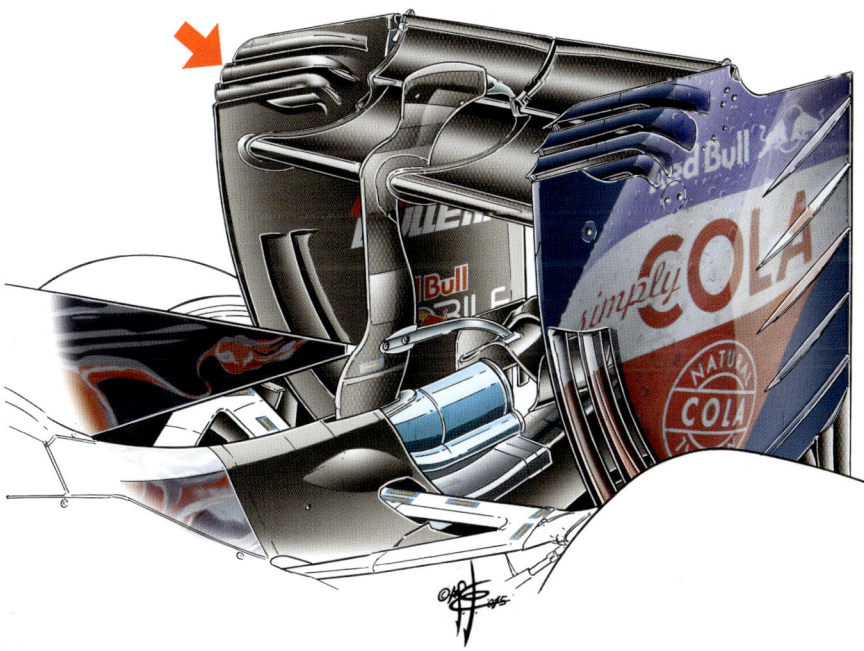

RED BULL

This long vertical slot starting from the horizontal one in the upper part of the end plates was new and descended to merge with the earlier slot (in the circle) seen on the RB12. The two designs were alternated throughout the season.

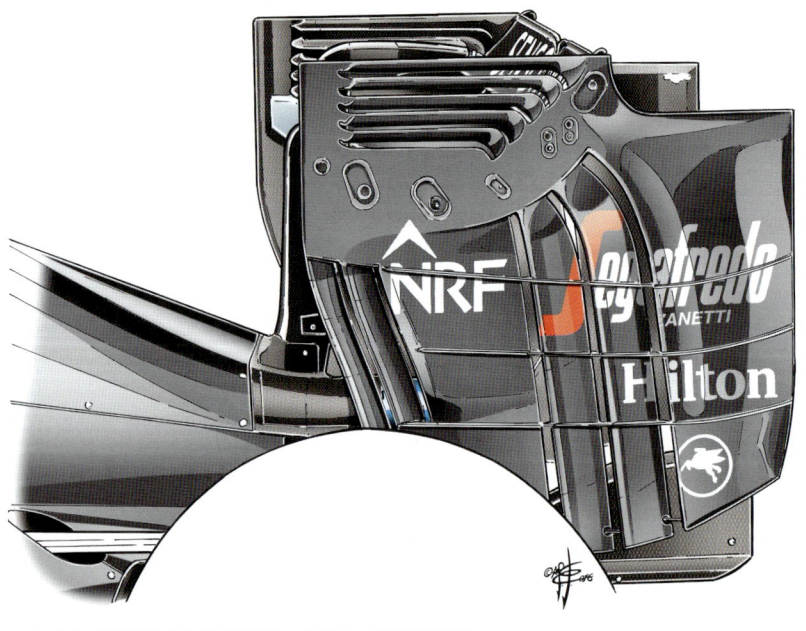

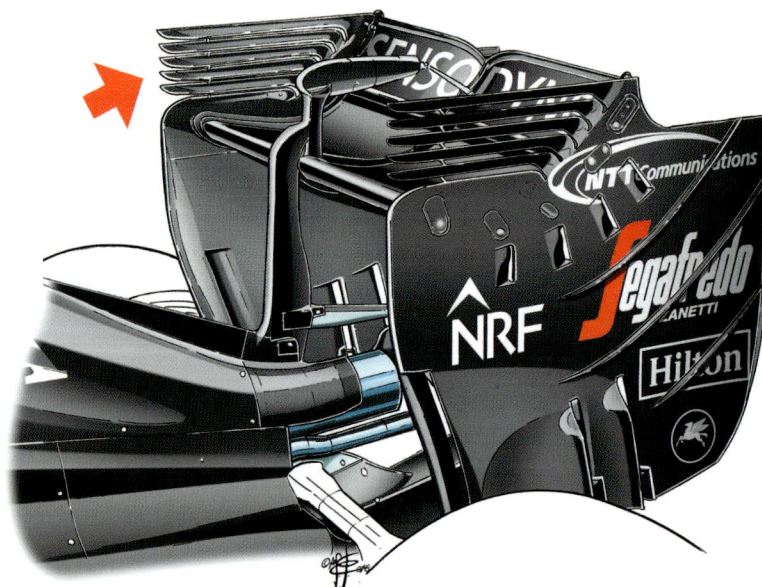

McLAREN ZELTWEG AND SEPANG

On the Friday morning at Zeltweg, Alonso tested these new and extreme end plates, characterised by long vertical slots. This configuration was almost immediately abandoned as it created loading variations.

At Sepang instead, McLaren, following Sauber and Mercedes, added open horizontal slots (red arrow) to the upper part of its sophisticated end plates, a feature introduced by Toro Rosso from the first laps of winter testing.

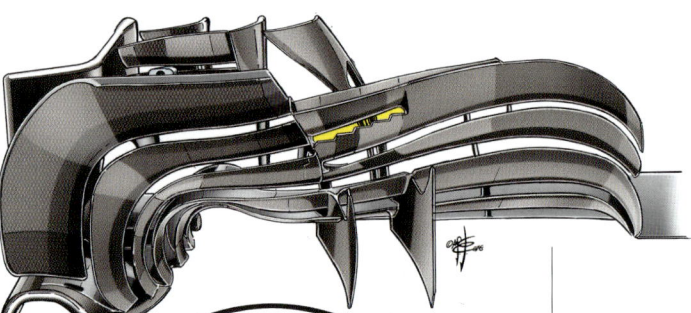

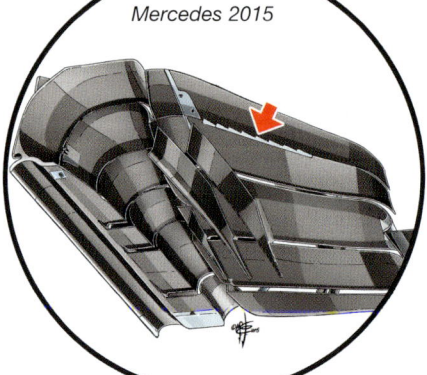

Mercedes 2015

RED BULL SHANGHAI

Red Bull also introduced an aerodynamic modification that proved to be a failure: the sawtooth trailing edge (highlighted with the arrow) of the penultimate flap, tested on the Friday morning in China but never seen again on the RB11.

SERRATED MERCEDES WING

The rear wing with a serrated trailing edge to the main profile was seen only in winter testing and was based on experiments conducted during the last race in 2015 at Abu Dhabi when sawtooth adhesive tape was applied to the main profile.

A similar feature had been introduced in Sochi, applied to the front wing (see the Mercedes chapter), with the experience and data gained being transferred to the rear. The serrated trailing edge on the main profile was intended to allow the flow to reattach when the DRS closed, improving stability under braking.

WILLIAMS

At Sepang, Williams introduced a small serrated portion on the flap, as indicated in yellow in the drawing. A feature that recalled one used by Mercedes in 2015, in that case applied to the trailing edge of the penultimate flap..

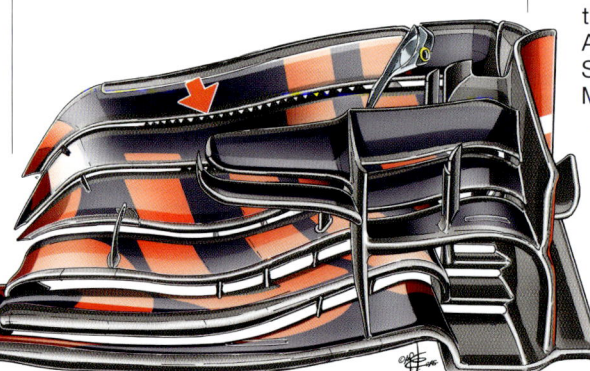

GIORGIO PIOLA

FORCE INDIA

Force India was the pleasant surprise of the 2016 season. Despite a limited budget it managed to finish 4th in the Constructors' Championship. The most important step was seen in Spain when to all intents and purposes a "B" version was fielded. The most conspicuous new feature was the front wing, completely different to its predecessor. The most important difference was the more pronounced "kink" (1) in the main profile, with a wider flat area (2) at the end plate mount, following the lead set by Mercedes the previous season. In the upper part, a fin (3) was added to divert the flow to the outside, together with a further two (4) directly linked to the end plate, immediately after the upper flaps. A wing therefore designed to channel as much of the flow as possible to the outside of the front wheels.

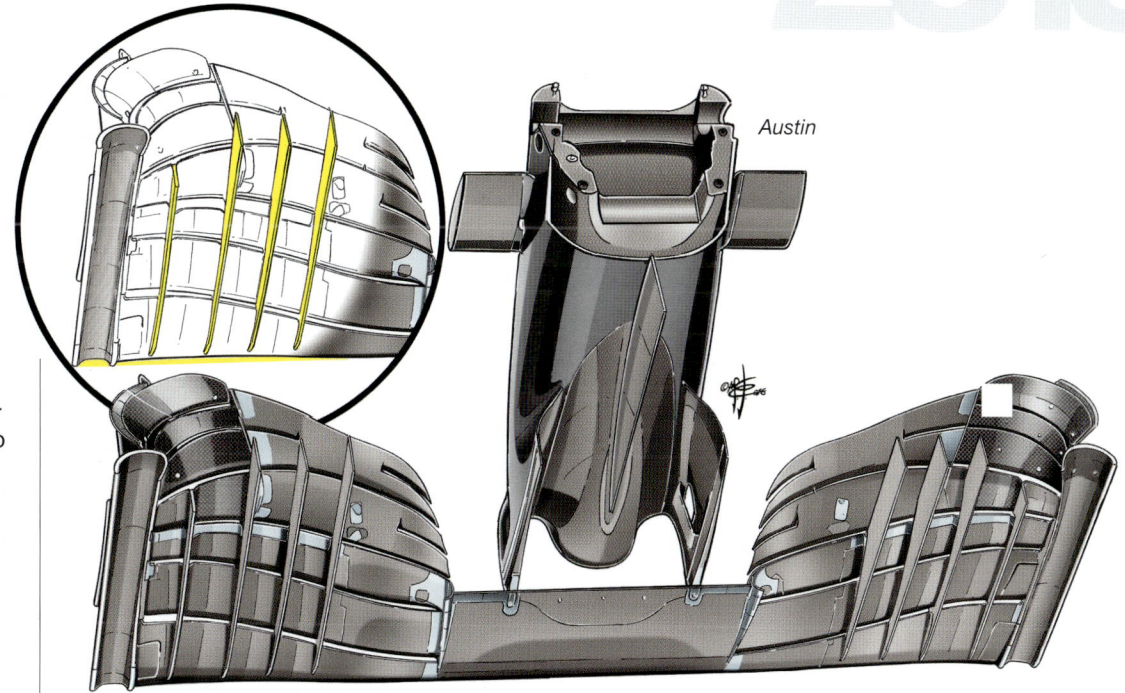

Austin

McLAREN SUZUKA E AUSTIN

At the end of the 2016 season, McLaren intensified development with an eye on 2017, firstly with a new front wing for Suzuka, characterised by a few wide flat area adjacent to the end plate and slots between the various very large profiles; at the following race in Austin, a new front wing made its first appearance, lacking the "kink" in the peripheral area and characterised by a straight profile with a tendency to curve down at the ends. The different configuration is highlighted by the yellow area in the two drawings. Note the presence of the four vertical mini-vanes in the lower part. This experiment was not followed-up in 2017.

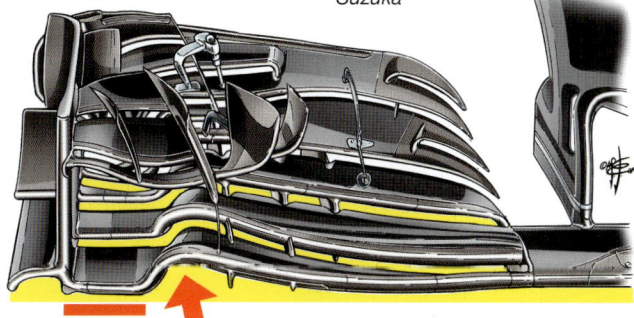

Suzuka

Austin

MERCEDES FRONT END PLATE

Mercedes also introduced a front wing in view of the 2017 season, with the end plates in particular presenting a new feature. Rather than being connected directly to the main profile they featured an ample vent. Having appeared at Sepange, it was tested again in Austin, but was not taken up for the following season.

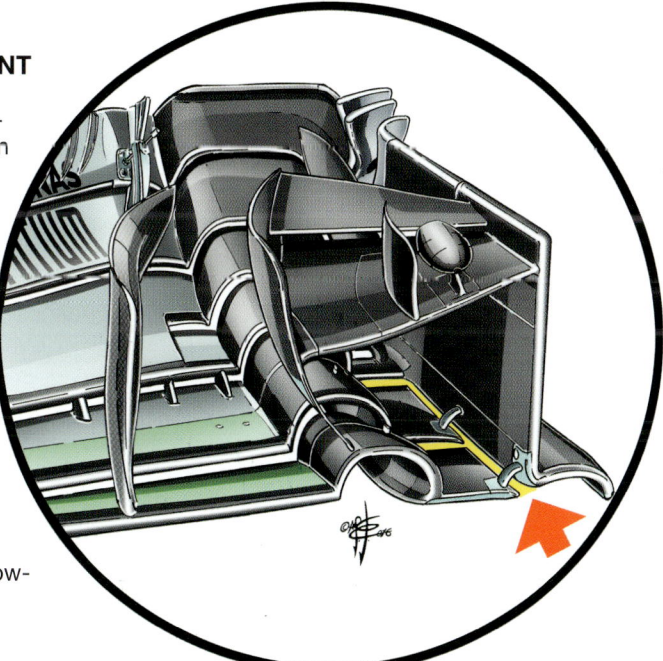

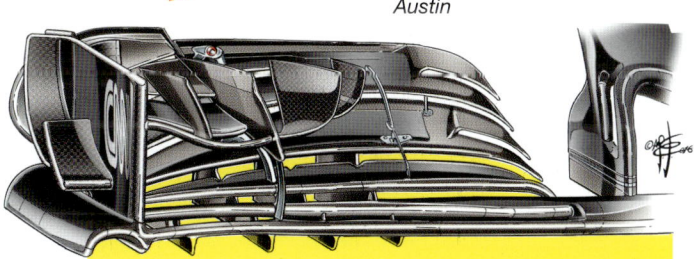

FERRARI DIFFUSER BUDAPEST

Considerable refinement work was conducted during the 2016 season on the delicate area of the diffuser between the rear wheels to neutralise the negative effects generated by the tyres.

As well as the usual mini-flaps (already seen in Austin in 2015) a conspicuous triangular fin was added to direct the air flow and increase downforce in this area.

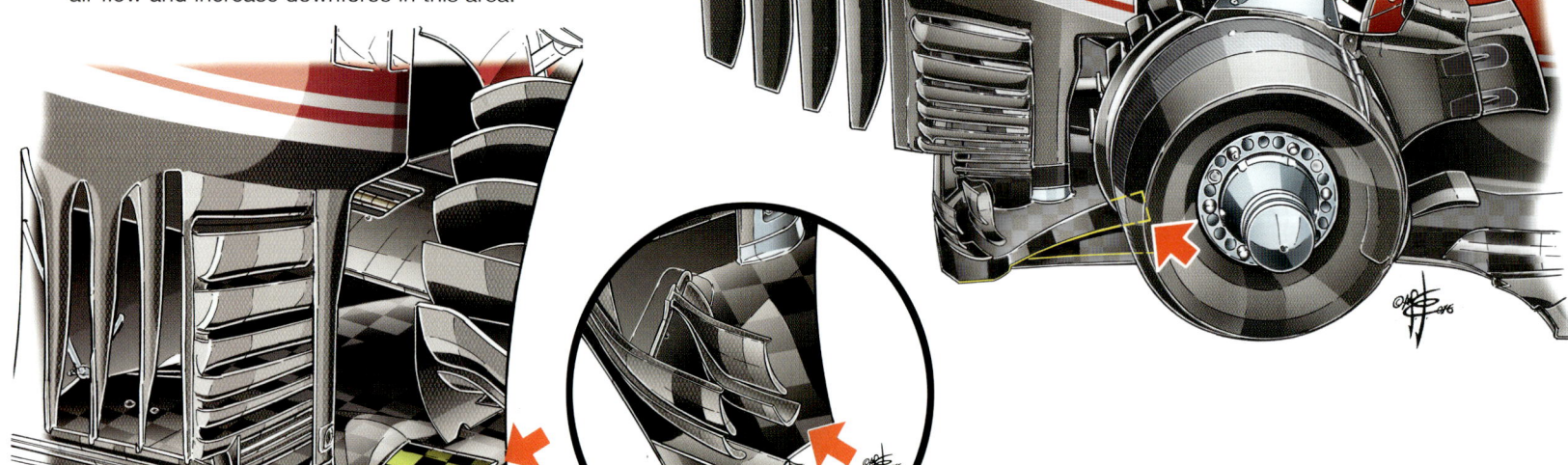

SAUBER

The Sauber's Y-shape rear wing pylon was new and replaced the two-piece support used in 2015. The flow in area of the link with the main profile was therefore unchanged, but the Toro Rosso idea from 2015 (copied by many teams) was adopted with the mono-pylon passing through the exhaust.

McLAREN FRONT SUSPENSION

The McLaren MP4-31 had its front suspension arms set very close together, almost as if to direct and control the flow of air towards the sidepods and the lower part of the car, therefore improving the quality of the flow in this area.

McLAREN CAMERAS

McLaren even exploited the possibility of inserting a small vent in the camera mount so as to improve the flow in this area; the feature was then copied on the Williams with the shorter nose.

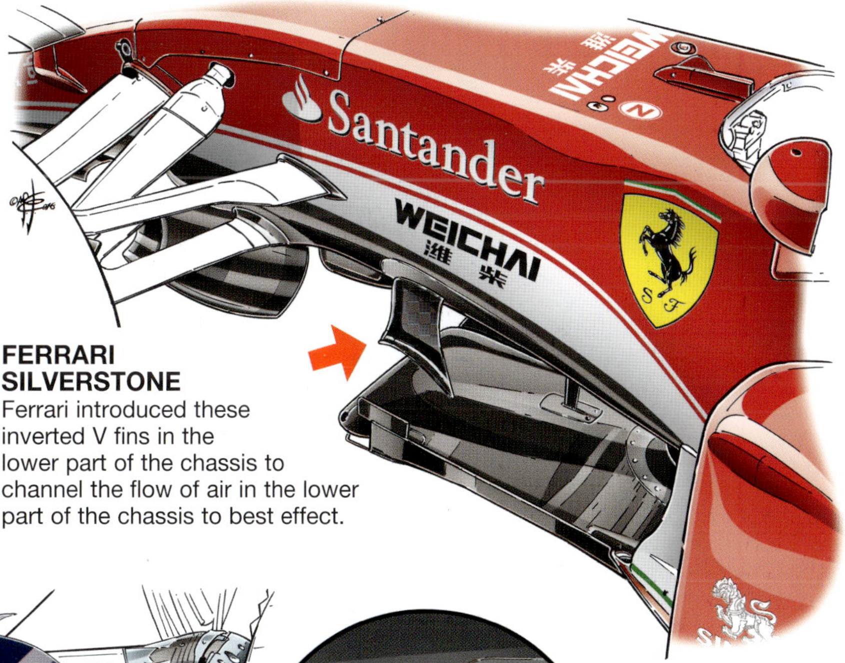

FERRARI SILVERSTONE

Ferrari introduced these inverted V fins in the lower part of the chassis to channel the flow of air in the lower part of the chassis to best effect.

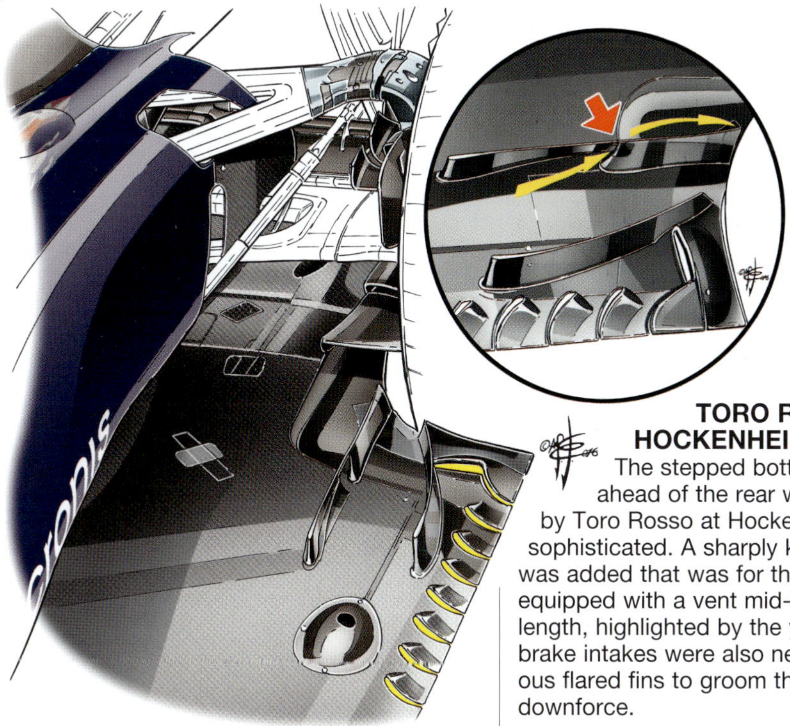

TORO ROSSO HOCKENHEIM

The stepped bottom in the area ahead of the rear wheels introduced by Toro Rosso at Hockenheim was very sophisticated. A sharply kinked vertical fin was added that was for the first time equipped with a vent mid-way along its length, highlighted by the yellow arrow. The brake intakes were also new with conspicuous flared fins to groom the flow and create downforce.

WILLIAMS

Another new feature was this shark-gill arrangement used by Williams on the "hot" tracks to disperse heat from the sidepods. The idea was then reprised by Mercedes in 2017.

The chapter dealing with cockpits in this edition is particularly rich in new features as in both seasons, the FIA introduced severe limitations to start line assistance, thereby stimulating the inventive minds of the designers. As a result, the steering wheels became even more differentiated, personalised and ergonomic, with the various teams trying to conceal their latest innovations.

For the 2016 season, in fact, the FIA imposed a drastic reduction on the number of electronic aids at the start, in particular banning the use of dual clutch management paddles. In recent years this feature had become a true replacement for the traction control sys-tems that had been banned again in 2008 after having been reintroduced in 2001 for the start only. Once the gear had been engaged with the first, the second paddle allowed the driver to modulate the power delivery on the basis of the clutch biting point, manually reproducing the effects of a traction control system. It should be remembered that the dual clutch control had been introduced as recovery with both hands in the case of an incident. However, from 2001, the second lever became more a kind of manual traction control to avoid wheelspin.

The steering wheel sector was enlivened by Ferrari which revolutionised the rear part of those on the SF16-H: in place of the dual paddles it introduced a long rocker pivoted not in the centre but rather to one side, providing a much greater range of movement. The Ferrari solution permitted greatly increased modularity but sparked a degree of controversy on its debut that was subsequently dampened by the FIA.

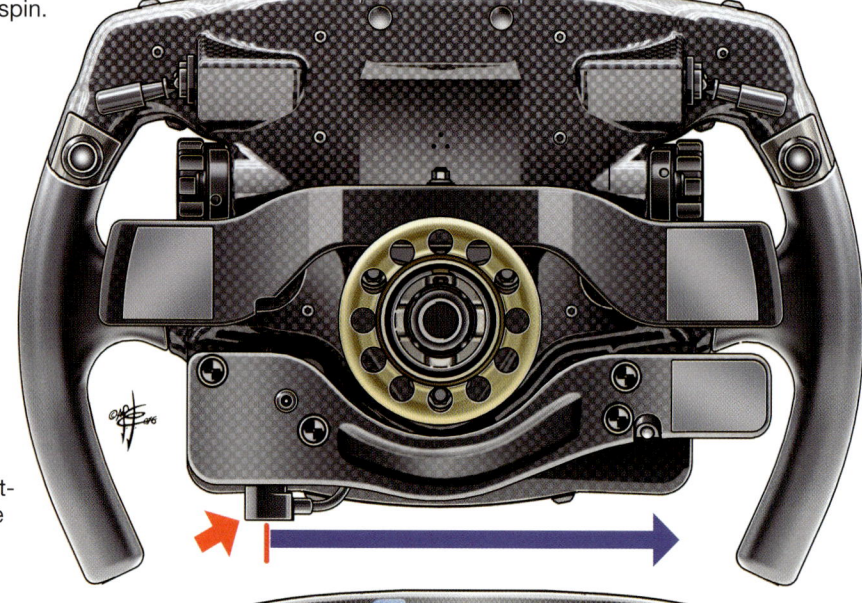

FERRARI STEERING WHEEL

In the comparison with the SF15-T steering wheel, it is clear that Ferrari opted for a new configuration; not only were the dual clutch paddles eliminated but also the two at the bottom , making way for a single rocker. The most important thing was, however, that this rocker was pivoted not in the centre as indicated on its unveiling in Australia (red arrow), but rather on one side only. There were a total of six paddles on the SF15-T in 2015.

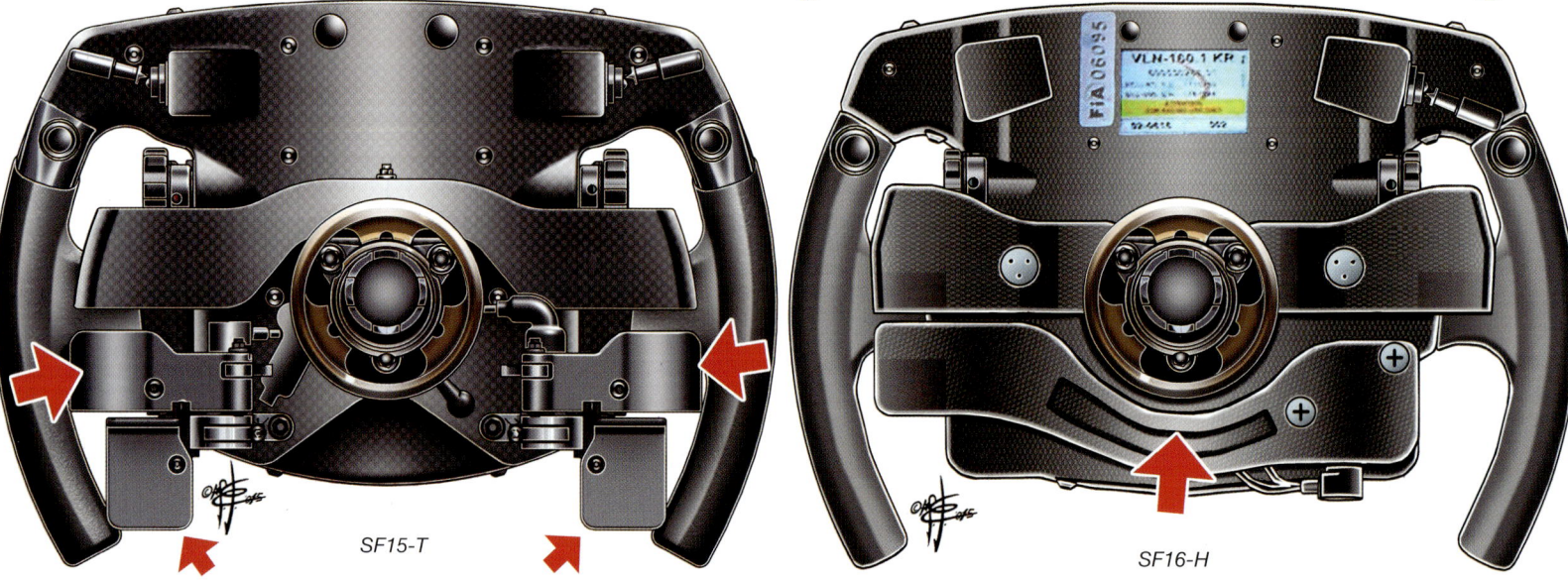

SF15-T

SF16-H

OFFSET ROCKER

This drawing "captured" at the start in Melbourne clearly shows the strongly offset pivoting point of the long clutch paddle behind the steering wheel; this feature provided much greater travel with respect to the previous short dual paddles. The Ferrari drivers could therefore count on improved modularity of the clutch at the start. The result was that the Ferrari car camera was the most closely observed by the rival teams early in the season.

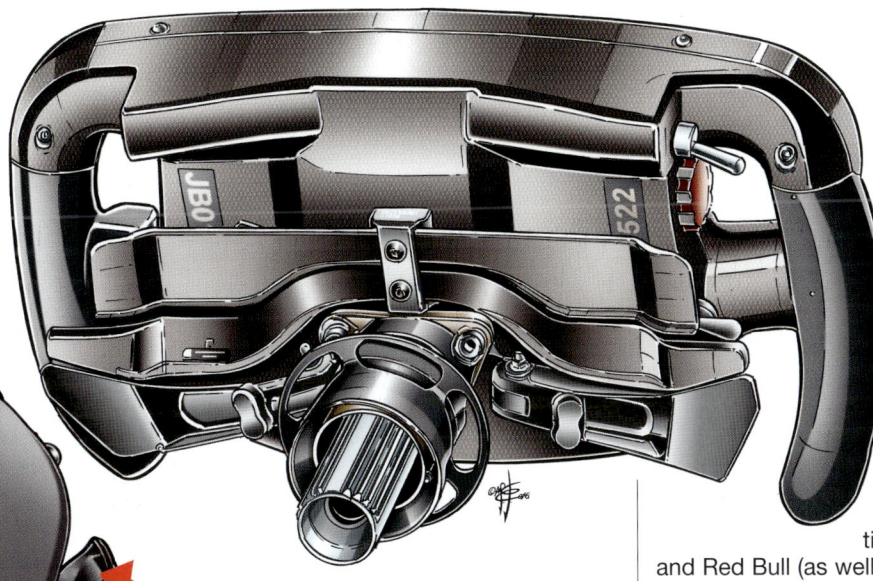

Räikkönen 2016

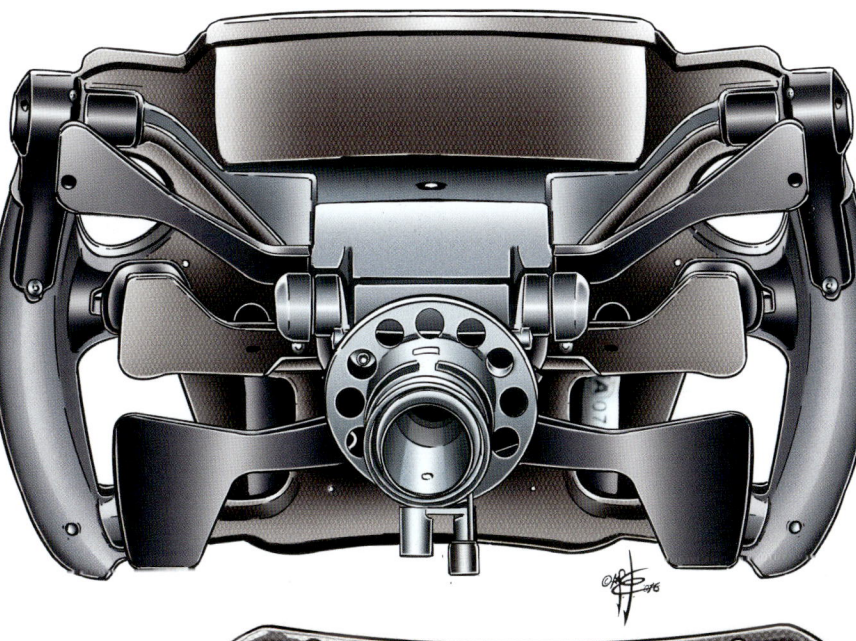

MERCEDES AND RED BULL

Mercedes, Red Bull and McLaren all retained the same steering wheel configuration as in 2015, as shown in the comparison of Rosberg's wheel with paddles only slightly modified with regard to their ergonomic functioning. Both McLaren and Red Bull (as well as Renault) retained the six-paddle configuration, with the two supplementary levers at the bottom, never previously seen on Mercedes steering wheels.

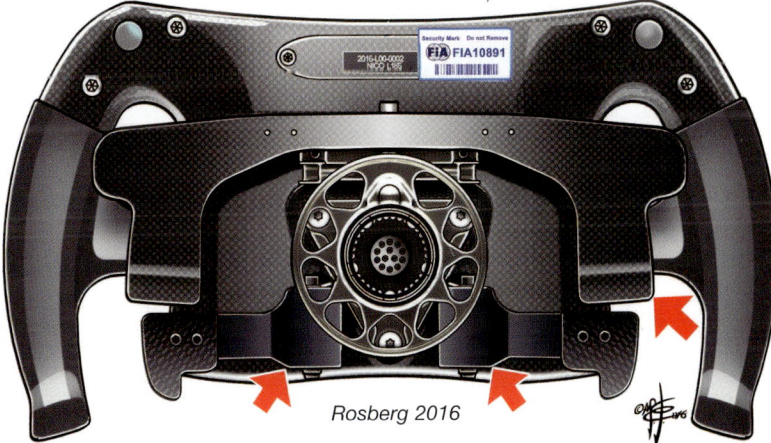

Rosberg 2016

ROSBERG STEERING WHEEL

For the 2016 season, as well as the ban on the second clutch paddle, the FIA introduced in Art. 21.1 of the Sporting Regulations very strict restrictions regarding communications between drivers and their pits. As a result, the steering wheels gained somewhat cryptic post-its with further instructions for the drivers, as seen on Rosberg's wheel (which even had questions). From the German GP, the severe restrictions on radio transmissions were lifted, to the satisfaction of the enthusiasts.

FORCE INDIA

With the season already underway, Force India introduced a steering wheel with a large incorporated display and to same time and money used the prototyping system, as seen from the colour of the material, leaving Williams as the only team with the display on the dashboard. This feature was retained by the Didcot team in the 2017 season too.

HAMILTON BRAZIL

For the last two races of the season, on Hamilton's steering wheel extensions were added to both the gear change rocker and the twin clutch paddles, as indicated by the arrows.

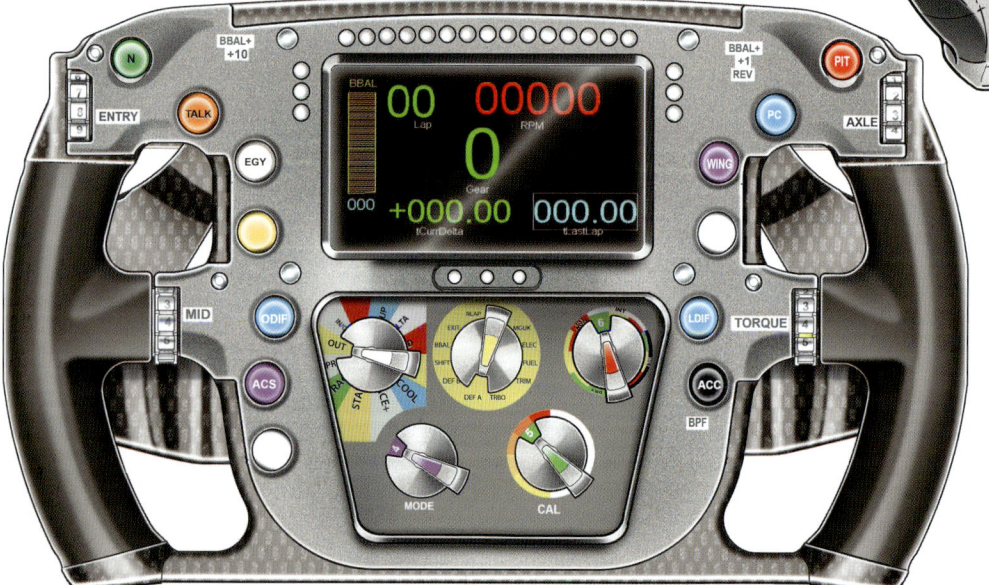

GIORGIO PIOLA

Vettel

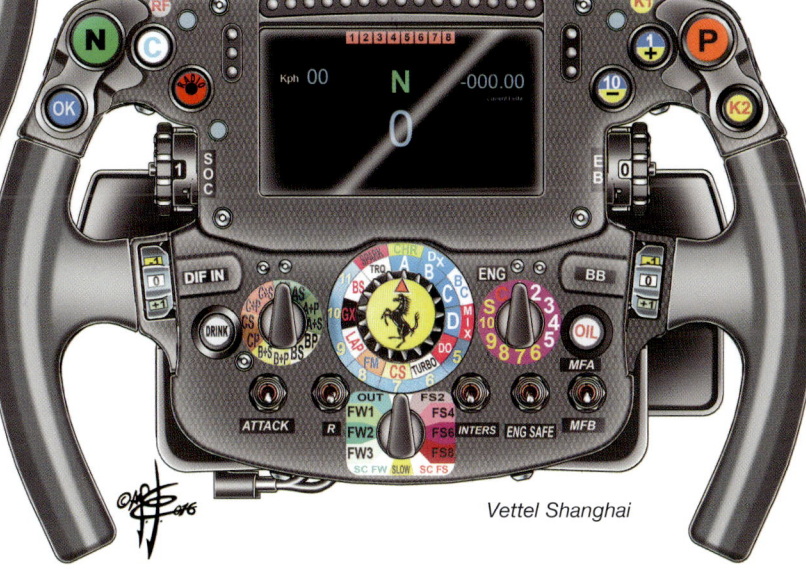

Vettel Shanghai

FERRARI STEERING WHEELS

Vettel and Räikkönen used two different steering wheels from the start of the season. At Shanghai both received a more highly developed version of the one supplied to the German driver that was significantly different with respect to the base version, with twin buttons at the top; it also featured a different lower central knob.

Räikkönen

Räikkönen Shanghai

Regulations 2016

Ahead of the major technical revolution announced for 2017, the 2016 season opened under the sign of technical stability, with a limited number of modifications designed to improve safety, the sound of the Power Units and to bring about a drastic reduction in the electronic clutch management aids at the start. In recent seasons, such aids had become a kind of "substitute" traction control, which had been banned again in 2008 after having been reintroduced in 2001with its use restricted to the start alone. Lastly, Pirelli introduced a new Ultrasoft compound distinguished by a violet band and designed to guarantee greater freedom in strategy management with the possibility of having three rather than two compounds available to the teams at every race, while retaining the limit of 13 sets assigned for the race weekend.

In order to enhance protection of the drivers' head, in 2015 the FIA had commissioned Mercedes to design a kind of cage (Halo), while for 2016 it introduced an increment of 20 mm in the protection structures either side of the driver's head while, more importantly, making the crash test values even more severe. The value was in fact more than tripled from 15 N to 50 N, applied for three seconds. This test proved much more difficult to pass that the frontal crash test.

In order to improve the sound of the Power Units, the wastegate valve exhaust no longer vented into the principal exhaust but had to be kept separate. The teams were now permitted to adopt two small exhausts, opening the way for different configurations, but they had to be symmetrical and set within 10 mm of the principal exhaust.

All of the 2016 cars adopted the twin configuration with the exception of the Renault, while Red Bull fell in line with the twin exhausts. The number of engines available to the teams was unchanged at five, following an initial suggestion of reducing it to four, the increased number of races causing this decision to be postponed. The number of development tokens was also unchanged at 32 (in theory, it should have fallen to 25) for each engine maker, with the possibility of homologating an engine for the 2016 season for a client team (e.g. Toro Rosso with Ferrari). However, the restriction that had the greatest influence on the season was the use of a single clutch paddle at the start that, together with the restrictions on the team radio made the driver much more the focus at the start of the race.

Once the gear had been selected with the first paddle, the second had allowed the driver to modulate the handling of the power on the basis of the clutch biting point, thus reproducing a kind of manual traction control. It should be remembered, that the dual clutch had been introduced to permit rapid reaction to an incident with both hands, as illustrated in the 2015-2016 edition. However, from 2001, when traction control was tolerated, the second paddle had lost its "recovery" function in the case of a spin to become instead a kind of manual traction control, with the driver able to feather the clutch with the paddle to avoid wheelspin.

All teams left the disposition of the paddles unchanged, as shown in the drawings of the Mercedes and Red Bull steering wheels, while Ferrari revolutionised the rear part of the SF16-H wheel; in the place of the paired paddles was a long rocker pivoted not in the centre but on one side only, so as to achieve the movement of just one of the driver's hands much broader. This design permitted greatly increased modularity but sparked a degree of controversy that was immediately dampened by the FIA.

15kg (2015)
50kg (2016)

+20mm

+20mm

PROTECTIONS

In order to improve protection of the driver's head in the case of one car mounting another, the safety structures on either side were raised by 20 mm and above all had to pass a test in which the load applied was more than tripled from 15 N to 50 N and applied for three seconds. This test proved much more difficult to pass that the frontal crash test.

GIORGIO PIOLA

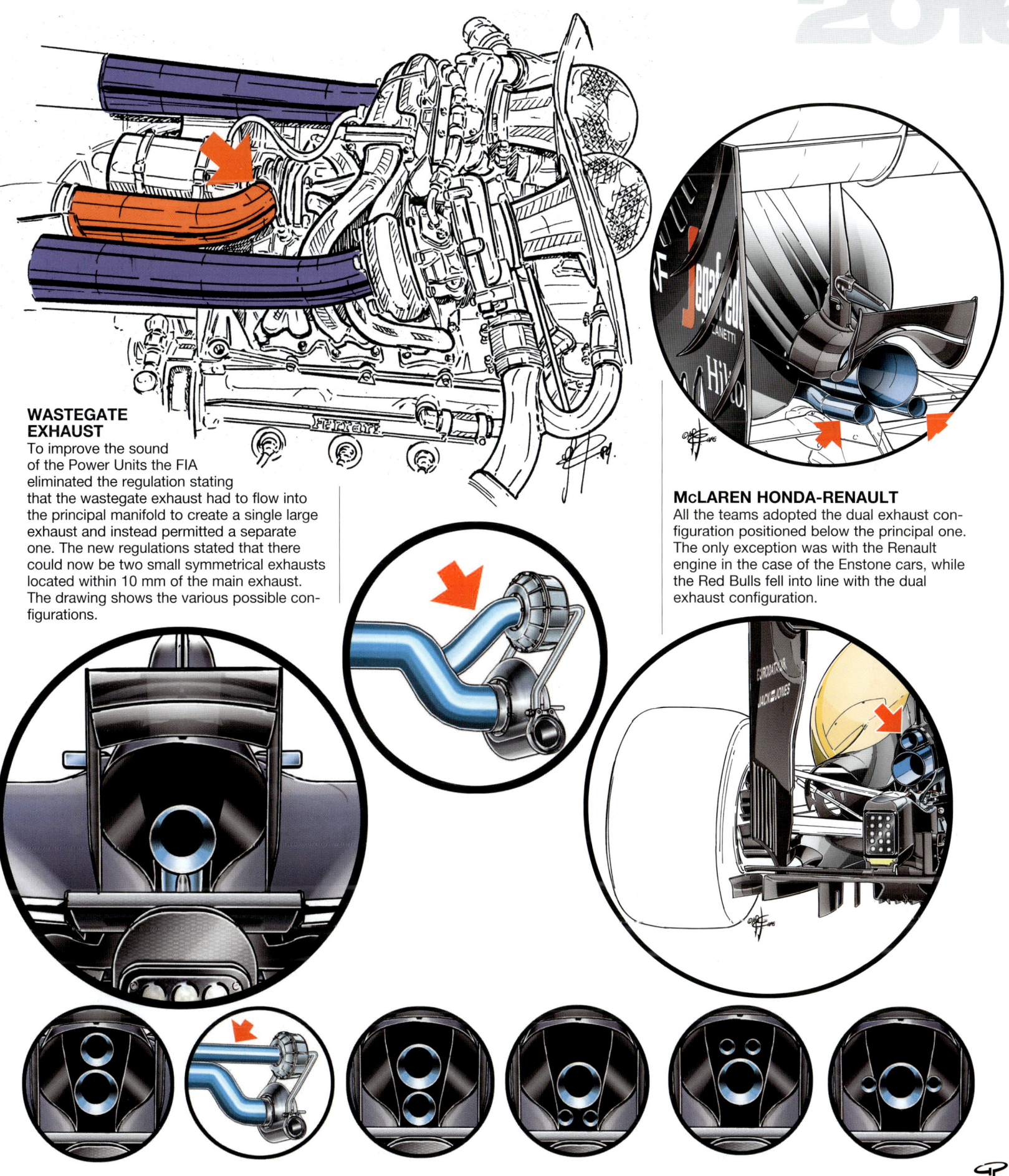

WASTEGATE EXHAUST

To improve the sound of the Power Units the FIA eliminated the regulation stating that the wastegate exhaust had to flow into the principal manifold to create a single large exhaust and instead permitted a separate one. The new regulations stated that there could now be two small symmetrical exhausts located within 10 mm of the main exhaust. The drawing shows the various possible configurations.

McLAREN HONDA-RENAULT

All the teams adopted the dual exhaust configuration positioned below the principal one. The only exception was with the Renault engine in the case of the Enstone cars, while the Red Bulls fell into line with the dual exhaust configuration.

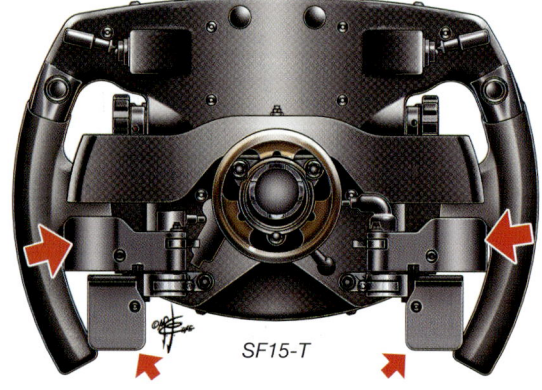

SF15-T

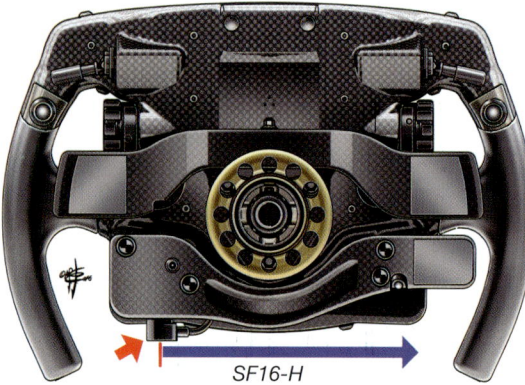

SF16-H

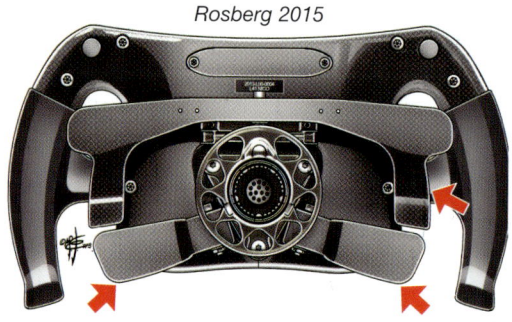

Rosberg 2015

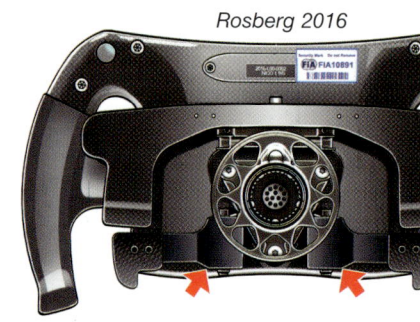

Rosberg 2016

FERRARI STEERING WHEEL

In 2016, Ferrari introduced a new clutch control feature on the SF16-H: a long rocker in place of the dual paddles for the clutch control that was to be operated by the driver using one hand only. The most important feature was however that the rocker was pivoted not in the centre but to one side. There were a total of six paddles on the SF15T in 2015.

OFFSET ROCKER

The offset pivoting point of the long clutch rocker on the Ferrari steering wheel allowed greater modulation of the clutch at the start. The result was that the Ferrari car camera was the most closely observed by the rival teams early in the season.

MERCEDES AND RED BULL

Both Mercedes and Red Bull retained the same steering wheel architecture as in 2015, as documented in the comparison with Rosberg's wheel with paddles modified only in terms of their ergonomic functioning. The RB12 retained the two supplementary paddles at the bottom, never previously seen on Mercedes steering wheels.

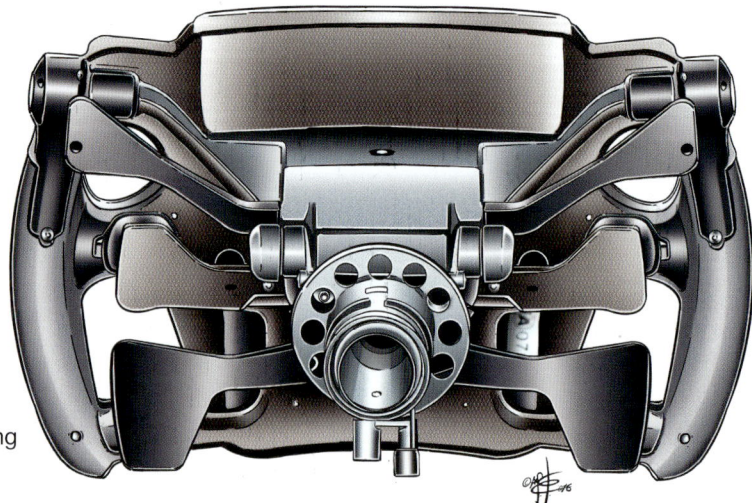

NEW ULTRASOFT TYRES

For the 2016 season Pirelli introduced an extra slick tyre, the Ultrasoft with a violet band, taking the number of wide slick tyres available to five, along with the usual two rain tyres.

The range of P Zero covers therefore became: Hard (orange), Medium (white), Soft (yellow), Supersoft (red) and the new Ultrasoft (violet).

The two rain tyres were the Cinturato full wets (blue) and the intermediates (green).

With the tyre off the car, the FIA checked that the minimum inflation pressure was in line with the indications provided by the sole supplier on a race by race basis, in accordance with the camber rates, while the temperature with the thermal covers removed could not exceed 110°. Since 2016, the Milan company has also indicated the maximum recommended distance (in laps) for each tyre in relation to the data harvested during free practice so as to avoid safety issues deriving from over use of the tyres in the race.

MANOR

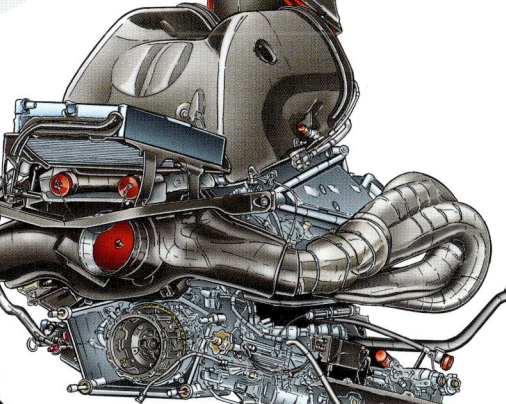

While one team made it into F1, Haas, after endless problems another had to throw in the towel at the end of 2016: Manor, the former Marussia, which dropped out at the end of what was actually its best season in which it came very close to achieving its objective of not finishing at the foot of the Constructors' Championship. The outfit missed out on the objective at the penultimate race, overtaken by Sauber, which until then had always been behind the little English team thanks to Felipe Nasr's 9th place in Brazil.

Key to rendering the MRT 05 more competitive was the switch from the Ferrari Power Unit, which contractually had been the previous season's, to the Mercedes standard version with gearbox and rear suspension supplied by Williams.

The designer of the MRT 05 was Luca Furbatto, the former Toro Rosso chief engineer, under the direction of John McQuilliam. The English team was reinforced with expert engineers such as Pat Fry, ex-McLaren and a former Ferrari technical director, and Nikolas Tombazis, the Greek aerodynamicist who for years had been head of aerodynamics and CFD at Maranello. Characterised in particular by a very long, flat nose, as this configuration allowed the crash test to be passed with the minimum of expense, the MRT 05 saw interesting developments throughout the season that focused mainly on aerodynamics, the new feature nonetheless failing to allow the team to achieve its objective of overtaking Sauber in the Constructors' Championship.

CHASSIS
As in the previous season, Manor adopted the stratagem of an addition at the front of the chassis to ensure the crash test was passed.

MERCEDES POWER UNIT
The Manor's strong suit was its Mercedes Power Unit and in this drawing the Stuttgart six-cylinder is seen installed in the two English cars.

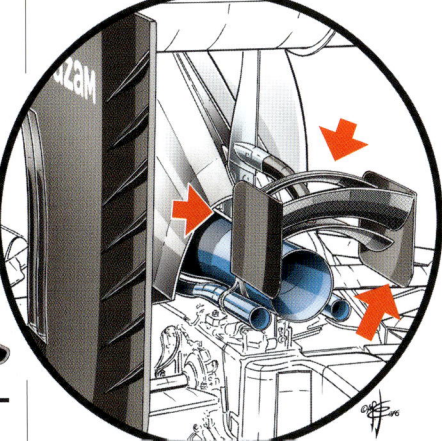

Manor MR03B
Brazil

Manor MRT05
Launch

Manor MRT05
Abu Dhabi

MONKEY SEAT
At Monaco the Manors ran this simple monkey seat with a single central support and an advanced profile on the leading edge, as well as a second equipped with a flap.

HAAS

The American team's return to F1 could hardly have gone better. Underlying Hass's success was the approach adopted by Gene Haas: no leaps in the dark, but rather a relationship of close collaboration with Ferrari, the supplier of virtually all of the mechanical organs and the use of its own wind tunnel, as well as with Dallara, the outfit responsible for the construction of the chassis and the assembly of the car. The result was a reliable, healthy car that caught the eye in the early races but which as the season progressed proved to have reached its limit compared with its rivals. Team principal duties were taken by Gunther Steiner, while technical direction was entrusted to Rob Taylor, with Ben Agathangelou as the chief aerodynamicist.

Satisfied with the points conquered in those early races, the powers that be at Haas decided to stop development on the car so as to focus on the more complex project for the 2017 car as early as possible. The only aerodynamic package introduced in the second half of the season was seen at the Japanese GP, four races before the end of the championship.

On that occasion the team introduced a new underbody and new brake intakes. However, Haas did have the merit of creating the only new low downforce feature for Monza: a rear wing with very sophisticated shapes.

The VF16's weak spot was the braking system even though it was identical to the one fitted to the Ferrari, both in terms of components (Brembo calipers and discs) and their location. Haas also retained the blown front hubs used on the cars from Maranello, naturally in closed form for the high speed tracks. The braking issues were a cause of concern above all for Grosjean who frequently complained of excessive locking of the front wheels despite a very precise rear end. In an attempt to overcome these problems, discs from the French manufacturer Carbon Industrie were tried in place of the Brembos, but this failed to resolve the issue.

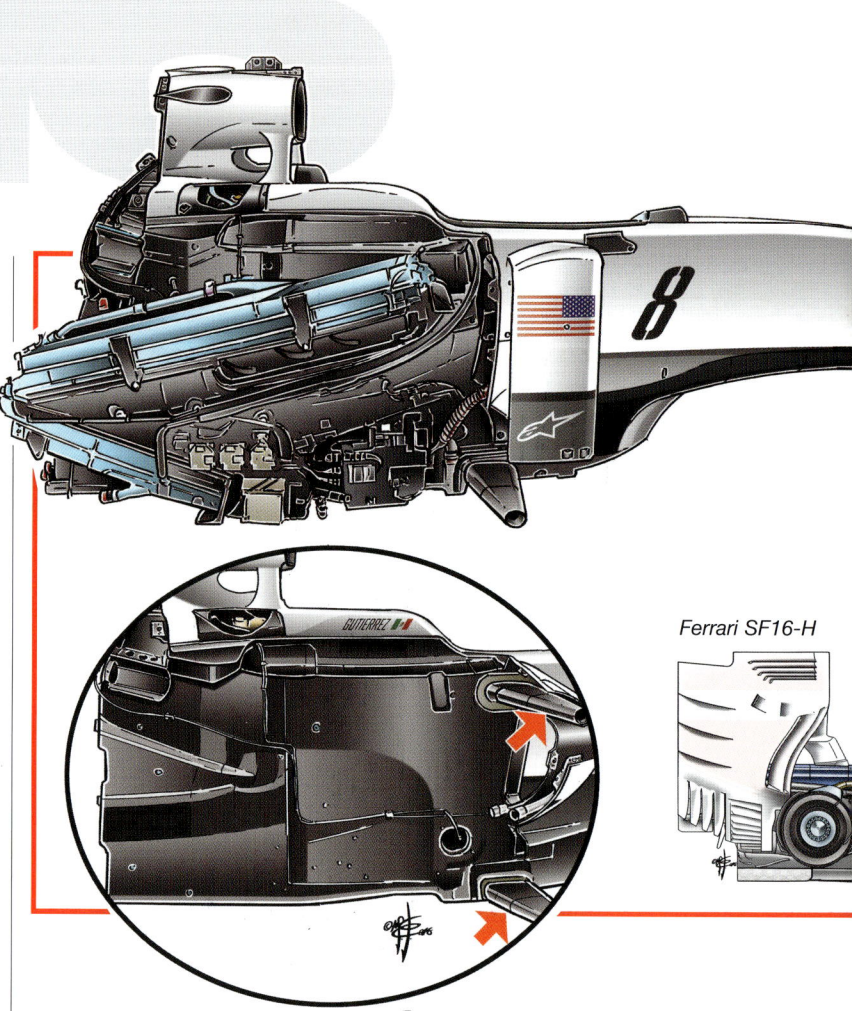

Ferrari SF16-H

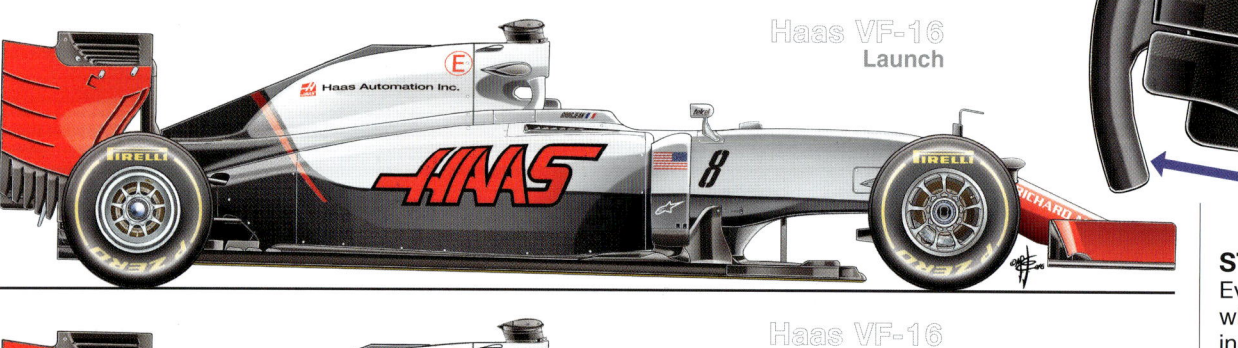

Haas VF-16
Launch

Haas VF-16
Suzuka

STEERING WHEEL
Even the Haas steering wheel was of Ferrari derivation, including the long offset rocker facilitating clutch control at the start by providing greater modularity thanks to its increased travel.

FRONT WING

The comparison with the Ferrari front wing from 2015 clearly reveals the family feelings of the Haas wing. This flared fin was added at Shanghai in order to better control the vortices that were created at the leading edge of the central neutral zone and the rest of the wing.

Ferrari 2015

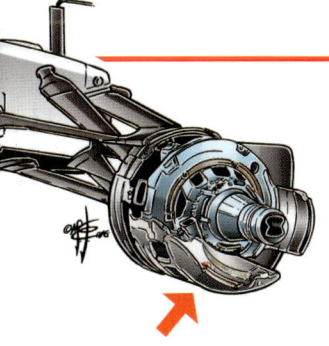

FERRARI MECHANICALS AND DALLARA CHASSIS

The engine, gearbox, radiator pack layout, braking system, suspension configuration and even the blown front hubs were all part of the Ferrari package, installed in a chassis built by Dallara, as shown in the drawings of the bare monocoque (in the oval) and with the radiators assembled in a horizontal V, as on the SF16-H. Also visible in the inset was the disposition of the safety structures, identical in terms of location, shape and technical characteristics on all the cars.

Renault Montréal 2010-2011

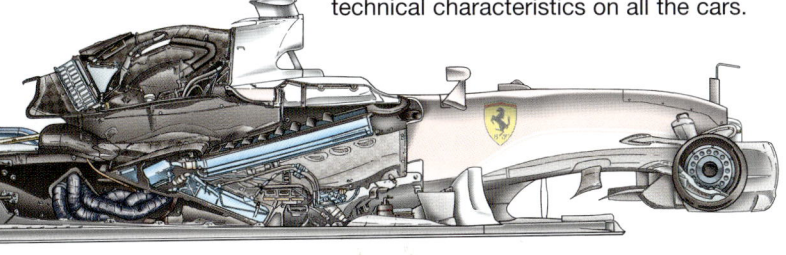

BRAKES

The Haas braking system with underslung Brembo calipers was identical to the Ferrari SF16-H's. Haas also adopted the blown front hubs, although this drawing shows the closed configuration used on the faster tracks. Grosjean frequently complained about excessive locking of the front wheels and requested a comparison test in Brazil with discs from the French firm Carbon Industrie.

Ferrari SF16-H

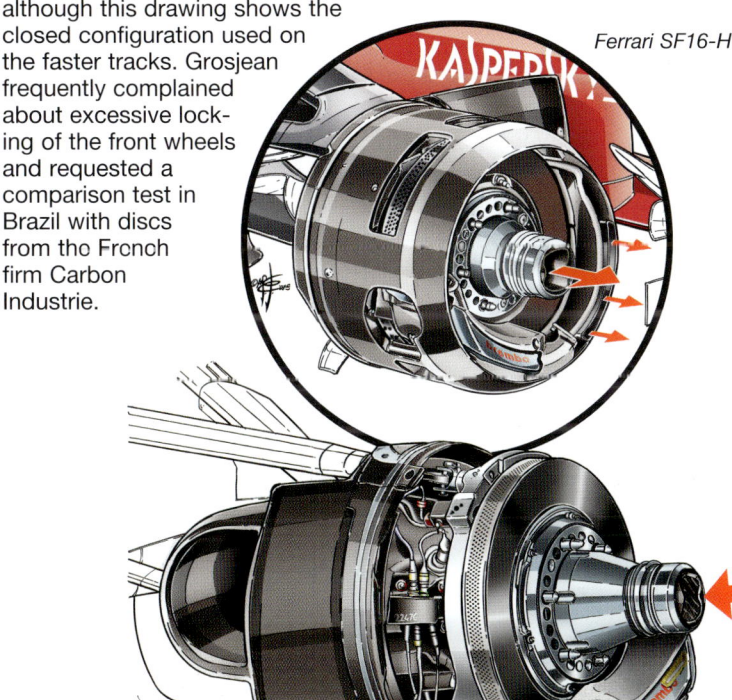

MONZA

Haas was the only team to field a feature designed specifically for Monza rather than an adaptation of the rear wing used in Montreal or at Spa. The very complex design referenced the configurations used in Canada by Renault in the 2010 and 2011 seasons.

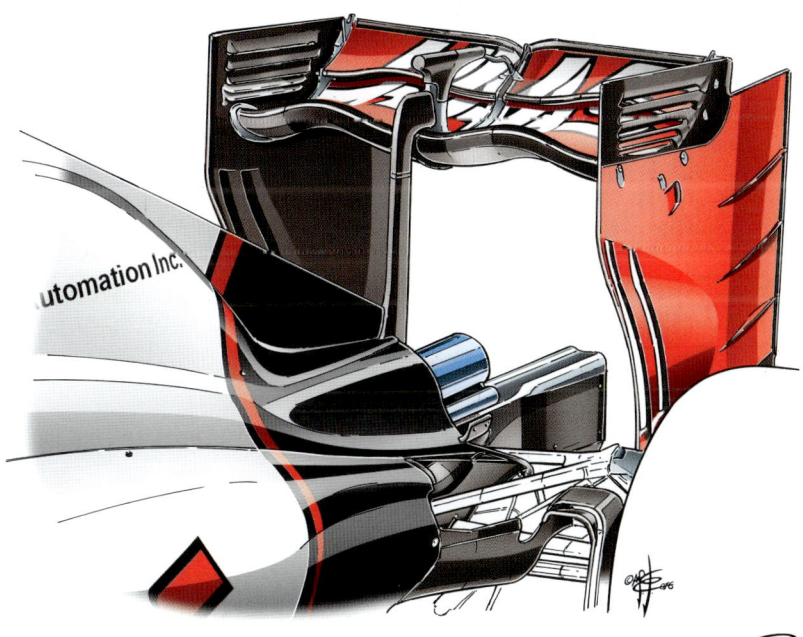

MERCEDES

CONSTRUCTORS' CLASSIFICATION			
	2015	2016	
Position	1°	1°	=
Points	703	765	+62 ▲

The Mercedes W07 totally dominated the 2016 season and brought an incredible haul: 19 wins from 21 races (10 for Hamilton and 9 for Rosberg). Red Bull did have the merit of taking advantage of the German outfit's only two false steps: the technical problem with Hamilton's engine when he was firmly in the lead of the Malaysian GP and the remarkable collision between the two Mercedes drivers on the first lap of the Spanish GP. In the first case vic-

tory went to Ricciardo, in the second to Verstappen.
The W07 proved from its presentation to be a perfect car and above all extreme, with an even more radical evolution with respect to what had been done during the winter break between 2014 and 2015.
Ferrari's three early wins under the new Marchione-Arrivabene management had convinced the powers-that-be at Mercedes to force the development pace with the objective of increasing their

advantage over their rivals. From the presentation extreme and innovative features were noted, including the most important, the splitting of the front part of the chassis, only discovered after the first few races and described in detail in the New Features 2016 chapter. Immediately evident were the conspicuous turning vanes in front of the sidepods, with no less than six separate elements and above all a kind of horizontal saw-tooth feature in the lower section (highlighted in

the oval), destined to create mini-vortices and groom the air flow towards to the underbody and the diffuser. This feature required lengthy study of the correlation between the vortices and the flows generated by the front wing and the turning vanes under the nose and the chassis. Designed by the team as the "W floor", it was defined by Paddy Lowe as the most important novelty on the German car and he expressed his surprise that no other team had tried to copy it.

Mercedes W06
Brazil

Mercedes W07
Launch

Mercedes W07
Abu Dhabi

The W07 proved to be perfect straight out of the box, with an aerodynamic configuration that had no need of the positive rake set-up introduced by Red Bull and then adopted by almost all the other cars in 2016. The merit for this was also due to the suspension which was capable of managing the car's stance as in the FRIC era, when the interconnection of the springing of the two axles was permitted. The engineers excelled in the development of the new PU106, using just 16 of the 32 tokens provided for by the regulations for development between one season and the next. The group led by Andy Cowell started out from the version raced at Monza in 2015, characterised by a larger turbine, introducing two further modifications during the course of the season, a modified MGU-H guaranteeing improved reliability despite wider use of the maximum power mode in Canada and a revised combustion chamber and turbine at Spa, achieving the incredible output of 1000 hp. A further version instead failed to pass homologation on the test bench and was never used on the track, in part because there was no need to take risks given the clear supremacy seen throughout the championship.

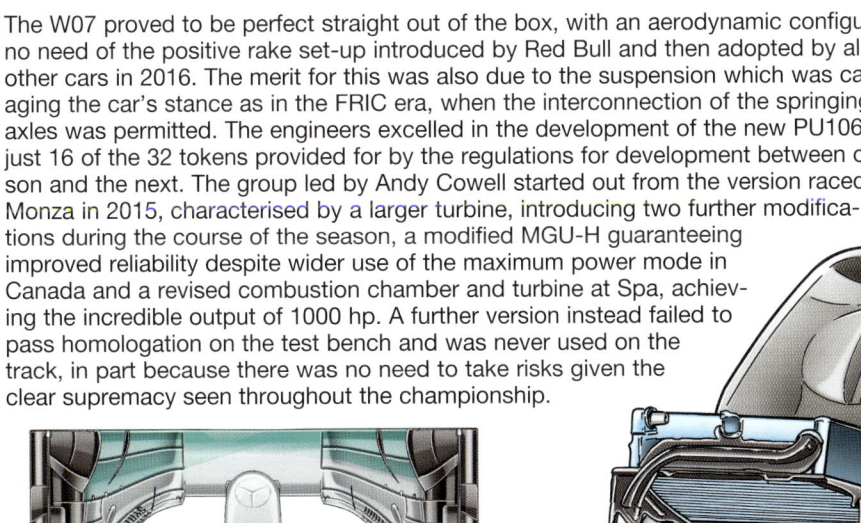

MERCEDES PU 106

The 2016 version of the 6-cylinder Mercedes Power Unit used just 19 of the 32 tokens provided for during the winter break and saw two major updates at the Canadian GP and at Spa where it reached the 1000 hp threshold. Note the exhausts leading to the larger turbo than the one used in 2015.

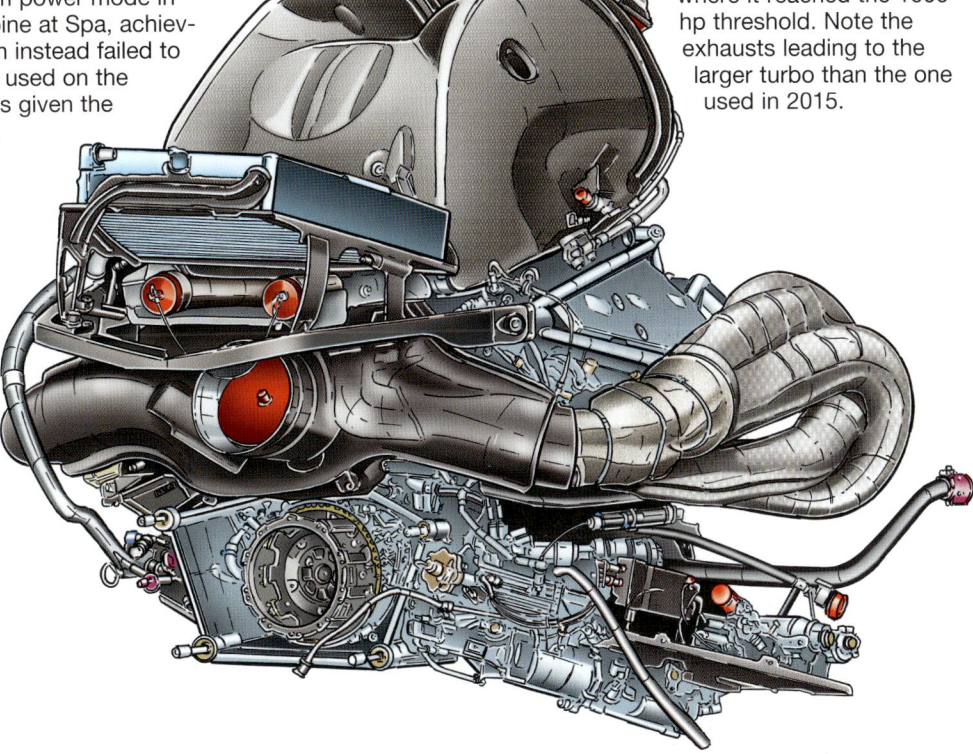

TURNING VANES

Paddy Lowe defined this kind of finger at the base of the bargeboard as the most important aerodynamic novelty of the year. They served to create mini-vortices that "interacted" with those generated by the front wing in order to optimize the flow towards the lower/rear part of the car. By the second week of pre-season testing, Mercedes had already introduced new vertical turning vanes in front of the sidepods, divided into two parts (insert). The lower one was connected to the floor so as to channel the air around the sidepods and presented a kind of ear that had been introduced by Sauber in the 2013 season. The upper part had a vertical section and a part curving towards the inside to better control the flows.

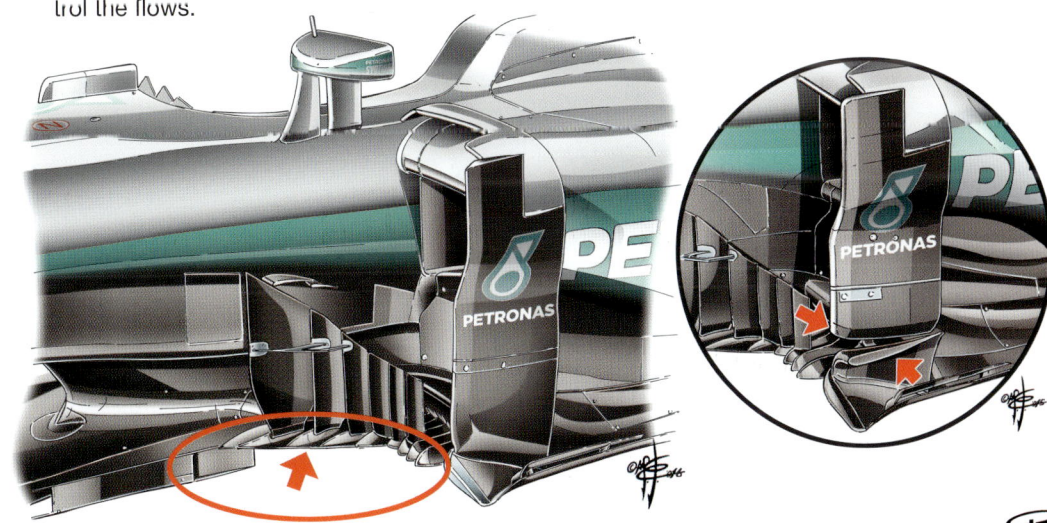

MERCEDES AIRBOX

With the W07, Mercedes introduced a very large oval engine air intake divided into three sections (monitored with sensors on its debut), each devoted to a component of the Power Unit. This feature followed the introduction at the 2015 Mexican GP of two "ears" compensating for the problems associated with the altitude of the circuit. This feature was then retained on the car for the following races.

It should be noted that Mercedes had already introduced novelties in this area in previous seasons, including a knife-edge roll-bar equipped with two conspicuous ears to reduce the section as much as possible and to improve the flow towards the rear wing. This configuration was then copied in 2011 by Lotus and Force India before being banned by the FIA.

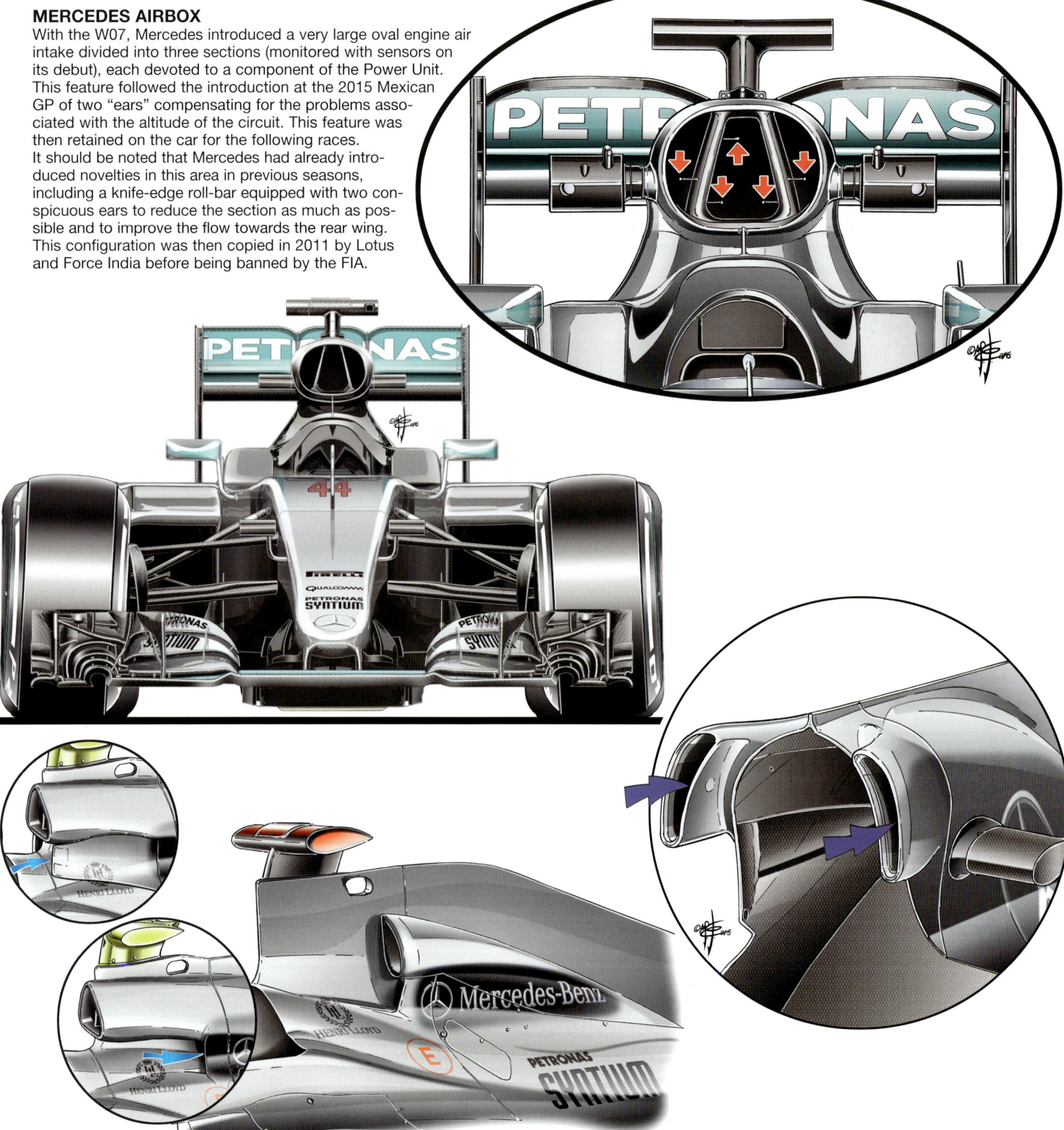

GIORGIO PIOLA

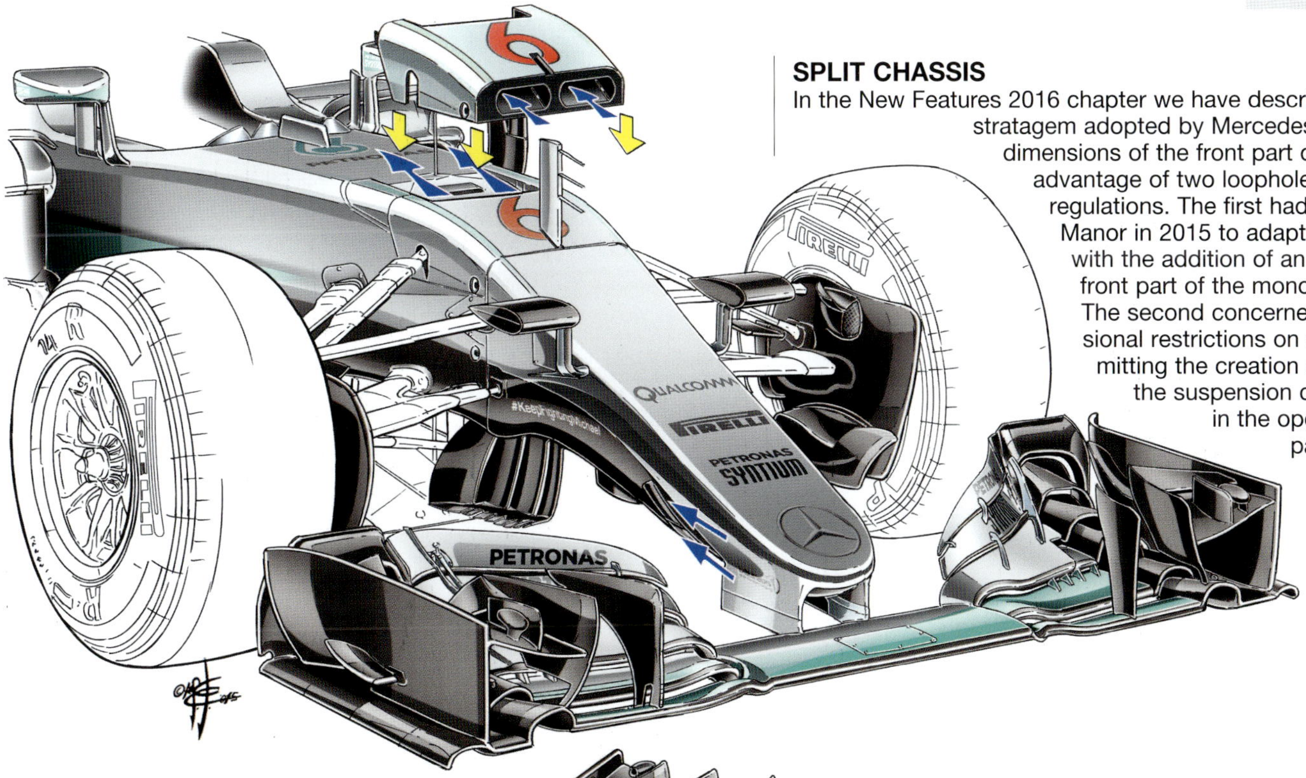

SPLIT CHASSIS

In the New Features 2016 chapter we have described in detail the stratagem adopted by Mercedes to reduced the dimensions of the front part of the chassis, taking advantage of two loopholes in the technical regulations. The first had been exploited by Manor in 2015 to adapt the 2014 chassis with the addition of an extension to the front part of the monocoque.

The second concerned the lack of dimensional restrictions on any apertures, permitting the creation of an area in which the suspension could be worked on in the open, using a vanity panel to satisfy the regulations. This drawing highlights the split.

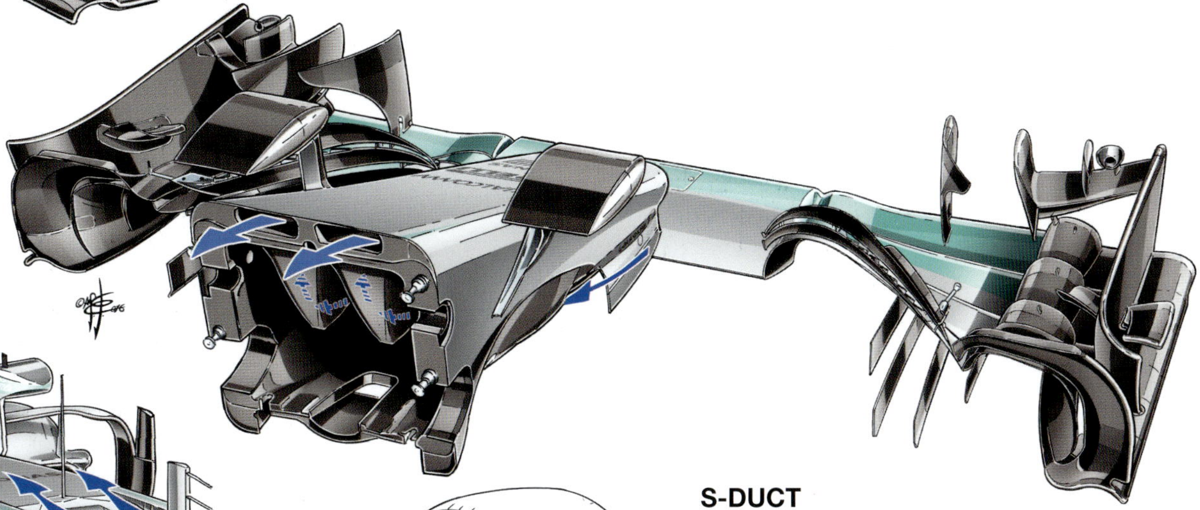

S-DUCT

These two drawings highlight the configuration used by Mercedes for the S-Duct, nicknamed Bruce after Nemo's shark and the conspicuous slot that served to feed the S-Duct system thanks to which it was possible to draw on the flow of air under the nose and transfer it to the upper part of the chassis with a dual opening as can be seen in the rear view of the nose.

FRONT SUSPENSION

Despite FRIC having been banned during the course of the 2014 season, Mercedes continued to use sophisticated hydraulic components to control the W07 dynamics, not only for improved driveability but also to ensure a more constant aerodynamic configuration; Mercedes was however the only one of the top teams not to exploit the positive rake set-up. The innovation? Along with the third hydraulic damper, the anti-roll bar was now controlled hydraulically.

SOCHI-BARCELONA

This new end-plate equipped with two vents for better control of the flow around the front wheels was taken to scrutineering at Sochi but not used on track. It was part of the aerodynamic package devised for the Spanish GP. The feature was easily identified as it presented a dual vent at the back of the end-plate, exaggerating the concept thanks to which it was possible to deviate the flow around the front tyres.

BARCELONA MONKEY SEAT

Mercedes decided to present at Barcelona a number of the technical features that had been kept in reserve for the Spanish GP, even though they had already been ready in China.
The first concerned the monkey seat, which in its latest form saw the addition of another small flap in the upper part in order to increase downforce at the rear, together with a high downforce wing. It can also be seen that the reflective heat insulation material that had been installed on the upper surface of the gearbox protection structure was applied to the inside face of the monkey seat too.

MONACO

For the Monaco GP the flow of cooling air had to be increased given the low average speeds attained on the street circuit and Mercedes introduced the 6th evolution of its front brake air intake.

The Brackley engineers devoted considerable attention to this detail after Nico Rosberg risked a retirement in the Australian GP when detritus had entered the unprotected intake causing the calliper to overheat.

For Monaco, the intake was larger and moved further forwards, beneath the flap that stood over the eight-segment carbonfibre grid that helped laminate the flow in an area of the car that was very sensitive. Both versions of the end-plates were subsequently used.

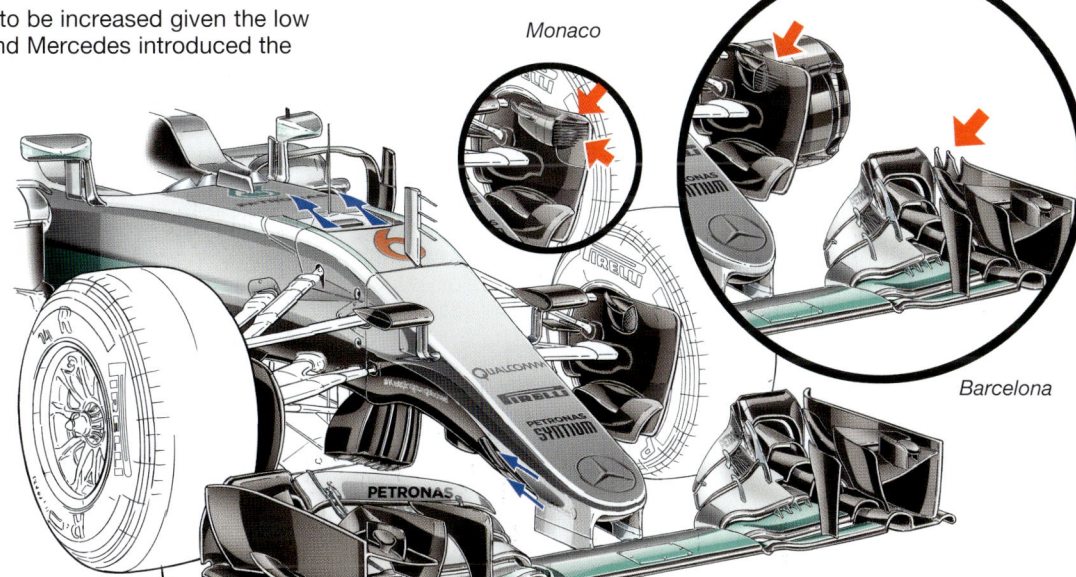

Monaco

Barcelona

BARCELONA

A new front wing in Barcelona and new brake shrouds, while the turning vanes under the chassis were also different.

BAKU

Mercedes reintroduced in Baku the low downforce rear wing that Nico Rosberg had tested for a couple of installation laps in free practice in Canada.

The dished profiles had a fairly short chord that allowed a good compromise between efficiency and aerodynamic loading to be found. In 2016, this configuration was favoured by a revision of the regulations that permitted greater freedom in the central section: the designers could go below the minimum height of 600 mm from the reference plane for an area of 200 mm from the centre of the car, while in 2015 this was limited to just 75 mm on either side of the mid-line.

It should be remembered that the dished design was nothing new for the Brackley team, which had already used the configuration at Spa and Monza. An unusual oval Monkey Seat also appeared at Baku.

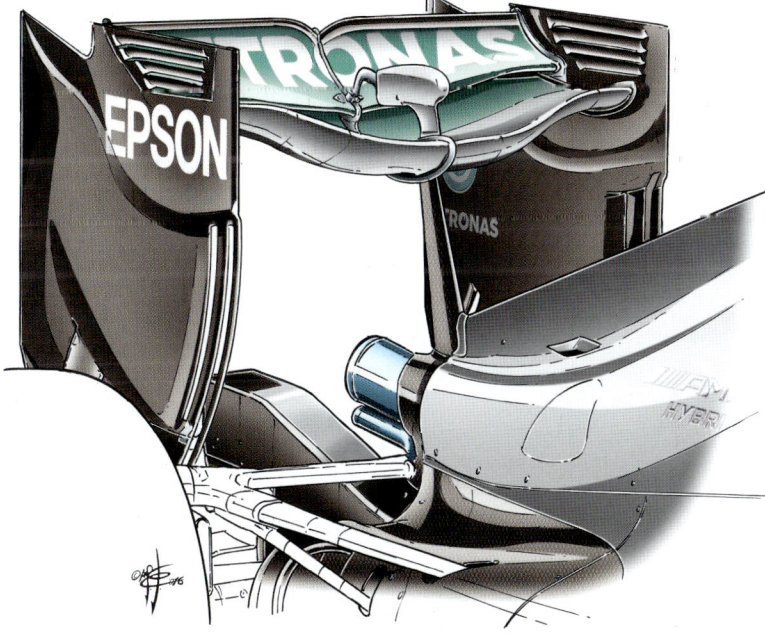

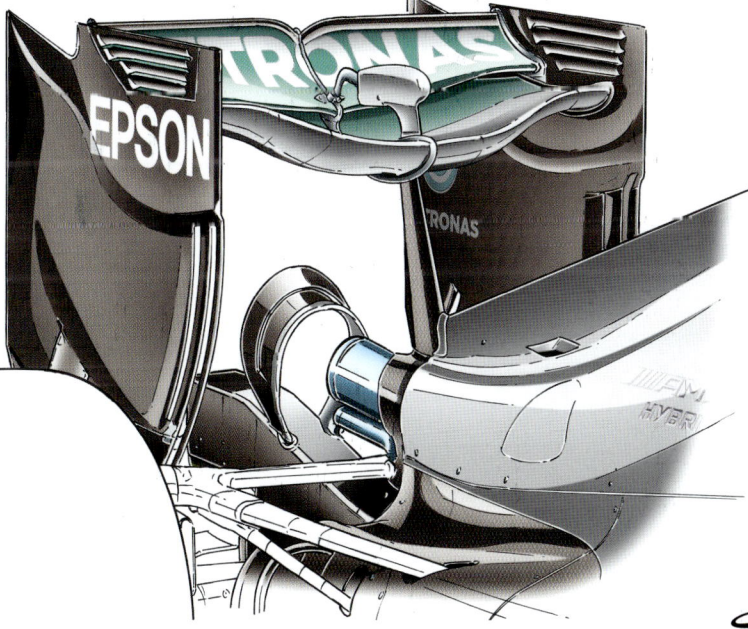

SILVERSTONE

The W07 complicated bargeboard was part of the car that was subjected to constant development by the aerodynamicists at Brackley. The comparison shows the winter test version and the one fielded in the Spanish GP.

For Silverstone, the six vertical vanes detached from the differentiated vents became five, while the blades, which previously numbered nine became eight, significantly modifying the entity of the flows. According to indiscretions gathered from the firm's engineers, at Brackley they had tried to generate a kind of "pneumatic side skirt" ahead of the slots either side of the floor, in front of the rear wheels.

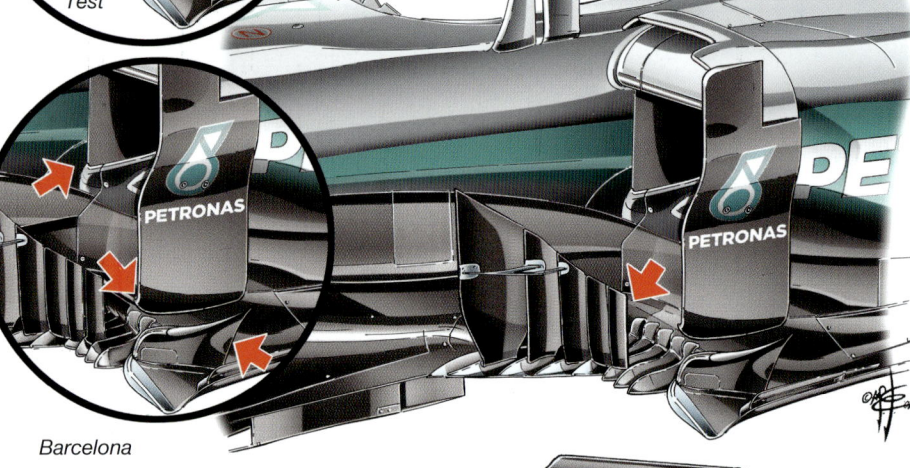

Silverstone

Test

Barcelona

REAR END-PLATE

Toro Rosso was the team that presented the most features that were subsequently reprised and copied by the others, including the top teams. The new rear wing introduced by Mercedes at Silverstone was in fact characterised by end-plates with the horizontal vents of the STR11 open at the leading edge.

The W07 presented no less than five horizontal slots: the smallest was part of the end-plate, while the others were linked to one another by small brackets to avoid the unwanted flexing that James Key's engineers had noted after the first track tests.

Toro Rosso

W07 Silverstone

SAWTOOTH PROFILE

Introduced in the second pre-season test, this new rear wing with the trailing edge of the main plane having a sawtooth configuration was never used in a race. It was based in tests conducted during the last race at Abu Dhabi in 2015, when sawtooth adhesive tape had been applied to the main plane.

A similar feature had been introduced at Sochi, applied to the front wing. The sawtooth trailing edge on the main plane allowed the flow to be reattached when the DRS closed, improving stability under braking.

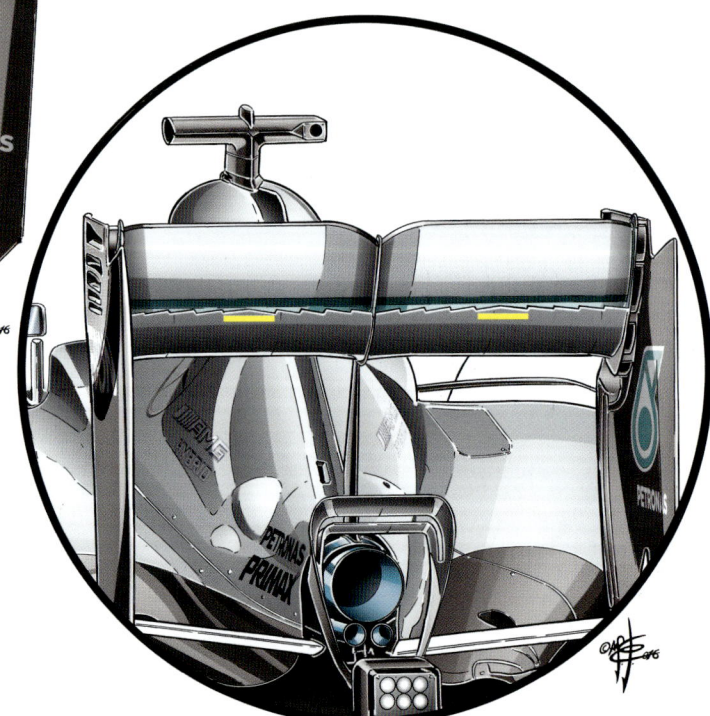

BUDAPEST

A high downforce rear wing for Budapest with the endplates not only featuring the Toro Rosso-style horizontal slots, but also the double vertical vents on the leading edge (which were more Red Bull), along with the four triangular turning vanes mounted in the second part of the flap.

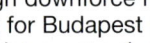

BUDAPEST

The hot air vents let into the bodywork either side of the cockpit were even more evident: rather than traditional louvres (with seven apertures) they were now true heat extractors, raised from the level of the sidepods and capable of generating a kind of chimney effect and carrying no less than eight small arched flaps that were useful in keeping the flow of hot air as low as possible, thus preventing it from interfering excessively with the streams destined for the rear wing.

SPA

A dished rear wing as used at Baku, this time combined with a Monkey Seat with a more square-cut shape and a vent.

FRONT BRAKES

In the last three races of the season, from the Italian GP onwards, Hamilton experimented in FP1 only with a new front brake configuration. The team's technical director himself admitted at the end of the season that this was a project from the R&D department in view of the 2017 season. In practice a form of casing or spacer in carbonfibre was added to the sides of the Carbon Industrie discs, the assembly stiffened with a pair of metal supports.

RED BULL

Red Bull RB11
Abu Dhabi

In the season in which Mercedes literally dominated the scene and Ferrari was left without a win, Red Bull was the only team to dent the German hegemony in two races, with Verstappen in Spain and Ricciardo in Malaysia, thanks to the collision between Hamilton and Rosberg in Barcelona and the engine failure suffered by Hamilton at Sepang.

From its debut, the RB12 appeared to a logical evolution of the previous year's RB11 of which it retained the design principals, starting with the nose and the overall aerodynamic configuration. In 2016, the RB12 was considered to be one of the cars able to generate the most downforce thanks to the positive rake config-

uration and the body of the car, thus being able to use wings with less resistance to penetration. There was fairly intensive development of the front end with significant modifications to the chassis, suspension and aerodynamics. On the RB12 Red Bull eliminated the S-Duct it had instead used on the three previous cars.

This choice was made as a result of the shortening of the nose that retained the shape of the RB11. The development of the perforated hub at the front continued while the suspension was radically modified, leading to a revision of the front section of the chassis; this permitted a more vertical inclination of the push-rod strut to be adopted thanks to a chas-

sis mount that was higher compared to that of the RB11. The steering rack was set lower so as to reduce the centre of gravity while the wishbones became true aerodynamic devices.

The lower wishbone adopted the configuration introduced by Mercedes in 2014, characterised by a narrow base with an element faired with a wide profile; an even more extreme design than that of Mercedes, Ferrari and Force India which had already copied the configuration in 2015. This modification inevitably entailed considerable work to enhance the torsional rigidity of the chassis.

The possibility of obtaining the same effects as the FRICS system (the connection between the front and rear suspension, banned in the summer of 2014)

was crucial in that it guaranteed an optimum stance and allowed the positive rake configuration to be exploited to the full. The sophisticated third damper on the RB12 was combined with a Belville cupped spring washer, in contrast with the Mercedes' fully hydraulic damper.

With regards to the engine, Renault managed to reduce the gap on its two rivals Mercedes and Ferrari in terms of pure power, with the possibility of reducing the radiator packs to the benefit of aerodynamic efficiency with the sidepods sections further reduced and less need for a large vent in the end section. The RB12 consistently ran with lower downforce on the wings that its rivals, a sign that much was obtained from the car's underbody aerodynamics. Also of note was the new shape of the tapering sidepods, with less volume in the upper and no cutaway in the lower part.

The management of the flow towards the rear end was also totally different that that of all the other cars.

Red Bull RB12
Launch

Red Bull RB12
Melbourne

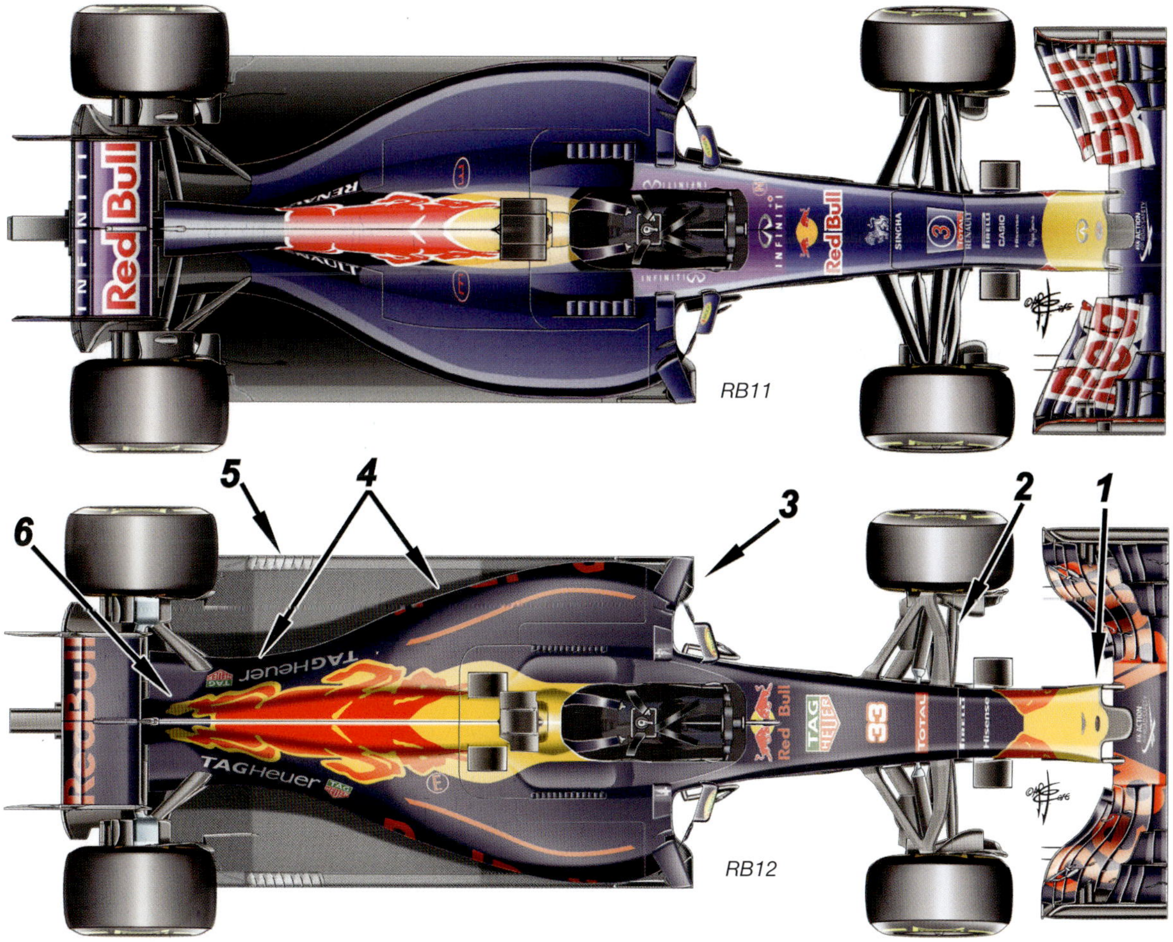

RB11

RB12

OVERHEAD COMPARISON

(1) The nose of the RB12 retained the same shape although it was shorter than that of the RB11. (2) A great deall of work went into the suspension, with the lower wishbone having a tuning fork shape with a narrow base and the aerodynamics of the various element being exploited more. (3) The shape of the front part of the sidepods and the relative aerodynamic appendages was almost unchanged. (4) The shape and section of the sidepods was very different from the mid-point of the car back. (5) In the 2016 season, all the teams worked hard in this area, with significant cuts to reduce the toxic effects of turbulence from the wheels. (6) The greater aerodynamic efficiency is also shown by the reduction of the large central vent present on the RB11.

NOSES COMPARISON

The RB12 adopted the same shaped nose as the RB11, definitively abandoning the new vent that had characterised the RB10 in 2014. Note the absence of the S-Duct, sacrificed in favour of a slightly shorter nose that permitted improved flow management.

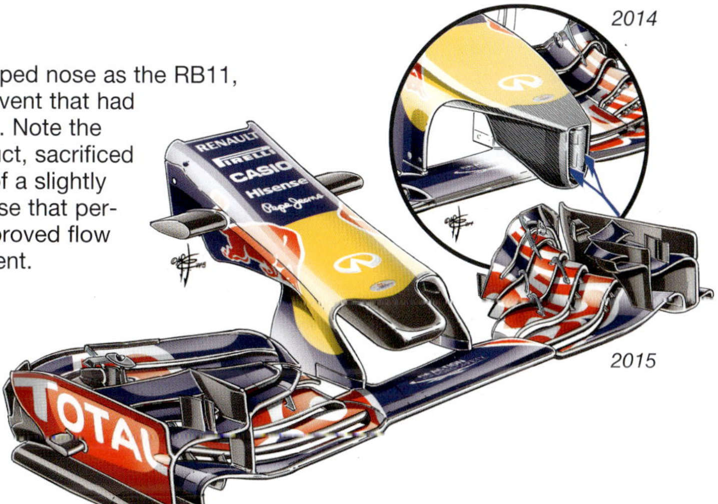

2014

2015

Red Bull RB12
Abu Dhabi

S-DUCT

After three seasons, Red Bull abandoned the S-Duct, the ramp of which passing from the lower to the upper section had been the object of particular attention on the part of the technicians (see the RB11/RB10 comparison in the small insert). The setting of the steering rack lower down on the level of the large lower wishbone was important. The flow vanes on the RB12 were very similar in shape to the last version of those on the RB11.

CHASSIS
FRONT SUSPENSION

Three teams (Mercedes, Ferrari and Red Bull) interpreted the regulations modifying the shape of the front section of the chassis, getting round the restrictions introduced by the FIA in 2014 (a strategy described in the New Features 2016 chapter). In the drawing, the Red Bull design, with a larger vanity panel with respect to the Ferrari, but smaller than the Mercedes'. The structure of the chassis corresponded perfectly to the spirit of the regulations. The third transverse element was exposed and permitted easy adjustment, but the shape of the aperture in the chassis respected the B-B and A-A section dimensions in full.

LOWER
WISHBONE

For the front suspension, Red Bull adopted the configuration introduced by Mercedes in 2014 and also used by Ferrari and Force India in 2015, that is the lower wishbone with very narrow base and a tuning fork shape, exploiting the 3.5:1 ratio between the thickness of the arms and chord. It did so in an extreme fashion, with a very wide wing profile so as to improve the quality of the flow in the critical zone ahead of the sidepods and the floor, an area critical to the overall efficiency of the car.

DUAL GEARBOX CASING

The RB12 retained the second carbonfibre shell that contained the true gearbox casing in titanium. The suspension mounts were modified to take into account the major changes made to the front suspension. The brake callipers, underslung horizontally, were again four-pot units rather than the 6-pot versions used at the front by all the teams.

POWER UNIT INSTALLATION

The layout of the cooling system for the various components of the Power Unit was unchanged, with the turbo heat exchanger (2) set below the radiator (1). The ducts cooling the hydraulics radiator (3) and the large ERS heat exchanger (5) were split either side of the principal duct (4).

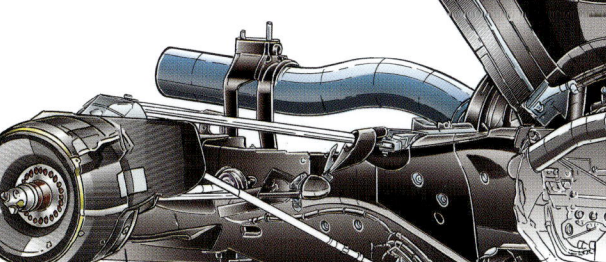

BLOWN HUB

The blown hub was retained in contrast to the S-Duct which was abandoned on the RB12. In the drawing, the legal version, with the hub blown via a static tube in the central hollow part. The basic shape of the large shrouds combined with the brake air intakes.

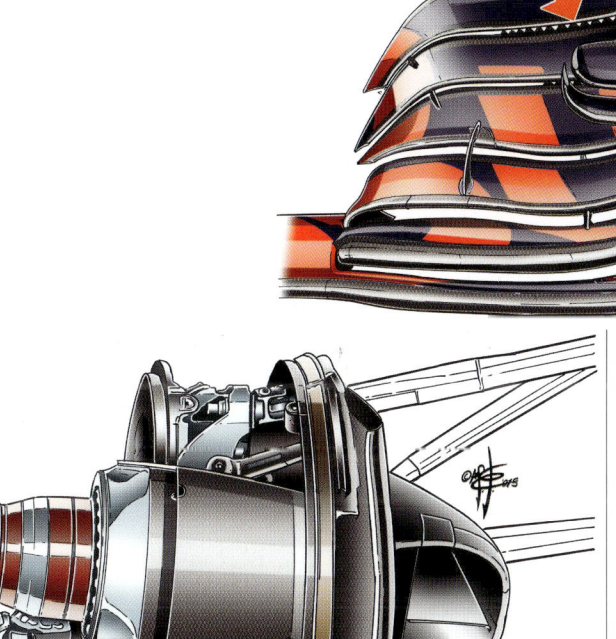

SHANGHAI

At the Chinese GP, Red Bull also introduced an aerodynamic modification that proved to be a failure: the saw-tooth trailing edge (highlighted with the arrow) of the penultimate flap, tested on the Friday morning but abandoned on the RB12.

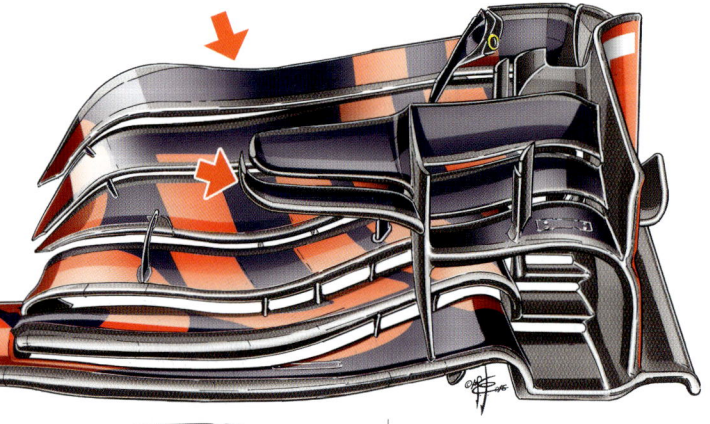

SPA

At Spa, Red Bull could count on two front wings: the one introduced in Germany with the simplified upper flaps, in particular with the last profile featuring a reduced chord, and with a more sinuous shape indicated by the arrow. The section of the upper flaps was also different, identical in the new version to those introduced in China with the rejected front wing.

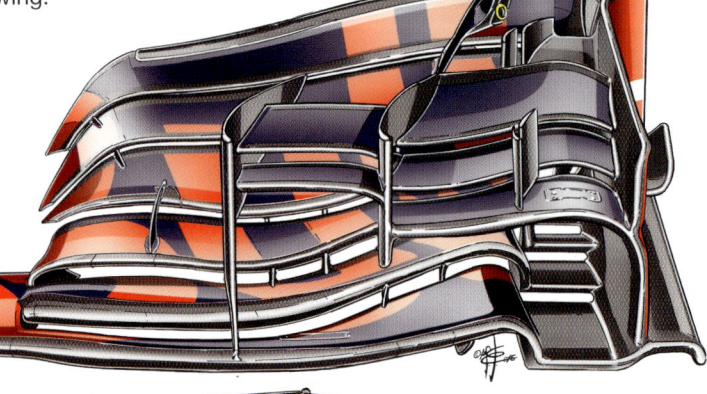

SOCHI

Red Bull brought a rear wing characterised by an end plate equipped with a long new vertical vent that started from a horizontal slot and descended to the height of the rear wheel. The were just three horizontal slots, instead of the four seen in Australia. Objective: lightening the rear end slightly to compensate for the lack of power from the Tag Heuer (Renault) engine, yet to reach the level of the Mercedes and Ferrari units.

Sochi

BUDAPEST

At Budapest, Red Bull began practice with two different rear wings: Ricciardo with the Monkey Seat and the less complex end plates, Verstappen without the Monkey Seat but with end plates equipped with fins flicked upwards at the tips. A combination of the two was adopted for the race.

Ricciardo Friday

Verstappen Friday

SPA

Red Bull was the team that tackled Spa with the least downforce, testing a low downforce wing with smooth end plates, before opting for a higher loading version for the race with end plates that retained both the horizontal slots at the top and the long vertical one, but without the oblique mini-fins on the trailing edge.

2014

MONZA

The rear wing used by Red Bull at Monza was a revised and modified version of the one adopted in 2015, both in terms of the profiles and the end plates without slots. The support pylon and the DRS control were different to those used in Belgium.

BAT WING

Another feature of the RB12 was the mounting of the Bat Wing on the T-Tray instead of the lower part of the chassis. This was associated with the difficulty in controlling the inclination of the floor in order to exploit the effects of the positive rake to the full.

FERRARI

CONSTRUCTORS' CLASSIFICATION			
	2015	*2016*	
Position	2°	3°	-1 ▼
Points	428	398	-30 ▼

Following the success of the 2015 season with the new Ferrari Marchionne-Arrivabene management structure (three wins and 2nd place in the Constructors' championship), 2016 was hugely disappointing, a season without a single victory (as had been the case in 2014) and the loss of a place behind Mercedes and Red Bull in the championship standings.
In the light of the success then enjoyed in the 2017 season, the SF16-H can be considered as a point of departure that served to lay the foundations for what in 2017 proved to be the most innovative car in the field.
Even the SF16-H was in its own way a revolution in technical terms with respect to 2015's SF15-T. There were numerous new features: the return of push-rod front suspension, the introduction of variable intake ducts with the consequent shifting of the intercooler from the V of the engine, the new location of the three heat exchangers and the moving of the MGU-K from the gearbox casing to the left-hand side, as on the Mercedes and Renault units, and the adoption of a new rear suspension layout. Underlying the SF16-H project was the objective of drastically reducing the volumes at the rear end in order to permit much more efficient aerodynamics, following the trail blazed by Ferrari's rivals, not to mention the abandonment of the long nose of the 2015 car that was practically unique in that year's field.
Great progress was also made with the power unit thanks to the deployment of all 32 tokens permitted by the regulations and the increase in 30 cc that took the maximum power output of the engine to 970 hp, around 16 hp down on the Mercedes unit that dominated the season.
The extreme sophistication of the rear end did, however, have negative effects on track.
The gearbox's lack of reliability penalised the SF1-6H on no less than four occasions, with a penalty of five grid positions being imposed. The clutch management system for the start was interesting with the long rocker offset with respect to its movement being retained for 2017 too (see the Cockpits chapter).

Ferrari SF15-T
Austin

Ferrari SF16-H
Launch

Ferrari SF16-H
End 2016

FRONT SUSPENSION

After four years of the pull-rod front suspension configuration, with the SF16-H Ferrari returned to the push-rod layout used on all the other cars. The narrow base, tuning fork-shaped lower wishbone was retained, while all the other suspension elements were radically modified. Note also the protections either side of the driver's head, 20 mm higher, as per the regulations.

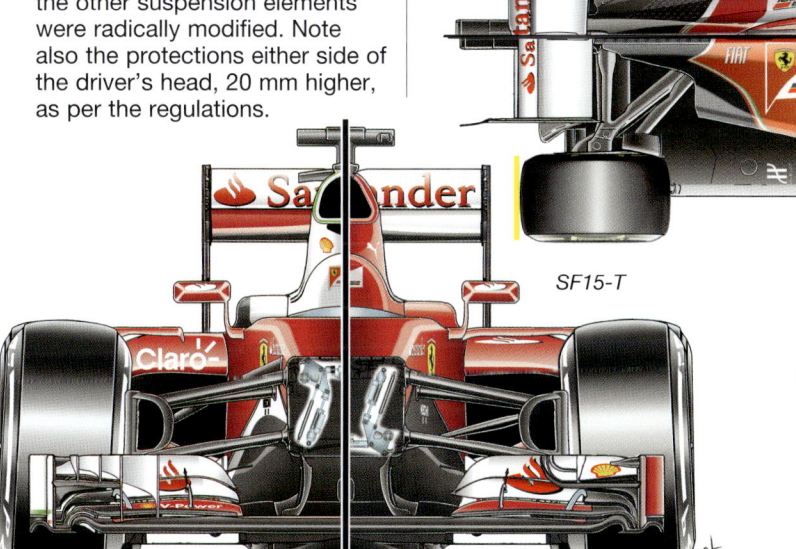

SF15-T

SF16-H

OVERHEAD COMPARISON
REAR SUSPENSION-GEARBOX

Underlying the SF16-H project was an attempt to drastically reduce the volumes at the rear end to achieve significantly more efficient aerodynamics, as can be seen in the overhead comparison.
In order to achieve a very narrow Coke bottle area, the gearbox casing was completely revised, the engine and its accessories and the the suspension elements were compacted towards the centre of the car and the protection structure was redesigned.

SF15-T

SF16-H

SIDE VIEW COMPARISON

The SF16-H nonetheless represented a starting point for Ferrari with the return to the push-rod front suspension layout and the introduction of variable length intake ducts to improve combustion, a modification that entailed the moving of the intercooler from the engine V (see insert). The greater cooling demands obliged the designers to install no less than three heat exchangers elsewhere (in green). The MGU-K was moved from the gearbox casing (in blue) to the left-hand side, as with the Mercedes, Renault and Red Bull engines, with McLaren and Force India later following suit. Modifications continued to be introduced at every pre-season test session.

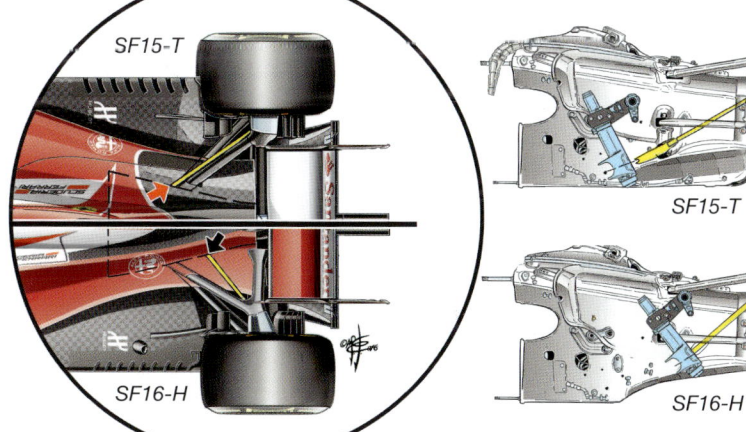

SF15-T

SF16-H

SF15-T

SF16-H

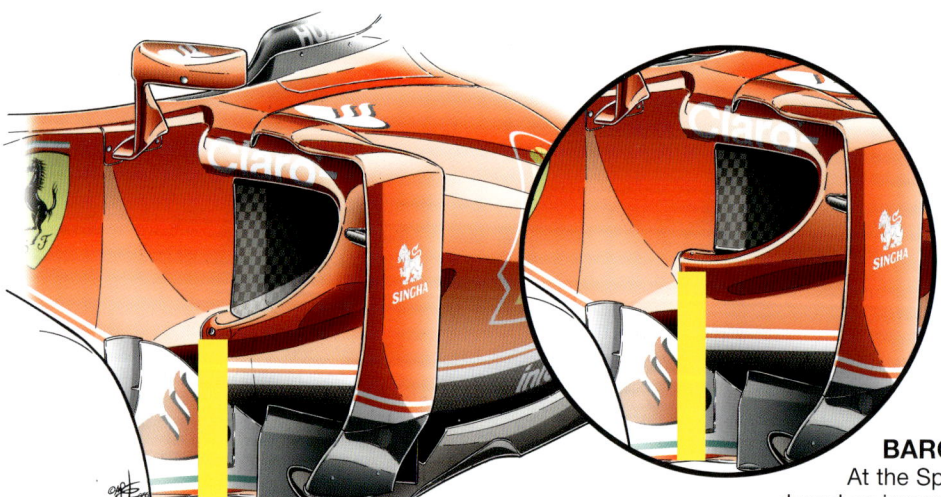

BARCELONA

At the Spanish GP Ferrari introduced an important modification at the front of the sidepods, with the intake mouth reduced in section and raised with respect to the ground. The smaller intake mouth also increased the quantity of air in the lower cut-away area to feed the flow towards the rear of the car.

BRAKE INTAKES BARCELONA/MONACO

In 2016, the Maranello firm took specific brake intakes to each circuit, with asymmetric apertures. Compared here are those used at Barcelona and Monaco. In Spain, the right-hand duct had the window at the top, the left-hand one just the oval apertures. In Spain the discs with more than 1,200 holes were not used but rather a version with around 840 holes in groups of four, aligned horizontally. The Monaco version had the maximum number of holes aligned obliquely in groups of six. A kind of window was also introduced, treated with a special paint to retain heat and direct it to warm up the front tyres, making them immediately ready to provide maximum efficiency.

Barcelona

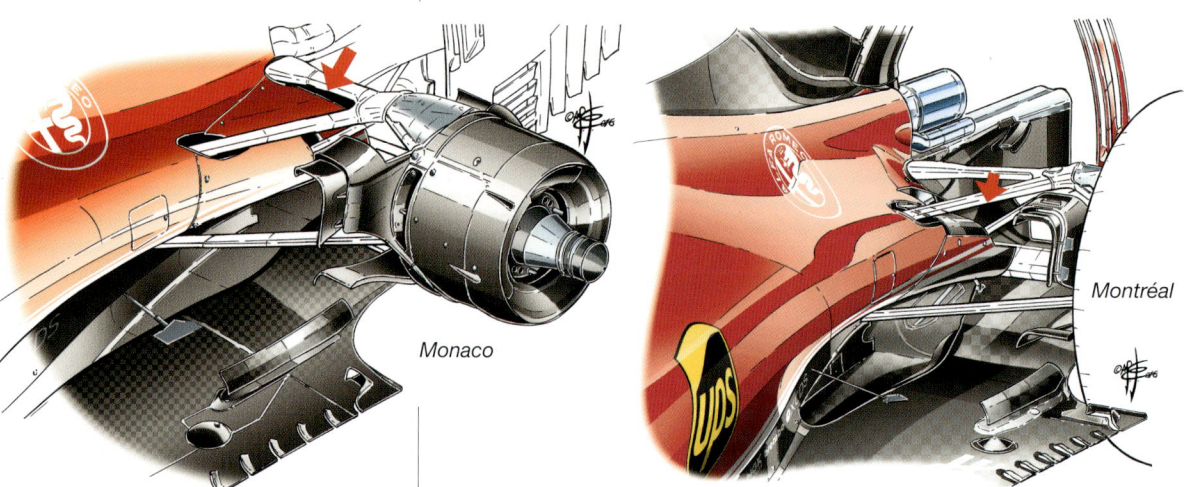

Monaco

Montréal

MONTRÉAL

Fast circuit bodywork for the end part of the SF16-H sidepods in Canada. In practice, the upper part was cut away, revealing the internal fairing of the hot air vent. A comparison between, left, the bodywork used up to Monaco with the upper part aligned with the upper suspension wishbone.

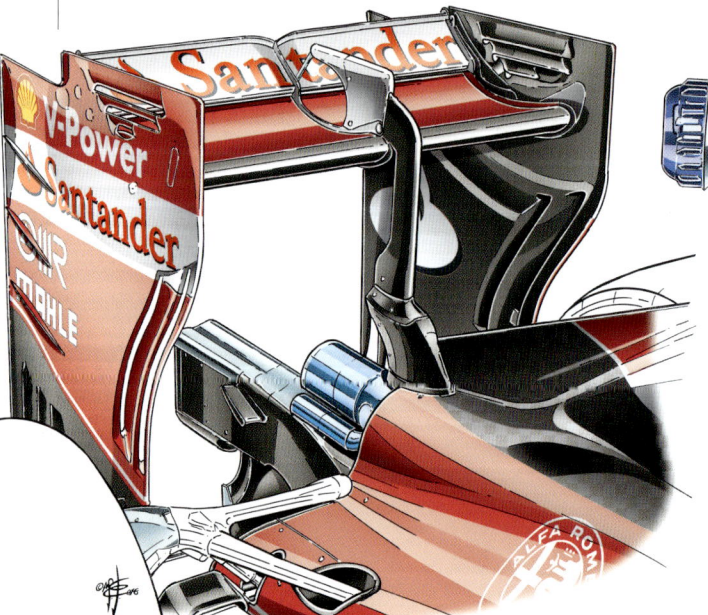

Monaco

BAKU

A new aerodynamic package for Baku complete with a new rear wing. The end-plates had only two slots in the upper part to pair with profiled with lower angles of incidence in order to privilege maximum speed. The earlier version was then chosen for the race.

Zeltweg

Silverstone

SILVERSTONE
New front wing end plates were introduced at Silverstone and were retained for the following races too. The venting towards the outside was increased with a cut and a flare similar to that used on the Mercedes end-plates.

BUDAPEST
Another zone subject to constant modification was the diffuser between the rear wheels. As well as the usual mini flaps (introduced at Austin 2015), a conspicuous triangular fin was added (indicated by the arrows) directing the air flow and increasing the downforce in this area.

SPA
Experiments ahead of Monza were also made to the rear end of the Ferrari that in Belgium had a low downforce wing (with just three slots and profiles with a lower angle of incidence), tested on the Friday before a high downforce version was fitted for qualifying and the race with endplates featuring four slots and higher downforce profiles. Note also the different DRS control.

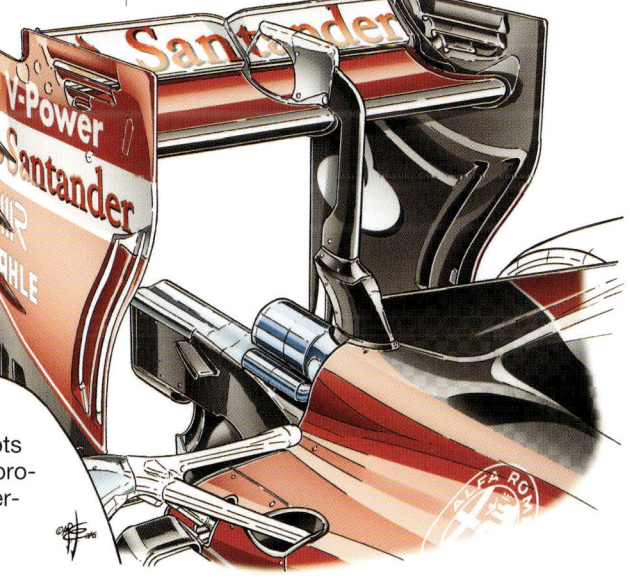

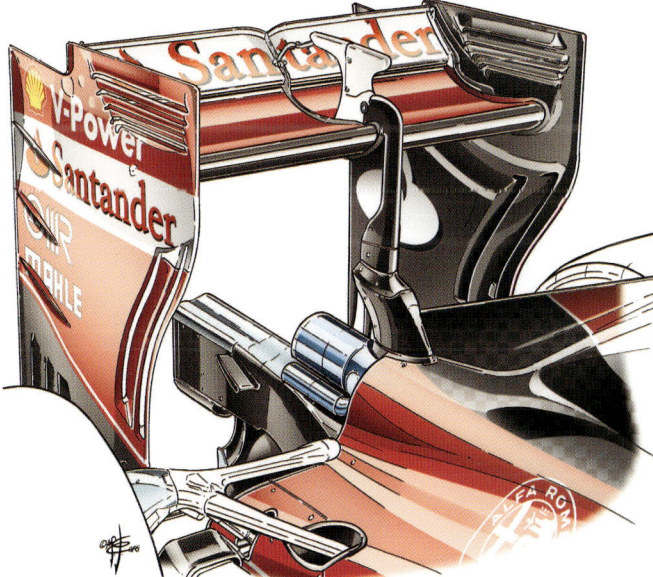

MONZA

At Monza Ferrari fielded a low downforce front wing with one fewer flap (red arrow) to gain in terms of penetration. The McLaren-style fin inside the endplates that had been used at Spa was instead eliminated.

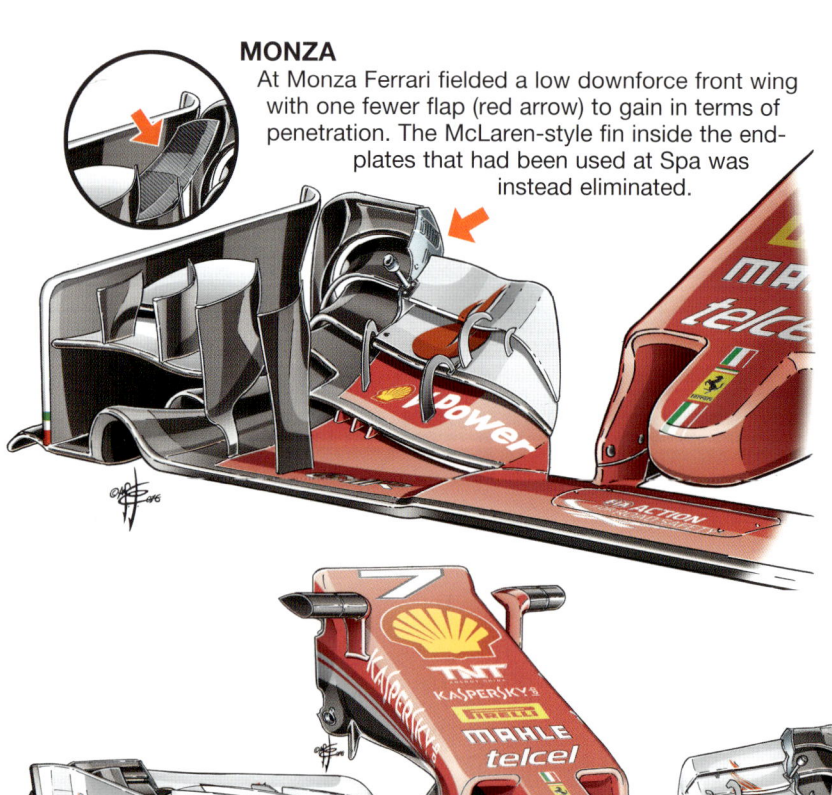

SINGAPORE

On the Friday morning, both Ferraris ran a new front wing characterised by a sawtooth trailing edge to the penultimate flap, as on the Mercedes from the 2015 Russian GP.

SUZUKA

There was a major aerodynamic package for Ferrari at Suzuka with a significantly modified nose and front wing assembly.
Two changes can be seen on the nose of the SF16-H: the rear part of the front wing support pylon was of a different design and was longer. Moreover, the front part was shaped to orient the air flows differently. You can also see an extra turning vane attached to the nose that is longer and more functional than the other two under the chassis of the rossa.

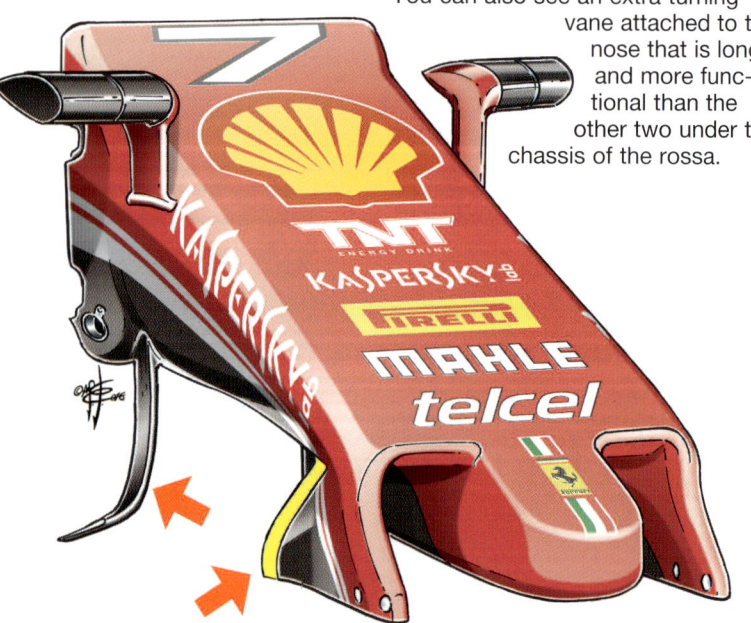

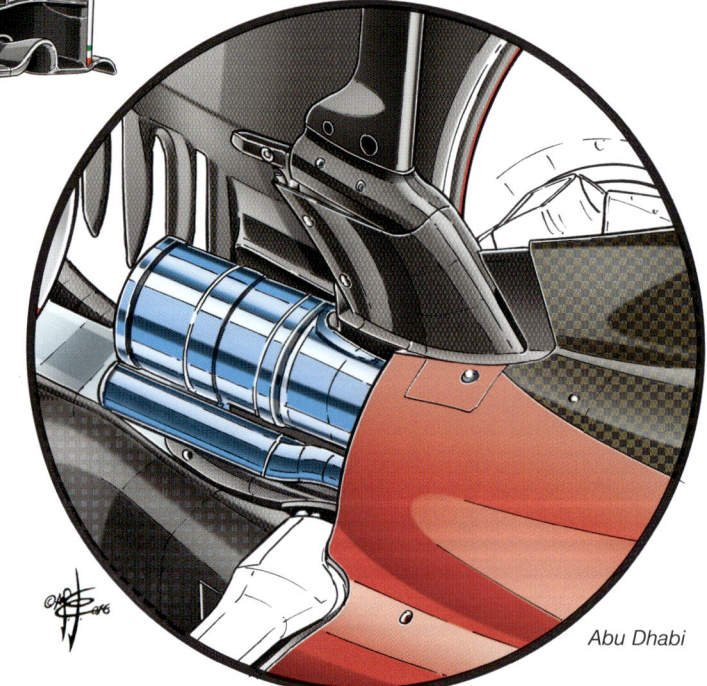

Abu Dhabi

EXHAUST

In the second half of the season, Ferrari introduced a novelty in the end section of the 6-cylinder 061/1 engine's exhaust. The Prancing Horse's engineers decided to adopt a component machined from billet. This was a very expensive path given that instead of working with a tube and welding, the metal is cut away from a solid block.
This feature did however provide two advantages: greater strength that in protecting against cracks guaranteed improved reliability along with a slightly lower weight as the pipe could be machined to lower thicknesses.

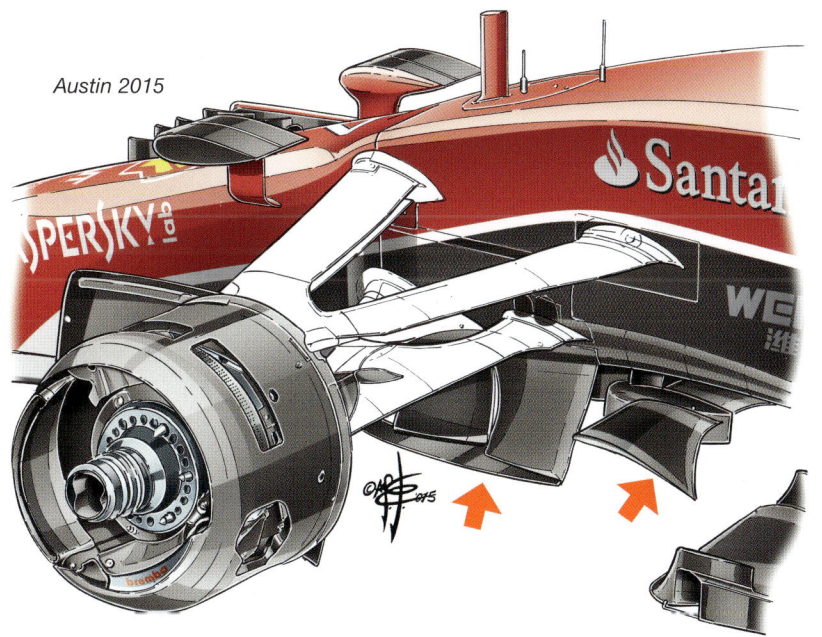

Austin 2015

TURNING VANES

At Austin at the end of the 2015 season Ferrari had already introduced a Red Bull-inspire "batman" horizontal profile, which was then abandoned at the British Grand Prix in favour of an inclined arrow-head turning vane designed to reattach those underbody air flows that were to be oriented towards the rear diffuser.

At Sepang the vertical vane that been seen during the course of the season at the edges of the T-Tray reappeared but in significantly elongated form so as to channel the air towards the turning vanes.

Lastly, in Mexico City a different version was tested of the bat wing that had detached at Austin, having been attached badly to the T-Tray.

The skid wing as the feature was nicknamed, can be recognised thanks to the dual profile that was designed to deviate the air flow downwards. It replaced those already used during the course of the season: the first profile under the chassis had been seen at Silverstone and subsequently had been replaced the bat wing that had appeared at Sepang. It can clearly be seen that the new wing design was very similar to the one McLaren had introduced at Hockenheim which itself was similar to the wing used by Red Bull from the start of the season.

Silverstone

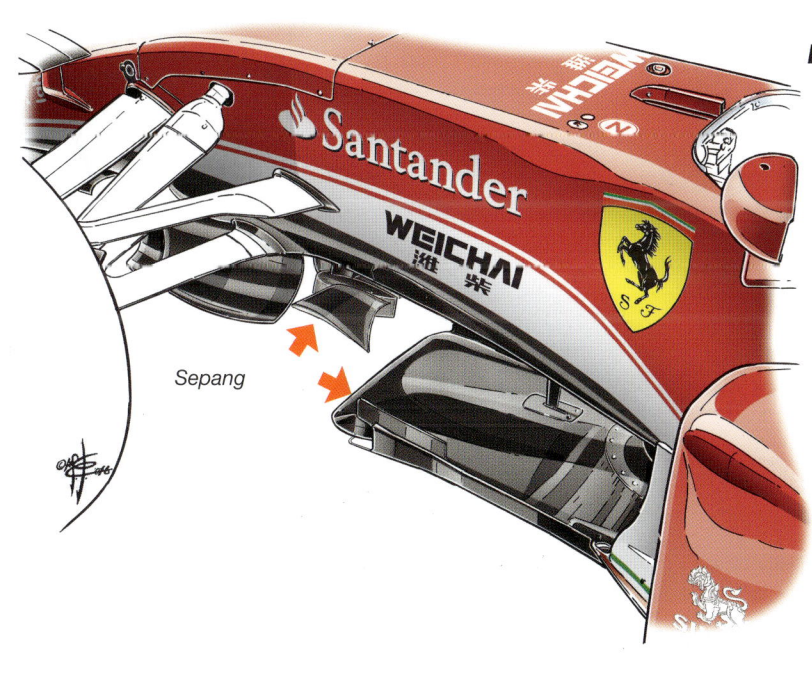

Sepang

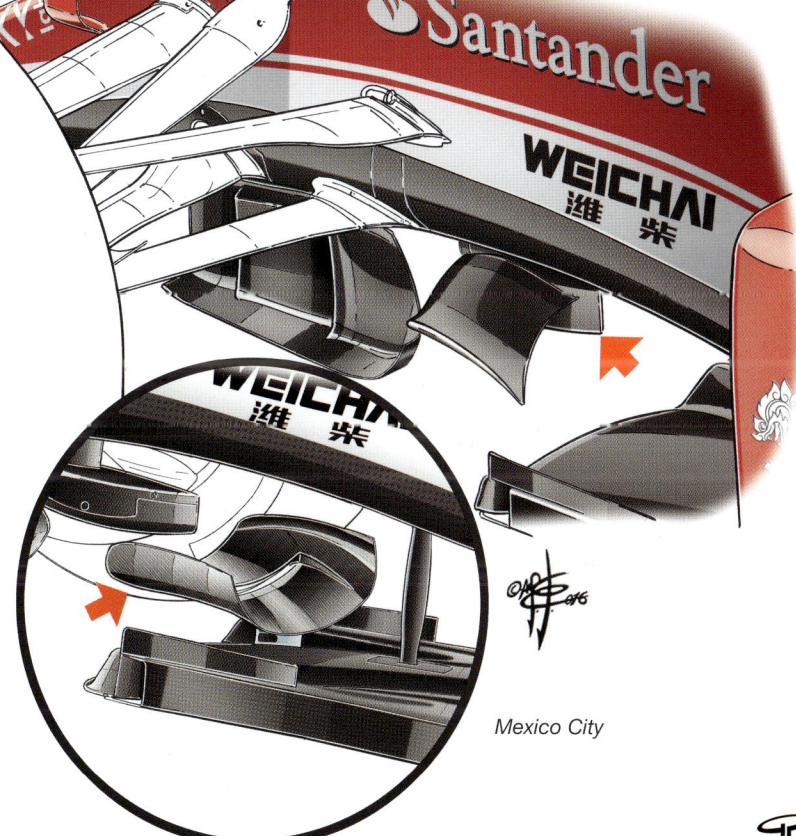

Mexico City

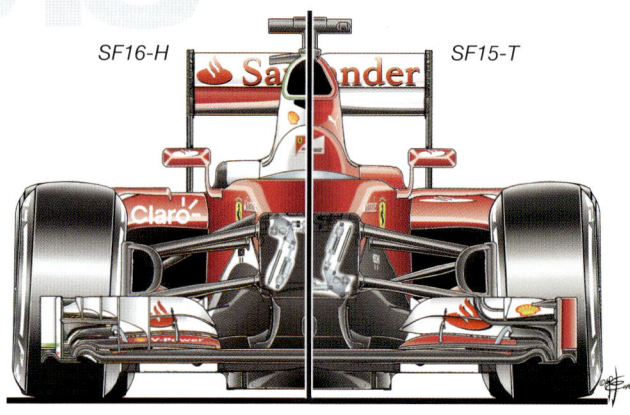

SF16-H   SF15-T

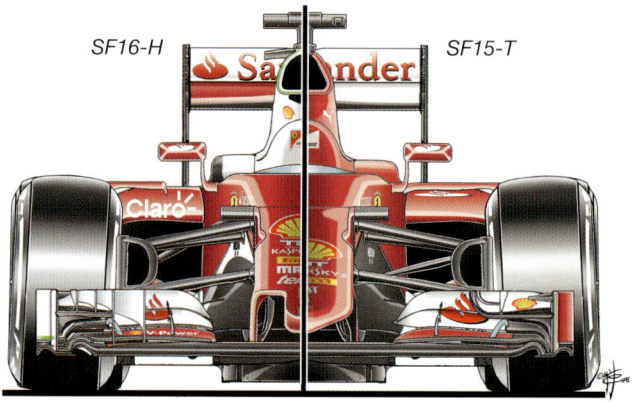

SF16-H   SF15-T

SIDE VIEW COMPARISON

Although relatively uncompetitive, the SF16-H was an important car in Ferrari's technical development, a starting point for the 2017 car. It in fact represented the abandonment after four seasons of the pull-rod front suspension system and the introduction of the short nose that featured on all the leading cars in the 2015 season. Below, the frontal comparison reveals the

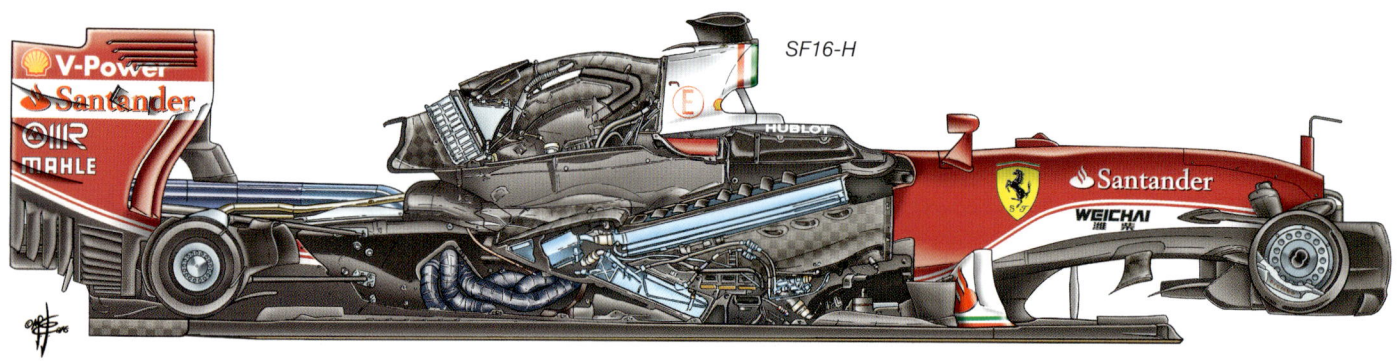

SF16-H

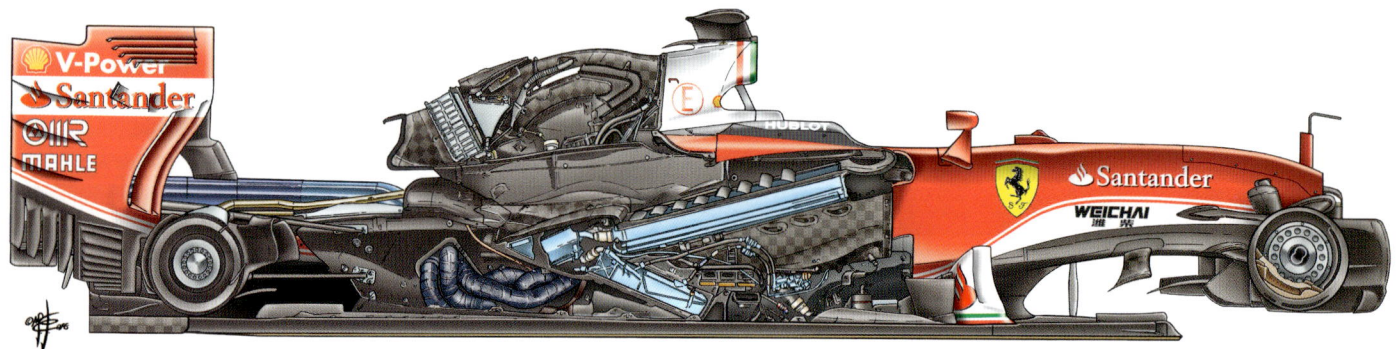

SF15-T

profile with a complete overhaul of the cooling system and a new arrangement of the radiators and heat exchangers in V-formation.

The sequence of six profiles shows the final phase of the assembly of the SF16-H in the pits. The division of the body-

work can be seen with the engine cover and the central part of the flanks forming a single piece.

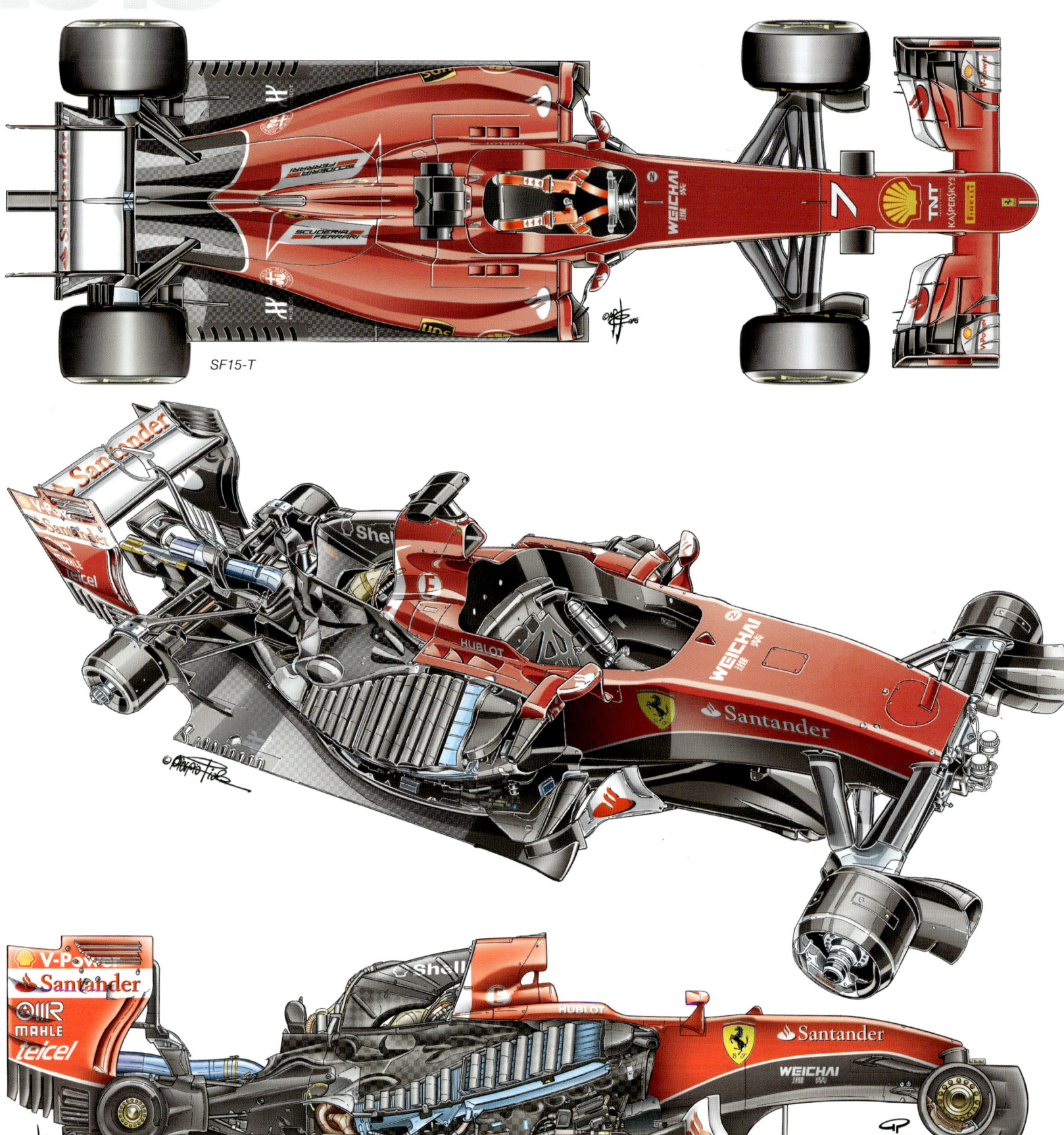

SF15-T

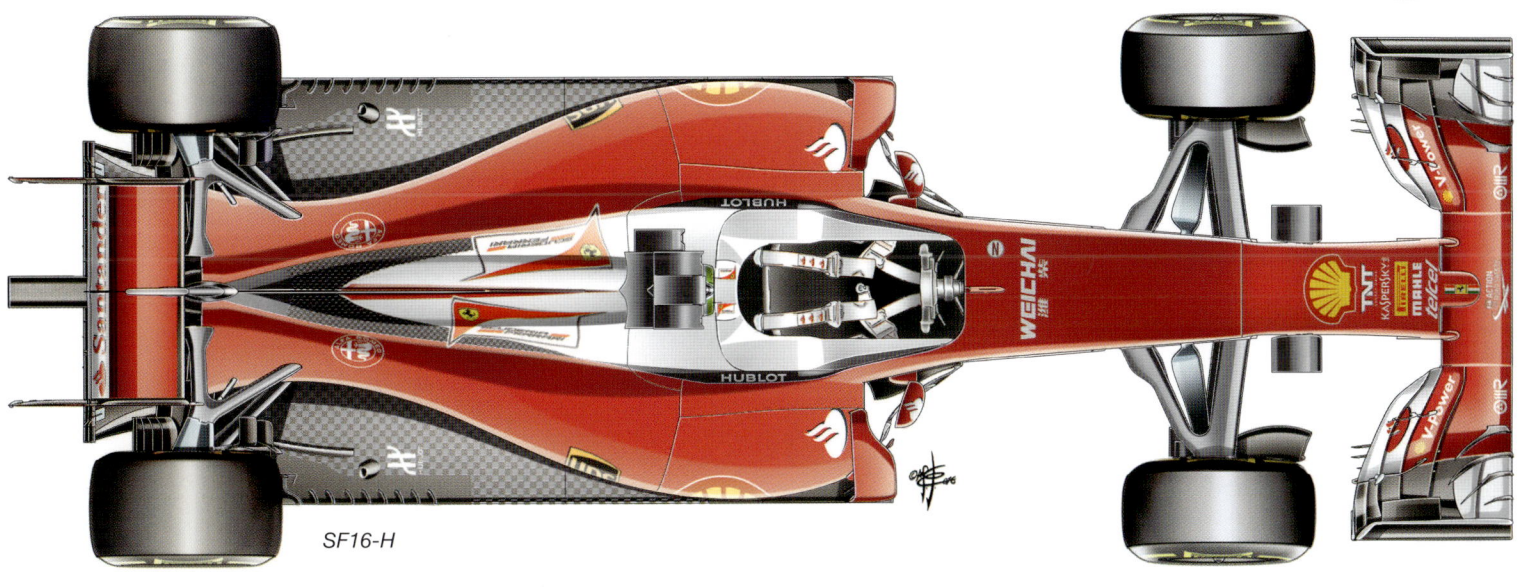

SF16-H

The extreme aerodynamic evolution of the SF16-H is particularly evident in the overhead view with the rear zone being narrower than ever thanks in part to the sophisticated integration of engine, gearbox and rear suspension. The side view instead presents the different configuration adopted for the radiators and heat exchangers.

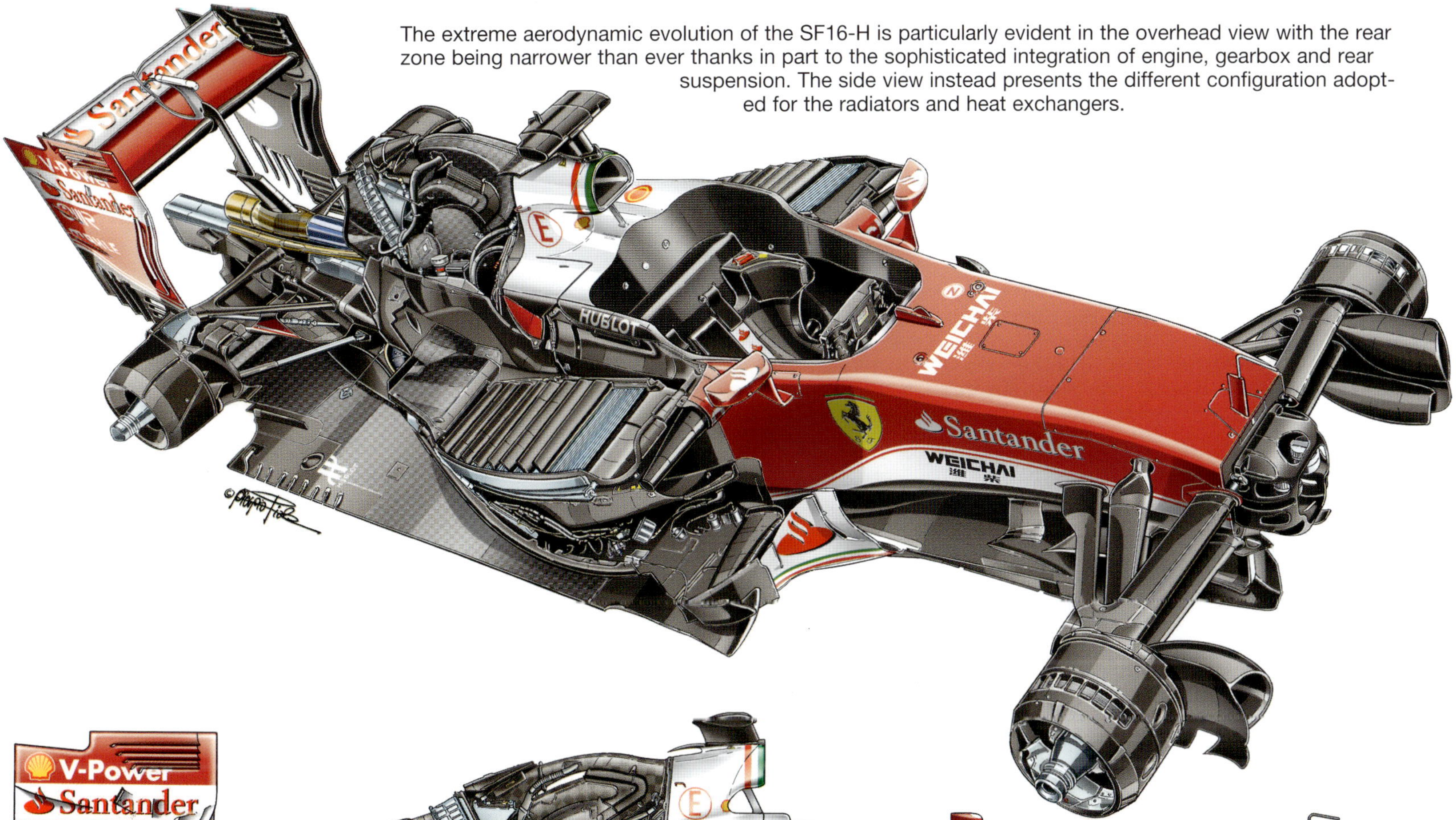

In the 2016 season, the last prior to the great regulatory revolution that led to the return of wide tyres, a brand-new "Ultrasoft" compound was added to the Pirelli range. This new tyre proved to be around half a second faster compared with the Supersoft that had up to the previous year been the softest compound. The coloured band around the sidewall was purple, chosen via a campaign launched by Pirelli on the social networks that asked the fans to express their preference. The Ultrasoft is a low working range tyre suitable for narrow, twisting circuits that exalt mechanical grip thanks to its capacity to reach maximum efficiency at relatively low temperatures (it is at its best in the window between 85 and 115 degrees centigrade) but has a rather limited duration.

The FIA also introduced new regulations regarding tyre management that allowed the teams to use three pre-selected compounds during the course of each race weekend rather than previous two.

The Federation's intentions, supported by the Milan-based tyre manufacturer, was clear: to make the teams' race strategies less predictable in an attempt to further enhance the entertainment. It was in fact no coincidence that no less than 866 passes were recorded in 21 races, with an average of over 41 at each Grand Prix.

In order to overcome the problem that emerged on Sebastian Vettel's Ferrari at Silverstone in 2016, the Milanese technicians tried a different tyre construction during testing at Abu Dhabi, reducing the tread depth by a few millimetres and introducing a very thin alert layer of rubber between the tread and the carcass. The compound was in fact harder than that of the tread and when the tyres wore down it provided the drivers with a warning that a pit stop was needed as performance dropped off significantly, avoiding the problems encountered the previous season. Max Verstappen, Red Bull Racing's young talent, recorded no less than 78 overtakes, the highest total since 1983, the first year the figure was recorded.

In the 2016 season Pirelli produced 42,782 Formula 1 tyres: 38,112 were supplied during the race weekends and the remaining 4,680 during the test sessions. 28,188 slicks took to the tracks while 14,604 rain and intermediate covers were also used.

In statistical terms the Soft tyres were the most used, with the compound covering 6,566 km with Sergio Perez (Force India), against the 4,598 km of the Supersoft used by Valtteri Bottas (Williams) and the 3,597 of the Medium also used by the Finn. The least used were instead the Ultrasofts (2,052 km for Nico Rosberg in the Mercedes) and the Hard (759 km with Daniel Ricciardo in the

WIDE TYRES

The tyres for the 2017 championship had a slighter larger circumference (10 mm) with respect to those used in the tests, while the width was increased by 25% with respect to 2016. The fronts now measured 305/670/13 and were fitted to 13.7x13" rims, while the rears were now 405/670/13 on 16.9x13" rims. There was therefore an increase of around 25% in the width of the tyres, which from the outset lent the cars a more aggressive appearance.

The new dimensions determined an increase in the weight of the tyres of 1 kg at the front and 1.5 kg at the rear, with the tyres therefore accounting for an overall increase in the weight of the car of 5 kg. The FIA and the teams requested different characteristics for the wide tyres with the objective of having tyres with a lower rate of decay with respect to those of 2016, aiming for a product less subject to overheating that would allow the drivers to push for longer during the individual stints. Result: in the 2017 season there were significantly fewer pit stops with a certain degree of freedom in terms of race strategies.

305 mm

245mm

2017

2016

Red Bull). The driver who covered the most ground with the Intermediate was Jenson Button (444 km in the McLaren), while Lewis Hamilton totalled 523 km with the Full Wets in the Mercedes W07.

It is also worth remembering that with the Pirelli P Zero Formula 1 tyres, Valtteri Bottas obtained the highest speed ever achieved in a GP, with 372.5 kph recorded in Mexico.

In the 2016 season, Pirelli developed the "wide" tyres for the 2017 season. Mercedes, Red Bull Racing and Ferrari were involved in the development using 2015 mule cars, with aerodynamic configurations that simulated the loadings provided for by the 2017 regulations: clearly, the suspension layouts were also modified and heavier wheels and tyres were adopted in this prototype guise.

The Bicocca-based tyre supplier, in agreement with the FIA and the teams, organized 10 test sessions totalling 24 working days (7 for each of the three cars involved, plus two cumulative days for the decisions taken at Abu Dhabi).

96 prototypes were tested by 11 drivers who alternated on track, covering a total distance of 12,148 kilometres (2,613 laps were completed in total on five different tracks: Abu Dhabi, Barcelona, Fiorano, Mugello and Paul Ricard).

The driver who covered most ground with the

experimental tyres was Pascal Wehrlein, who completed 3,248 kilometres in the Mercedes. Behind him came the red Bull tester, Pierre Gasly, with 2,494 km, while the first of the works drivers was Sebastian Vettel of Ferrari with 2,228 km. Lewis Hamilton instead avoided these sessions, limiting his involvement to just 50 km following the premature abandonment of the tests scheduled at Abu Dhabi.

At the behest of the promoter and the teams, Pirelli developed a range of tyres for the 2017 season that would suffer less decay and would provide a much wider operational window, allowing the drivers to push harder throughout each individual stint.

The 2017 tyres were 60 mm wider on the front axle, increasing from 245 to 305 mm, while the rears were 80 mm wider, reaching 405 mm compared with the previous 325 mm, that is to say, a 25% increase for the slick tyres as well as the intermediate and rain covers.

The wider tyres certainly lent the 2017 cars a more aggressive and more "muscular" appearance. Pole position times fell by an average of 2.450 seconds, with a peak of 4.1 seconds at Spa Francorchamps, while race lap times dell by 2.968 seconds and cornering speeds rose, with the cars reaching an

estimated loading of 2,700 kg on the Silverstone circuit at 300 kph. The increase was over 30 kph on the most important fast corners: peaks of 36 kph were recorded at corner No. 3 at Barcelona and in the Pouhon left-hander at Spa Francorchamps, while 30 kph higher speeds were seen at corner 9 at the Catalan circuit at Silverstone's Copse. These impressive figures give a measure of how the physical demands on the drivers have also increase given that the G forces generated have risen to peaks of 5.5- 6 G. In particular, the drivers have once again had to pay particular attention to their neck muscles, which are subjected to particular stress. The increase in performance encouraged Charlie Whiting to send a note to all the teams on the eve of the Austrian GP confirming that from the race at the Red Bull Ring onwards the tyre pressures would be checked before the wheels were fitted to the cars to avoid the possibility of teams increasing the pressure via the heat of the brake shrouds and then cool them to lower the pressure below the levels advised by Pirelli. This put an end to such "tricks".

The wide tyres provided more stable performance that allowed the drivers to push hard throughout the stint, extending their durability but reducing the opportunities for overtaking, which dropped with respect to 2016, above all due to the significant loss of front end downforce when one car found itself in the slipstream of another. 435 overtakes were recorded, with an average of 21.8 per Grand Prix.

In 2017, there were a total of 533 pit stops (of which six were drive through penalties and one stop and go). During each Grand Prix, there were on average 26.7 pit stops, the equivalent of 1.5 per driver, a clear reversal of the previous year's trend when there were 2.01 for a total of 933 and an average of 44.4 per race. It appears clear that the race strategies were simplified with respect to the previous season and that there were fewer opportunities to attempt more imaginative tactics than was the case in the past.

Pirelli supplied 38,788 tyres during the 2017 season (33,520 during the course of the race weekends and 5,268 in the various test sessions). 25,572 of that total were slick tyres, while the remaining 13,016 were wet weather intermediate and rain covers. A total of 12,920 tyres were actually during the race weekends, 11,532 slicks and 1,388 wets.

The numbers present a 2017 to remember for F1 and for Pirelli, with numerous lap records being broken after year, demonstrating that the leap in performance sought via the F1 regulations had been achieved with room to spare.

Franco Nugnes

white = medium

red = supersoft

orange = hard

yellow = soft

purple = extrasoft

TYRE SEQUENCE

Five compounds were again offered in the 2017 range: Ultrasoft, Supersoft, Soft, Medium and Hard with the possibility of using three each race weekend.

The P Zero Hard tyre had an orange band, the Medium white and the Soft yellow. With the softest compounds, the Supersoft had a red band and the Ultrasoft purple.

The two rain tyres were the Cinturato full wets (blue) and the intermediates (green).

With the tyres off the car, the FIA checked that the minimum inflation pressures were in line with the indications provided by the sole supplier, race by race, in accordance with the camber rates, while the temperature when the thermal covers removed could not exceed 110°. Since 2016, the Milan company has also indicated the maximum recommended distance (in laps) for each tyre in relation to the data harvested during free practice so as to avoid safety issues deriving from over-use of the tyres in the race.

MELBOURNE

Mercedes W08

Ferrari SF70H

Red Bull RB13

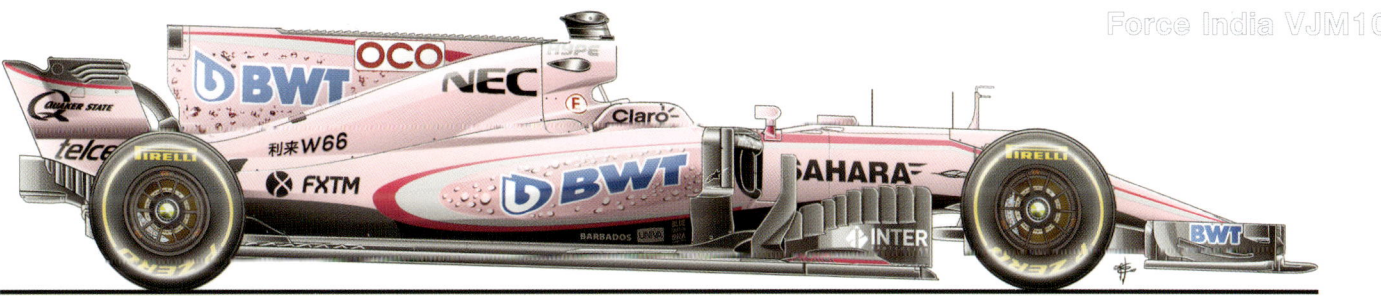

Force India VJM10

Williams FW40

ABU DHABI

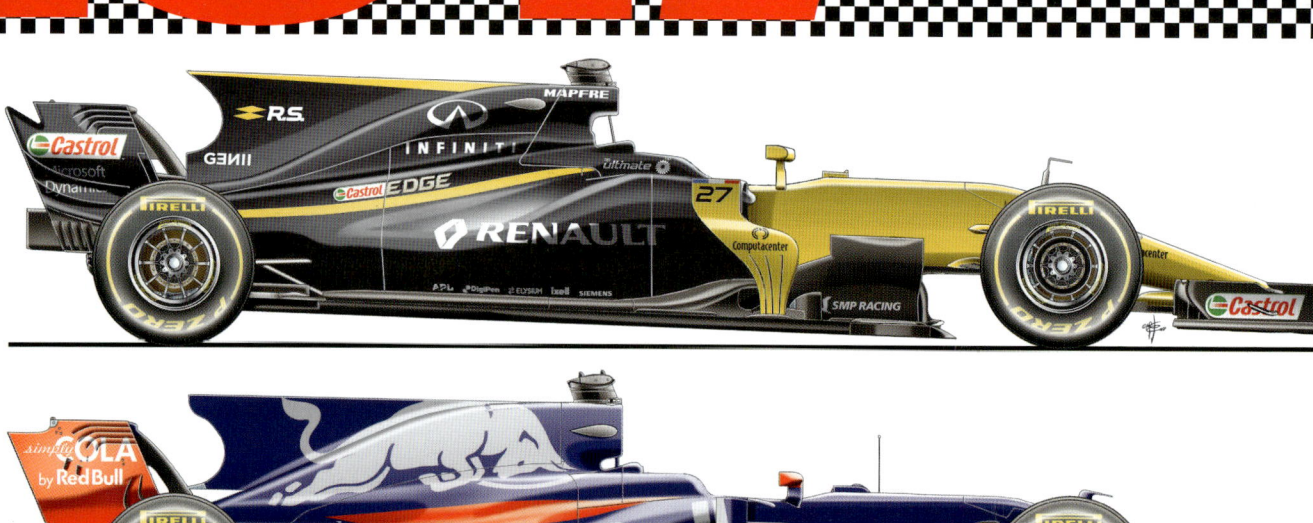

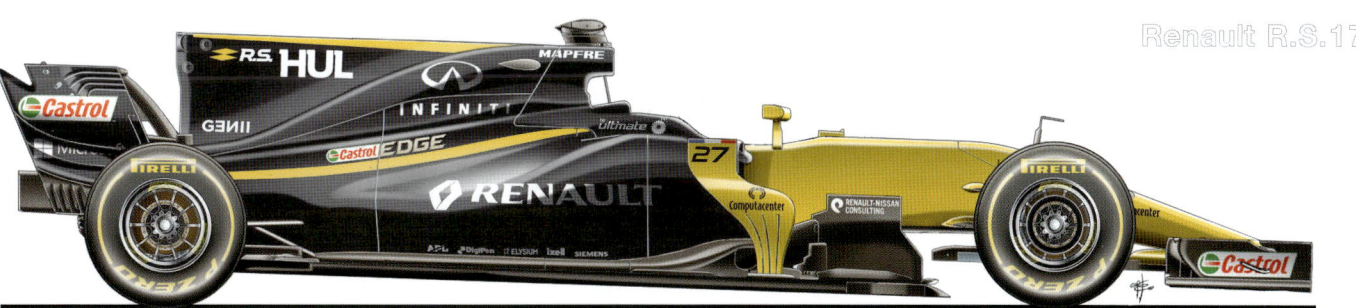

Renault R.S.17

Toro Rosso STR12

Haas VF-17

McLaren MCL32

Sauber C36

ABU DHABI

Car TABLE 2017

		44-77 MERCEDES	5-7 FERRARI	3-33 RED BULL	11-31 FORCE INDIA	
		W08	**SF70H**	**RB13**	**WJM10**	
CAR	Designers	James Allison Aldo Costa Andy Cowell	Mattia Binotto Simone Resta Enrico Cardile	Adrian Newey Rob Marshall Dan Fallows	Adrew Green Akio Haga	
	Race engineers	Andrew Showling Peter Bonington (44) Tony Ross (77)	Matteo Togninalli Riccardo Adami (5) Dave Greenwood (7)	Paul Monagham Simon Rennie (3) Giampiero Lambiase (33)	Tim Wright (11) Bradley Joice (31)	
	Chief mechanic	Mattew Deane	-	Chris Gent Lee Stevenson	Andy McLaren Will Wickery	
CHASSIS	Passo	3726 mm	3551 mm	3410 mm	3691 mm	
	Front suspension	Push-rod 2+1 dampers and torsion bars	Push-rod 2+1 dampers and torsion bars	Push-rod 2+1 dampers and torsion bars	Push-rod 2+1 dampers and torsion bars	
	Rear suspension	Pull-rod 2+1 dampers and torsion bars	Pull-rod 2+1 dampers and torsion bars	Pull-rod 2+1 dampers and torsion bars	Pull-rod 2+1 dampers and torsion bars	
	Dampers	Sachs	Sachs	Multimatic	Sachs	
	Brakes calipers	Brembo	Brembo	Brembo	A+P	
	Brakes discs	Brembo Carbon Industrie	Brembo CCR Carbon Industrie	Brembo	Hitco	
	Wheels	BBS	BBS	O.Z.	BBS	
	Radiators	Secan	Secan	Marston	Secan	
	Oil tank	middle position inside fuel tank	middle position inside fuel tank	middle position inside fuel tank	middle position inside fuel tank	
GEARBOX		Longitudinal carbon	Longitudinal carbon	Longitudinal carbon	Longitudinal carbon	
	Gear selection	Semiautomatic 8 gears	Semiautomatic 8 gears	Semiautomatic 8 gears	Semiautomatic 8 gears	
	Clutch	Sachs	Sachs	A+P	A+P	
	Pedals	2	2	2	2	
ENGINE		Mercedes AMG F1 M08EQ	Ferrari 062	RBR - TAG Heuer RB13 2017	Mercedes AMG F1 M08EQ	
	Total capacity	1600 cmc	1600 cmc	1600 cmc	1600 cmc	
	N° cylinders and V	6 - V90°	6 - V90°	6 - V90°	6 - V90°	
	Electronics	Mercedes	Magneti Marelli	Magneti Marelli	Mercedes	
	Fuel	Petronas	Shell	Total	Petronas	
	Oil	Petronas	Shell	Total	Petronas	
	Dashboard	Mercedes	Magneti Marelli	Red Bull	P.I.	

18-19 WILLIAMS	27-30 RENAULT	26-55-10-28 TORO ROSSO	8-20 HAAS	14-47-22 McLAREN	9-12-36 SAUBER
FW40	**R.S.17**	**STR12**	**VF-17**	**MCL32**	**C36**
Paddy Lowe	Nick Chester Martin Tolliday	James Key Phil Charles	Rod Taylor Ben Agathangelou	Timo Goss Matt Morris Peter Prodomou	Jorg Zander Eric Gandelin Luca Furbatto
James Urwin (18) Dave Robson (19)	Mark Slade (27) Karel Loos (30)	Marco Matassa (28-55) Pierre Hamelin (10-26)	Ajo Komatsu Gary Gannon (8) Giuliano Salvi (20)	-	Julien Simon Chautemps (9) Jorn Becker (12-36)
Mark Pattinson	Robert Cherry	Domiziano Facchinetti	Stuart Crump	-	Reto Camenzind
3545 mm	3630 mm	3550 mm	3551 mm	3584 mm	3551 mm
Push-rod 2+1 dampers and torsion bars	Push-rod 2+1 dampers and torsion bars	Push-rod 2+1 dampers and torsion bars	Push-rod 2+1 dampers and torsion bars	Push-rod 2+1 dampers and torsion bars	Push-rod 2+1 dampers and torsion bars
Pull-rod 2+1 dampers and torsion bars	Pull-rod 2+1 dampers and torsion bars	Pull-rod 2+1 dampers and torsion bars	Pull-rod 2+1 dampers and torsion bars	Pull-rod 2+1 dampers and torsion bars	Pull-rod 2+1 dampers and torsion bars
Williams	Penske	Koni	Sachs	McLaren	Sachs
A+P	A+P	Brembo	Brembo	Akebono	Brembo
Carbon Industrie	Hitco	Brembo	Brembo	Carbon Industrie Brembo	Brembo
O.Z.	AVUS	O.Z.	O.Z.	Enkey	O.Z.
IMI Marston	Marston	Marston	Calsonic	Calsonic - IMI	Calsonic
middle position inside fuel tank	middle position inside fuel tank	middle position inside fuel tank	middle position inside fuel tank	middle position inside fuel tank	middle position inside fuel tank
Longitudinal titanium	Longitudinal titanium	Longitudinal carbon	Longitudinal carbon	Longitudinal carbon	Longitudinal carbon
Semiautomatic 8 gears	Semiautomatic 8 gears	Semiautomatic 8 gears	Semiautomatic 8 gears	Semiautomatic 8 gears	Semiautomatic 8 gears
A+P	A+P	A+P	A+P	A+P	A+P
2	2	2	2	2	2
Mercedes AMG F1 M08EQ	Renault RS17	Renault RS17	Ferrari 062	Honda RA617H	Ferrari 062
1600 cmc	1600 cmc	1600 cmc	1600 cmc	1600 cmc	1600 cmc
6 - V90°	6 - V90°	6 - V90°	6 - V90°	6 - V90°	6 - V90°
Mercedes	Magneti Marelli	Magneti Marelli	Magneti Marelli	McLaren el.sys.	Magneti Marelli
Total	Total	Total	Shell	Mobil	Shell
Total	Total	Total	Shell	Mobil	Shell
Williams	Renault F1	Toro Rosso	Ferrari	McLaren	Magneti Marelli

2017 **REGULATIONS**

A major upheaval to the technical regulations was introduced by the FIA for the 2017 cars with a radical turning of the page with respect to the past seasons. Previous changes had in fact always been made to cap sudden increases in performance and to prevent certain safety limits from being exceeded. Instead, the for the first time the modifications introduced with the 2017 regulations were actually designed to increase performance: around five seconds per lap less than the times obtained at the start of 2016, thanks to the new wider tyres supplied by Pirelli and the significantly improved aerodynamics. The principal inspiration behind the new regulations was that of providing for wider tyres so as to guarantee greater mechanical grip, to which was added more aggressive aerodynamics thanks to less restrictions in this area.

The 2017 regulations therefore opened up whole new scenarios with the birth of cars capable of generating unheard of levels of downforce and with margins for development that were much greater than in the past.

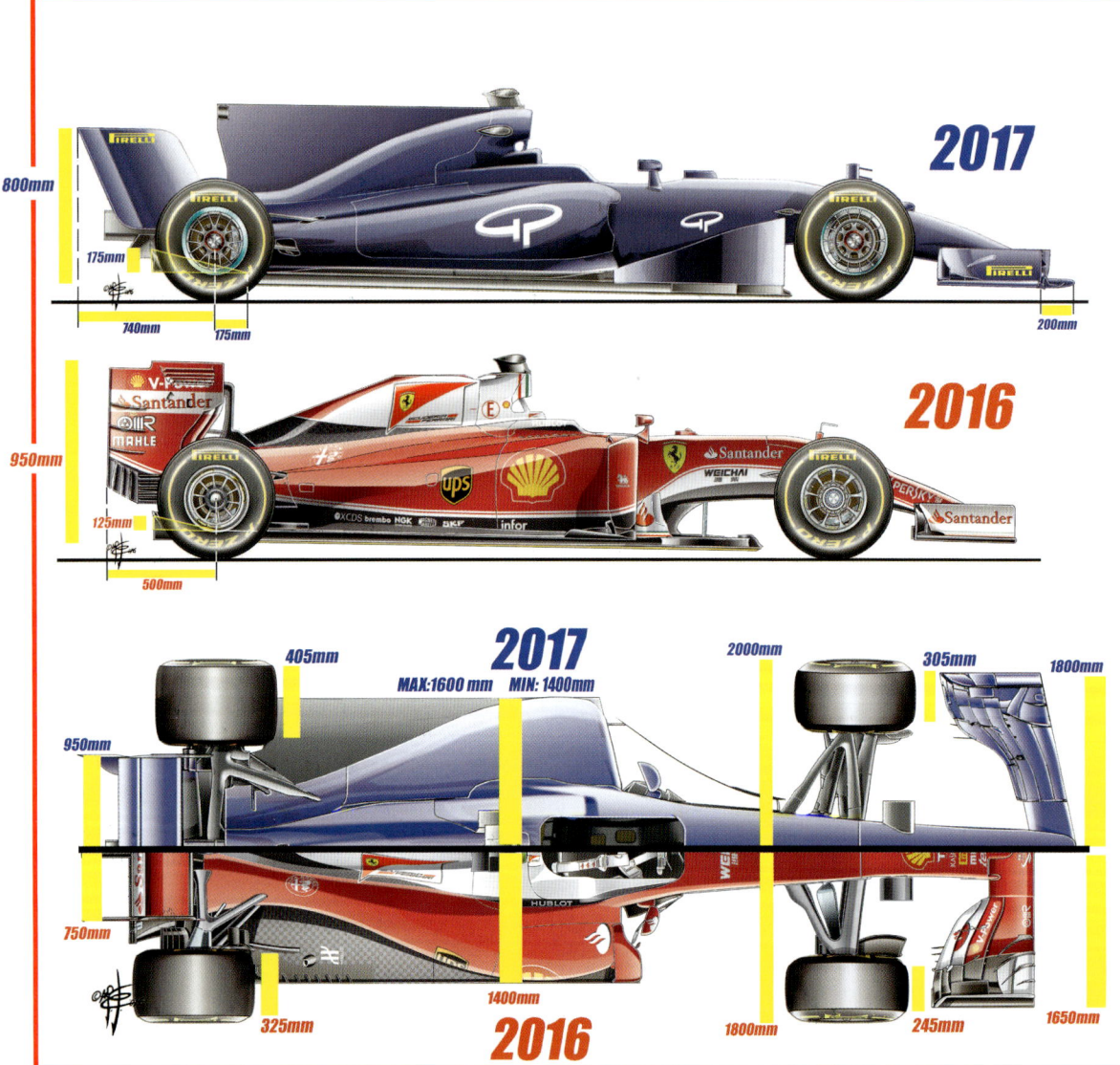

REGULATIONS 2017

Early in the summer of 2016 the FIA officially confirmed the regulations that would come into force in 2017 and be retained through to 2020 after having examined the proposals suggested by the various teams. In the end, it was decided to adopt the project presented by McLaren that, with respect to the configuration of the 2016 cars, provided for an increase of 25% downforce; the Red Bull proposal, seen as being more extreme, instead entailed an even greater increase in aerodynamic loading.

These drawings summarize the details that were modified with respect to the proposals illustrated in the 2015 edition of Technical Analysis. The maximum length was increased from 1800 to 2000 mm; the front wing was to have an obligatory delta shape and a maximum width increased from 1650 to a full 1800 mm. The bodywork and therefore also the underbody expanded from a maximum width of 1400 to 1600 mm with no cuts or indents required. The rear wing was also wider, extending from 750 to 950 mm, but lower at 800 rather than 950 mm.

Lastly, the diffuser was more powerful being higher at 175 rather than 125 mm and 50 mm wider, with a larger 675 rather than 500 mm ramp.

PIRELLI TYES 2017

There was a notable increase in the dimensions of the tyres underlying the technical revolution for the 2017 season, as seen in the comparison with the 2016 covers. The width was increased by 25%. The front tyre measured 305/670/13 and was fitted to 13.7x13" rims; the rears were now 405/670/13 fitted to 16.9x13" rims, while the diameter was increased by just 10 mm.

The new dimensions guaranteed notable mechanical grip, but clearly also entailed an increase in weight for each tyre: 1 kg for the fronts and 1.5 kg for the rears. This meant that the tyres accounted for an increase in the unsprung weight and a rise in the minimum weight of the cars to 628 kg. There was a consequent and notable increase in the width of the rims which permitted the designing of even more sophisticated brake ducts.

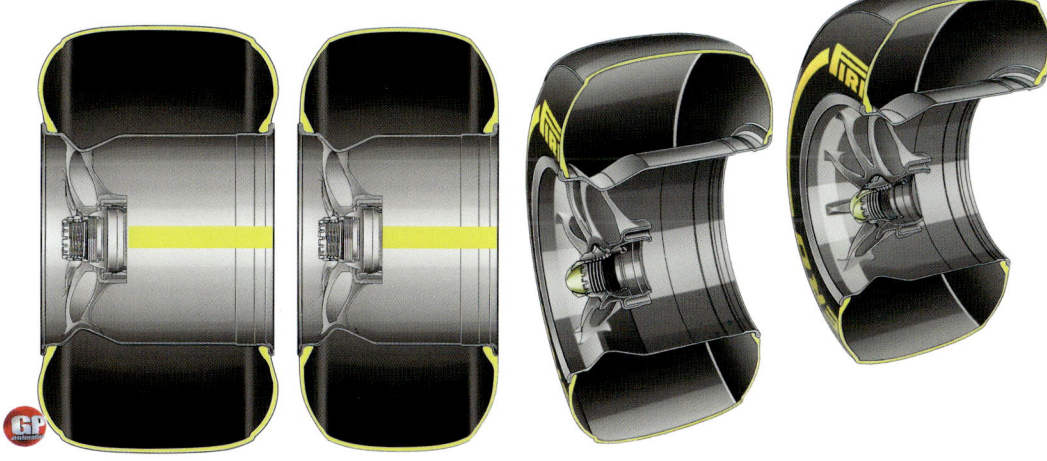

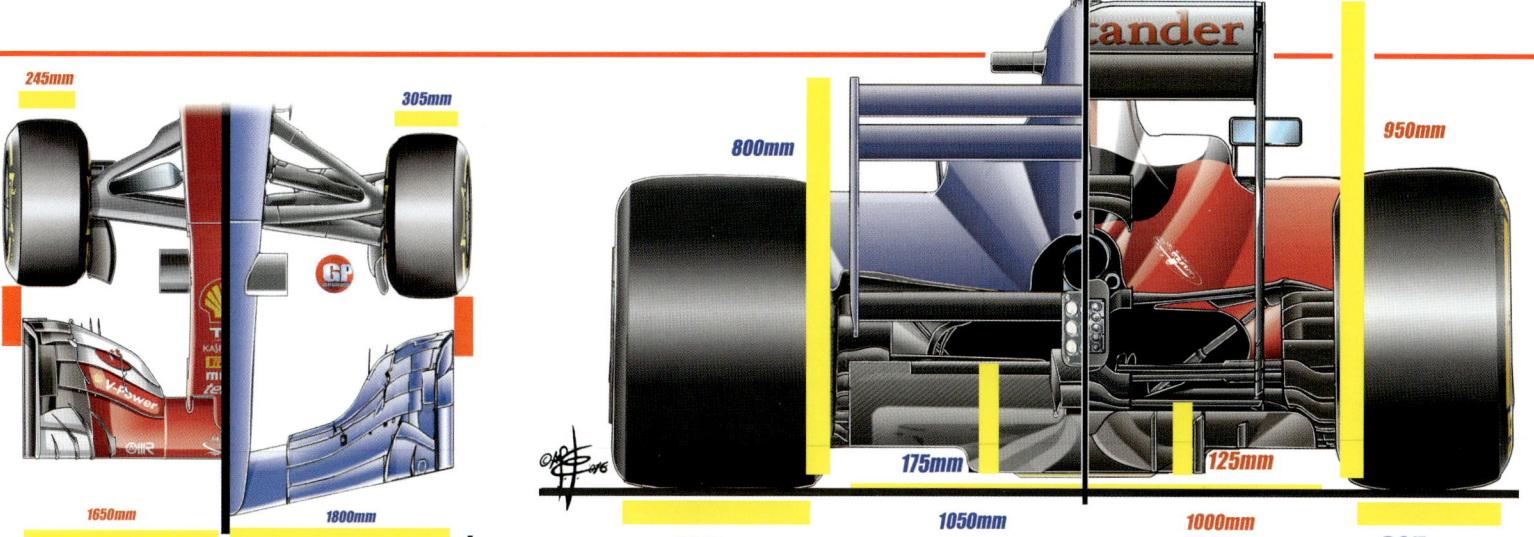

245mm   305mm

1650mm   1800mm

800mm   950mm

175mm   125mm

1050mm   1000mm

405mm   **305mm**

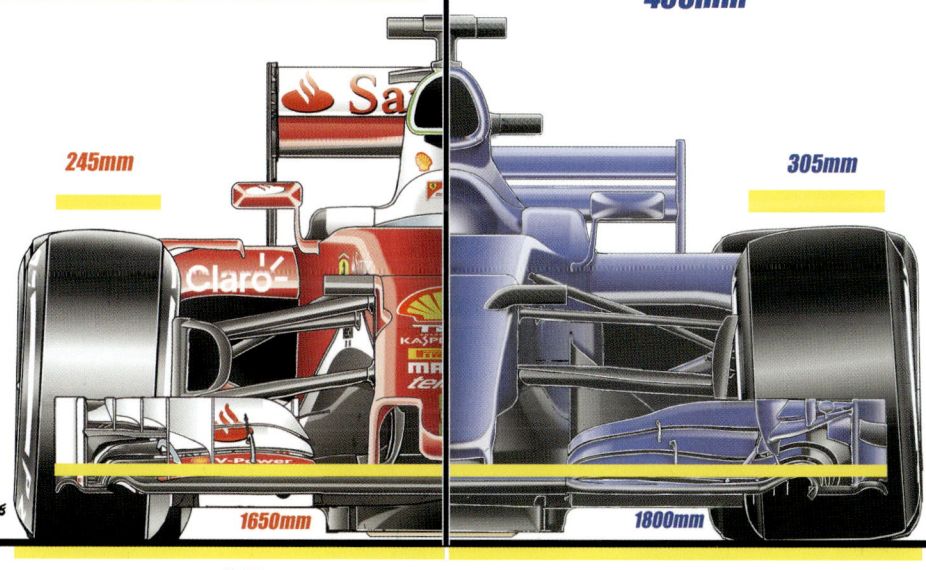

245mm   305mm

1650mm   1800mm

1800mm   2000mm

1998 CAR

Curiously, in the 2017 season, the opposite occurred to the situation in 1998 when grooved tyres were introduced and the maximum width of the cars was reduced from 2000 to 1800 mm, a dimension that remained unchanged through to 2017. At that time, however, the bodywork and the stepped bottom remained unchanged (1400 mm). In 2017, all the width dimensions were increased by 200 mm..

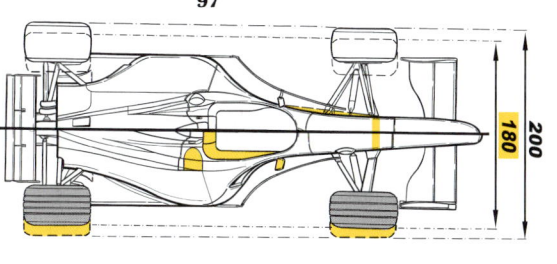

97

200   180

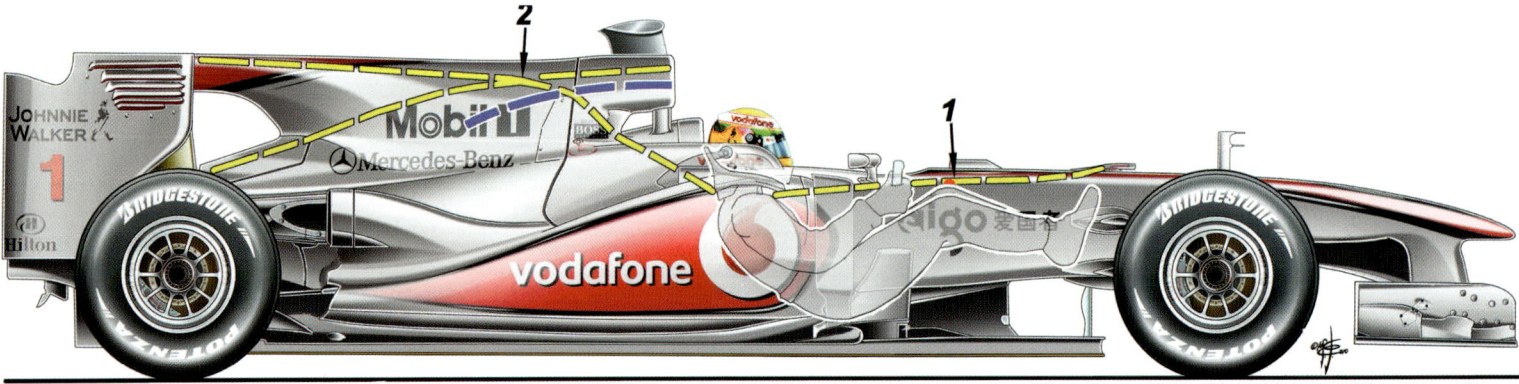

ENGINE COVER FIN

The engine cover fins returned after being banned in 2011 following a season that had been distinguished by the introduction of the F-Duct, firstly on the McLaren in 2010 and adopted on all the cars. In practice, the space indicated in yellow was to be left free so as to avoid the link between the engine cover and the rear wing (which acted as the F-Duct), the configuration of which is highlighted in the McLaren profile.

RED BULL 2008

In 2008, Adrian Newey had introduced a conspicuous fin on the Red Bull from the first race in Melbourne, a feature adopted by the rest of the field with the exceptions of McLaren, BMW and Williams. The substantial difference lay in the fact that on the 2017 cars it was made obligatory not so much to improve the efficiency of the cars as to create space for the race number and possibly sponsors.

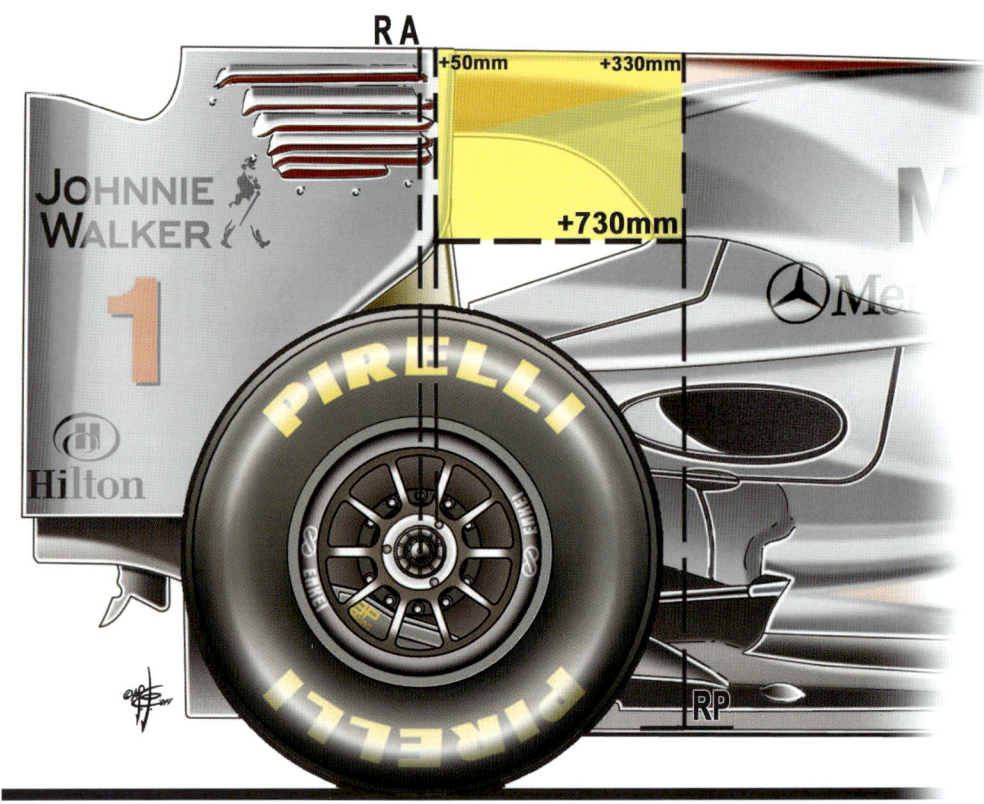

GIORGIO PIOLA

FERRARI 312 B2 (1972) AND McLAREN MP4/10 (1995)

The airbox sail on the 2017 car actually had a very distant in the 1972 Ferrari 312 B2 designed by Mauro Forghieri and another from 1995 with the McLaren MP4-10, when the sail acted as a support for a winglet.

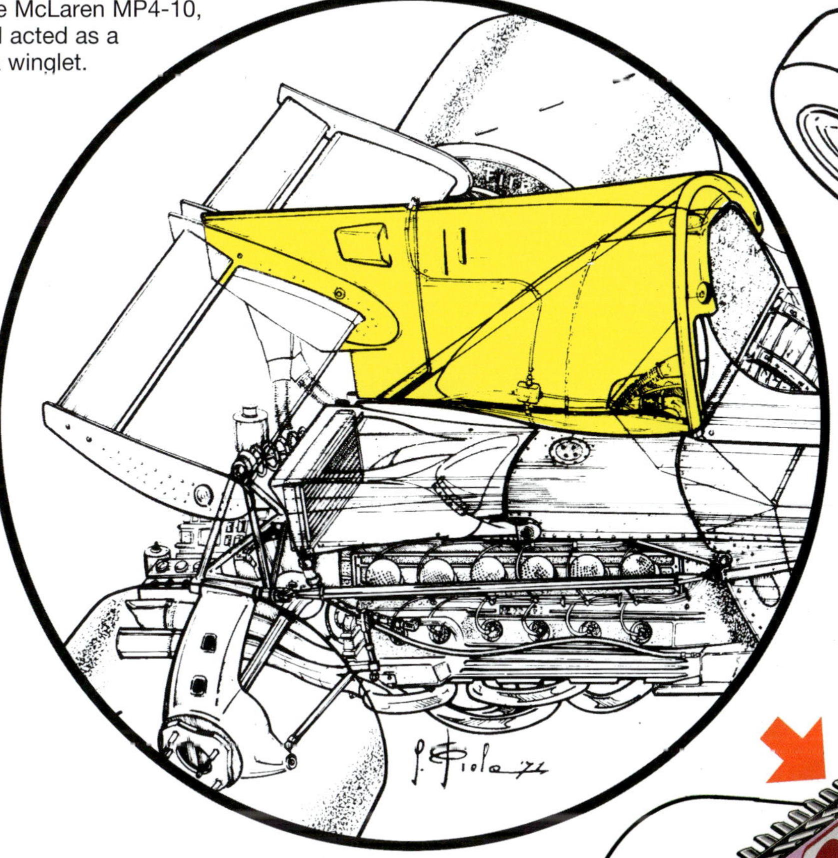

FORCE INDIA

The notional prize for the most innovative fin went to Force India, which at Singapore added this series of winglets in a kind of cascade so as to create mini-vortices and render more efficient the flow towards the triplane T-wing.

T-WING

The strict restrictions on the bodywork and the possible aerodynamic appendages presented a "loophole", a small X-zone that escaped any restriction, as highlighted in the areas in red. Obviously, from the launch of the various cars, all the teams worked to exploit this concession, fitting a 750 mm-wide winglet. The feature became the object of continual modifications with versions being copied to a greater or lesser extent, as with the triplane version used by Force India and Renault. Williams came up with a unique design, a biplane configuration with the two profiles set far apart. The biplane version with close-set profiles was more popular, being introduced by Mercedes and used by, among others, Ferrari and McLaren.

3.5.2c
3.5.2b
3.5.2a

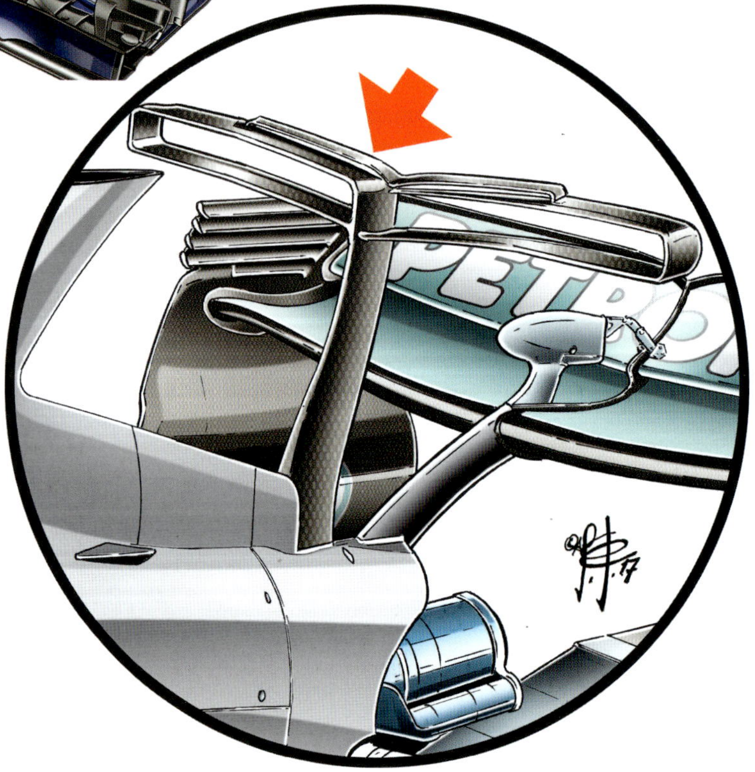

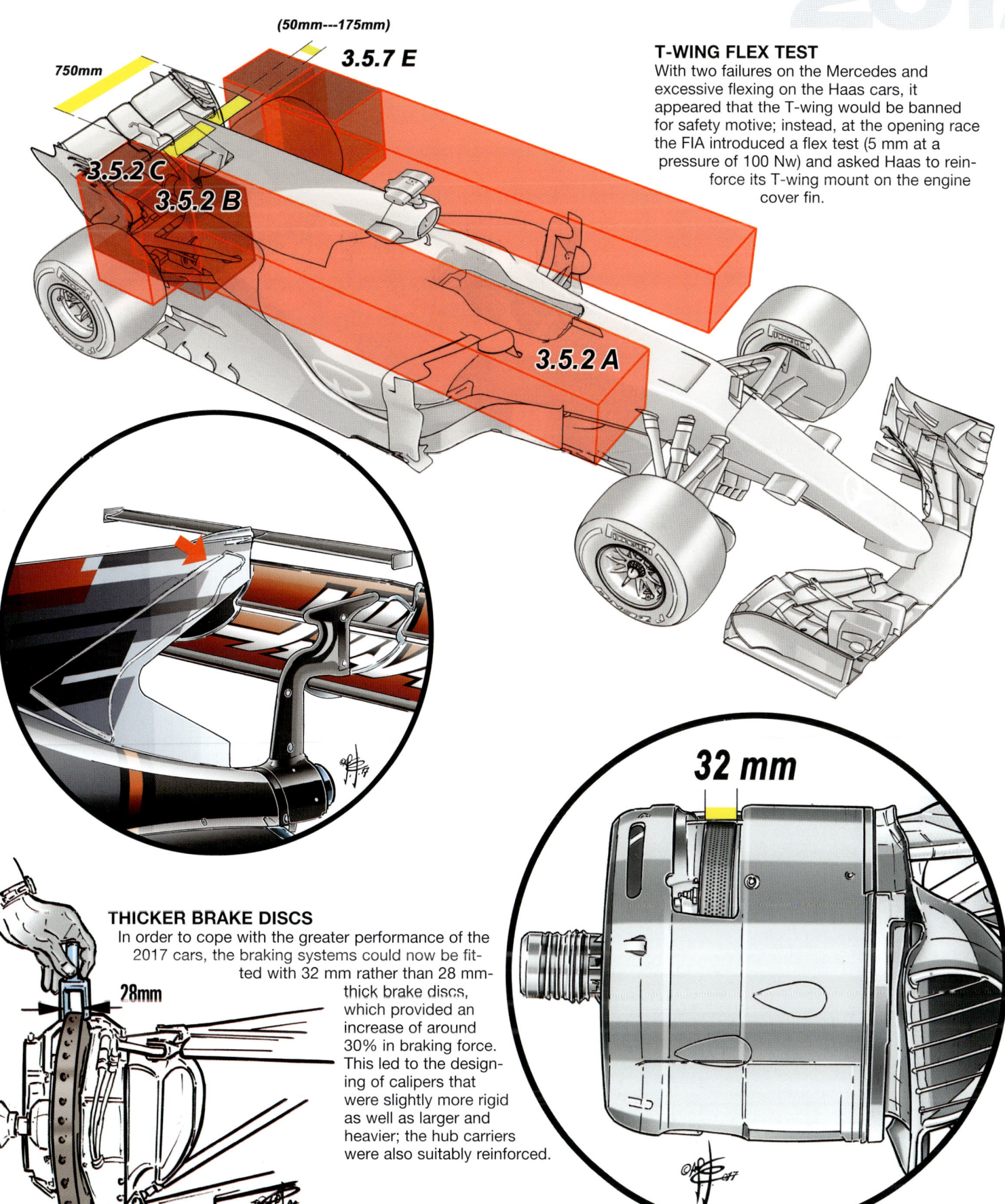

750mm

(50mm---175mm)

3.5.7 E

3.5.2 C

3.5.2 B

3.5.2 A

T-WING FLEX TEST

With two failures on the Mercedes and excessive flexing on the Haas cars, it appeared that the T-wing would be banned for safety motive; instead, at the opening race the FIA introduced a flex test (5 mm at a pressure of 100 Nw) and asked Haas to reinforce its T-wing mount on the engine cover fin.

THICKER BRAKE DISCS

In order to cope with the greater performance of the 2017 cars, the braking systems could now be fitted with 32 mm rather than 28 mm-thick brake discs, which provided an increase of around 30% in braking force. This led to the designing of calipers that were slightly more rigid as well as larger and heavier; the hub carriers were also suitably reinforced.

28mm

32 mm

New **DEVELOPMENTS 2017**

The 2017 saw modern Formula 1's greatest-ever technical revolution. Generally when starting from scratch it had always been Adrian Newey leading the way with stunning new features. This time, however, Ferrari came up with the most innovative car in the field with particularly extreme features.

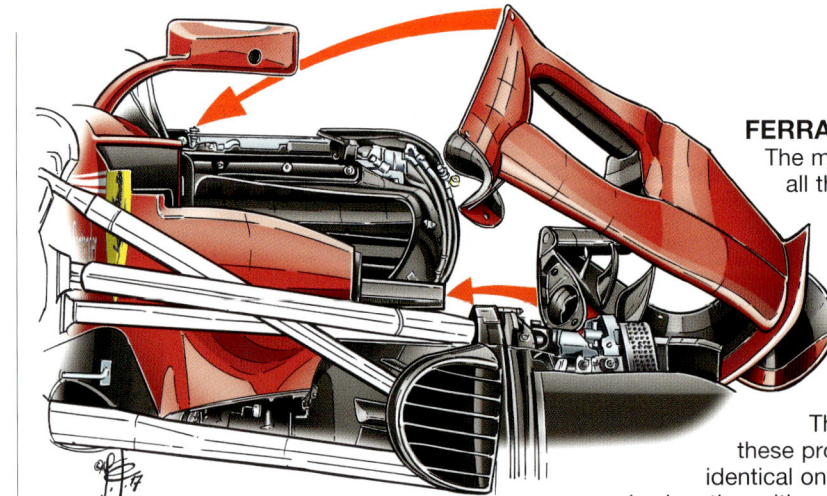

FERRARI

The most interesting of all the SF70H's hidden innovations was its extreme interpretation of the regulation governing the layout of the chassis safety structures.

The rules stated that these protections should be identical on all cars and have a precise location with respect to the cockpit. Ferrari instead placed them not at the edge of the sidepod intake opening, but rather within an aerodynamic profile that effectively allowed the sidepods to be set considerably further back. This solution had the effect of shifting the negative turbulence generated by the front wheels away from the sidepods, providing considerable benefits in terms of aerodynamic and cooling efficiency. The detail drawing shows the upper deformable structure incorporated in the horizontal plane, placed ahead of the sidepod mouths. In the comparison between the top view of the Ferrari SF70H (top) and the SF16-H (bottom) the bigger gap from the front axis of the sidepods is highlighted in yellow.

MERCEDES

The other all-new feature was introduced by Mercedes but not until the Spanish GP. In place of the usual vertical turning vanes, the team introduced a new kind of scoop that had never previously been seen on a Formula 1 car, at least not in this form. The feature was retained throughout the championship, although it was subjected to various modifications.

BRAWN GP 2009-WILLIAMS 2010

In Formula 1, the teams frequently make recourse to technical features that with introduction of new regulations come back into fashion after having been abandoned for years. This was the case with the Mercedes splitter that, in embryonic form, had been seen in 2009 on the Brawn GP, the team from which the current Mercedes outfit emerged.

The main drawing and the detail show how originally the splitter was very small and linked to the car's T-tray, a configuration that was copied by Williams for the following season.

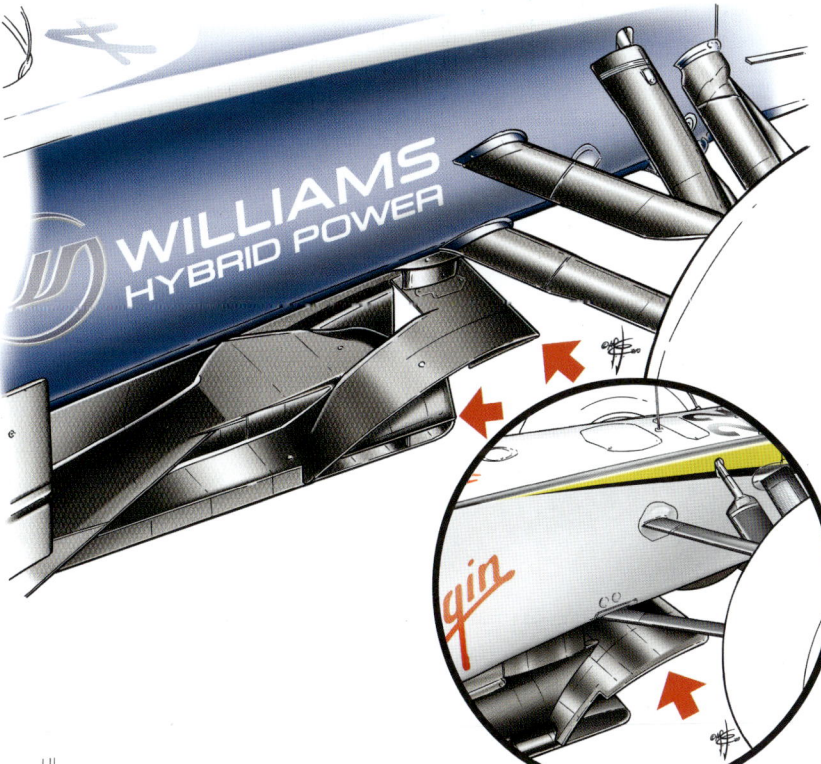

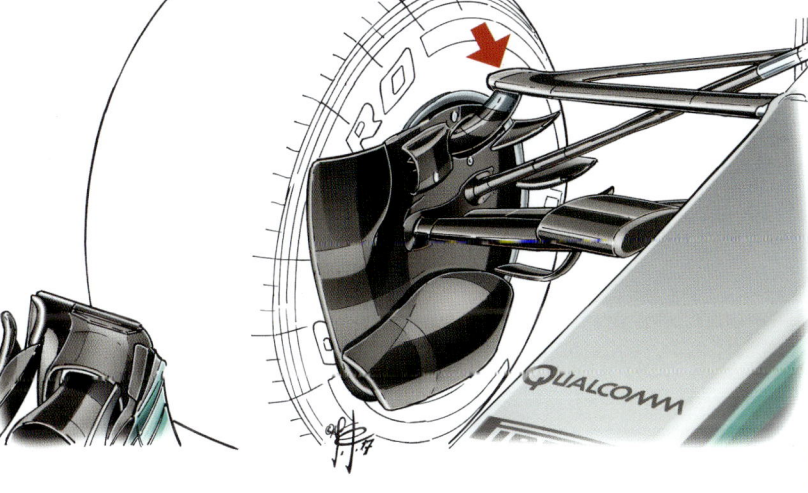

FRONT SUSPENSION

Curiously, and not without arousing a degree of controversy, two teams introduced an identical new technical feature. On the launch of their respective new cars, Mercedes and Toro Rosso both aroused great interest with this new upper wishbone mount that was set extremely high and protruded above the top of the wheel.

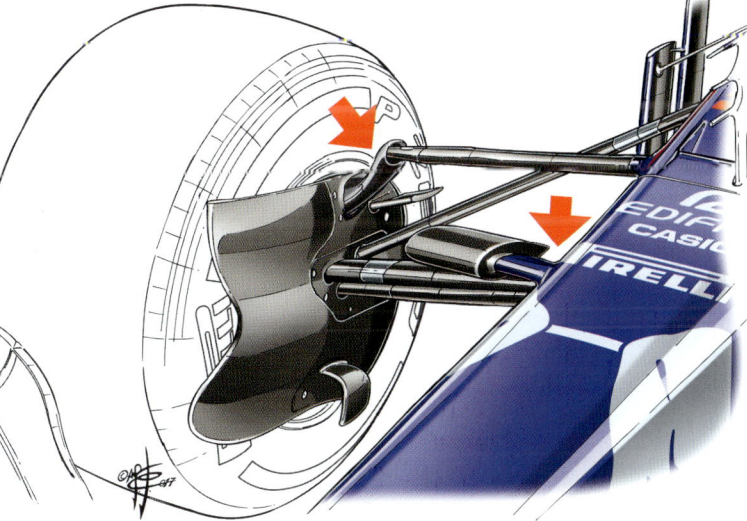

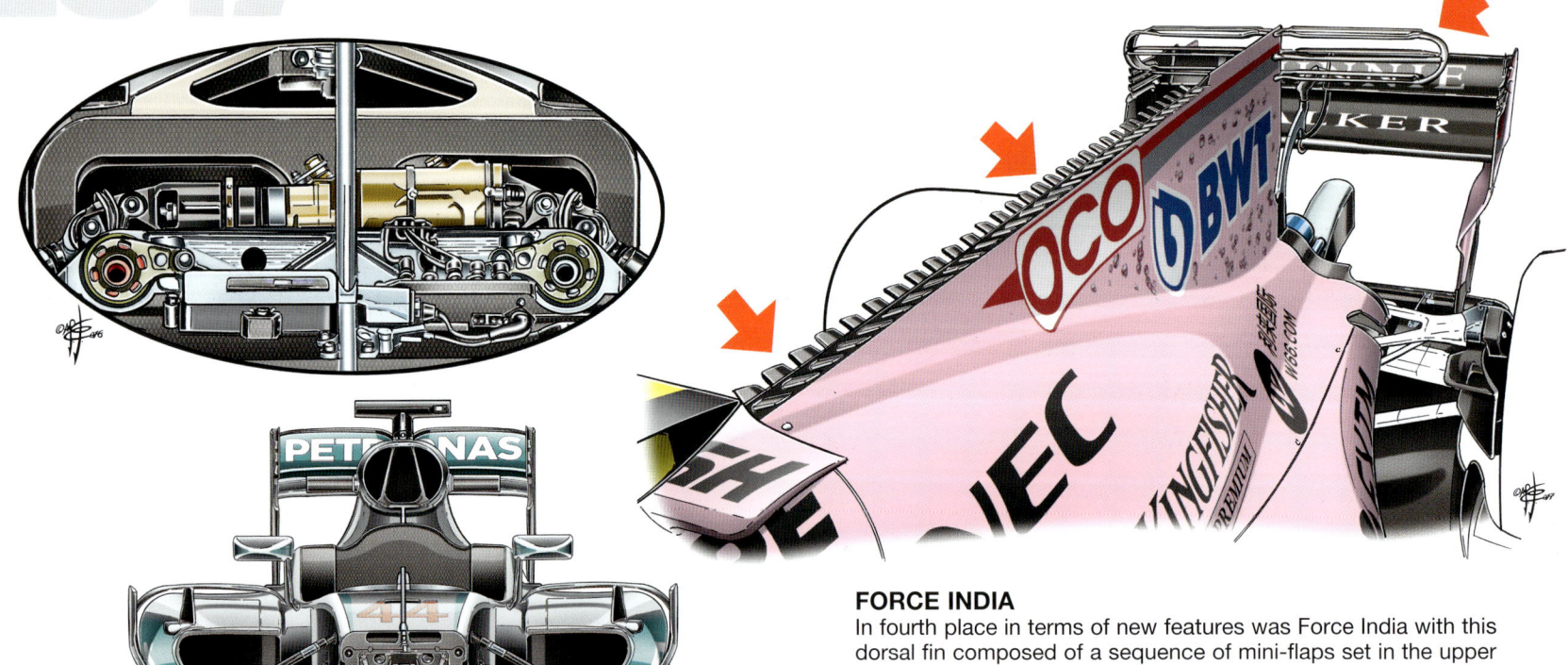

FORCE INDIA

In fourth place in terms of new features was Force India with this dorsal fin composed of a sequence of mini-flaps set in the upper part of the engine cover. Introduced in Singapore and clearly designed to try to recover downforce, this feature was never copied by any other team.

The drawing also shows the triple T-wing introduced by the British team.

MERCEDES FRONT SUSPENSION

Despite the banning of FRIC during the course of the 2014 season, Mercedes continued to use "clever" hydraulic components to control the dynamic ride height of both 2016's W07 (in the drawing) and the W08 from 2017, not only to improve driveability but also to ensure a more consistent aerodynamic set-up. This was actually one of the reasons why Mercedes did not adopt the Red Bull rake set-up that was instead used by Force India, even though the car had a wheelbase practically identical to that of its German rival. In the inset, the rational installation of the front suspension in the open cradle of the W08 chassis, exploiting the same stratagem that had been such a surprise on the previous W07 (see the New Features 2016 chapter).

FERRARI FRONT SUSPENSION

Ferrari had the merit of creating the front suspension installation that was most easily accessible for the mechanics. In practice, all major adjustment operations were performed in the open, with all the elements at the mechanics' disposal without having to work inside the chassis.
(1) anti-roll bar replaceable in less than two minutes.
(2) Torsion bars easily removable.
(3) Potentiometers for the damper travel.
(4) Third horizontal damper, also replaceable in a very short time.
(5) Steering rack removable from outside the chassis.

FERRARI-RED BULL

From the Belgian GP, Ferrari repeatedly used the Friday morning sessions, especially with Räikkönen, a new front suspension configuration with a bracket of a different length for anchoring the strut. The solution was subsequently tested by Red Bull with the involvement of the steering angle in the management of ride height, with the entire steering system being replaced when switching from one configuration to the next. Neither team ever used this suspension layout in qualifying or a race.

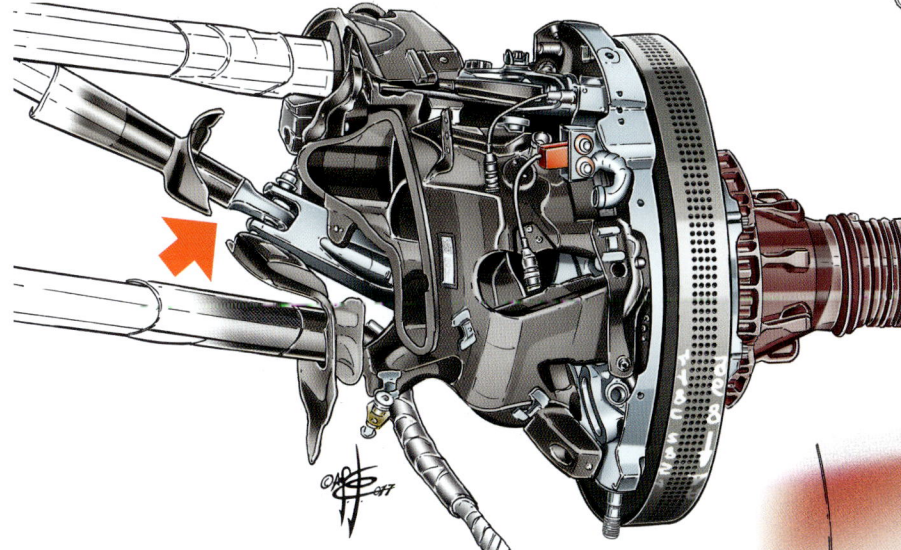

RETURN TO THE PAST

The greater freedom in creating aerodynamic devices in the area between the front wheels and the start of the sidepods revived features that had been seen on cars in 2008: large turning vanes in the upper part, to the sides of the sidepods, complicated barge boards ahead of the sidepods and arrow-head turning vanes in the upper part, as seen in these examples from Ferrari and Renault.

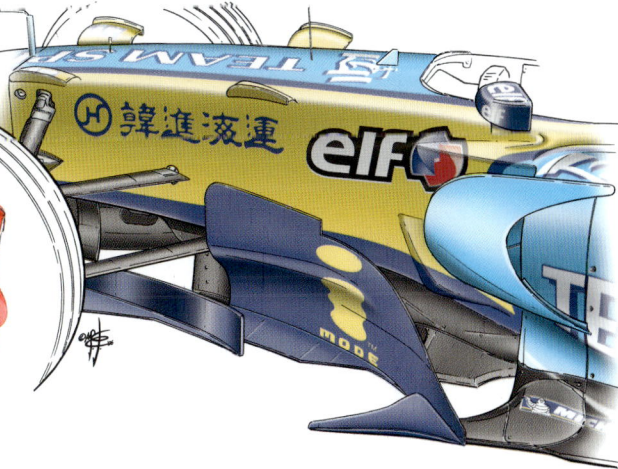

FERRARI

The aerodynamics between the front wheels and sidepods of the SF 70H was extremely sophisticated with conspicuous turning vanes and very complex bargeboards, especially in the lower area indicated by the arrow with configurations that were all-new.

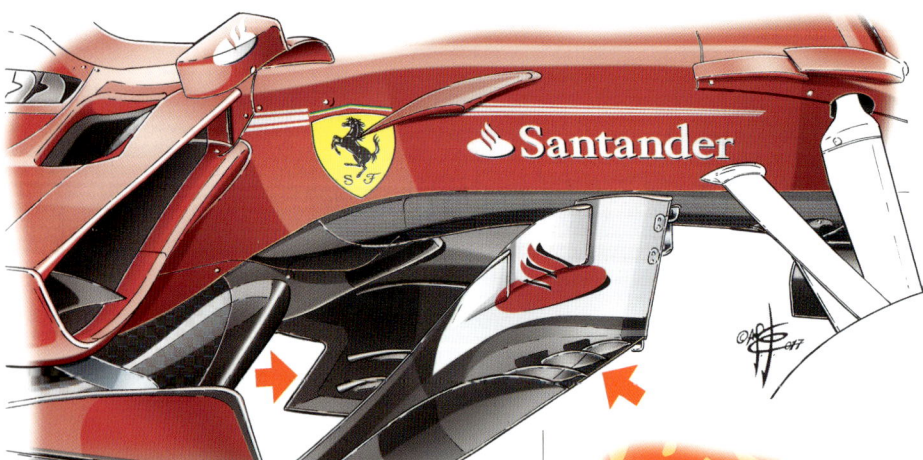

MERCEDES

Compared with the 2016 car, the 2017 Mercedes took to extremes the sophisticated venting and the longitudinal fingers in the lower part of the bargeboards. While not a new feature (Renault 2008), the boomerang turning vane in the upper part was very elaborate; it was then copied by Williams from the Austrian GP.

RED BULL

This new cut in the floor of the Red Bull was present on the 2017 car from pre-season testing in Barcelona; the feature was then dropped from the RB13 in favour of the longitudinal cuts made in this area b by Ferrari, Mercedes and one by one all the others.

RENAULT

Another example of continuous development of the bargeboard area with the introduction of new features was represented by Renault, which at almost every Grand Prix fine-tuned this area of the car, as seen in the drawing that illustrates the evolution of this specific zone.

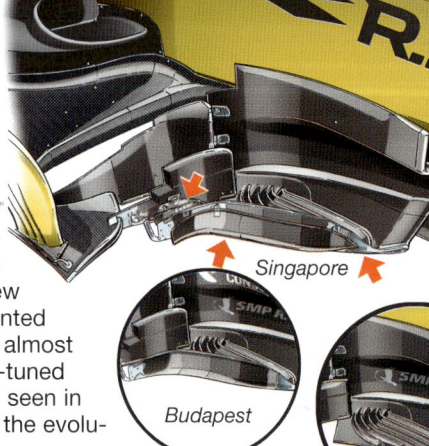

Singapore

Budapest

Zeltweg

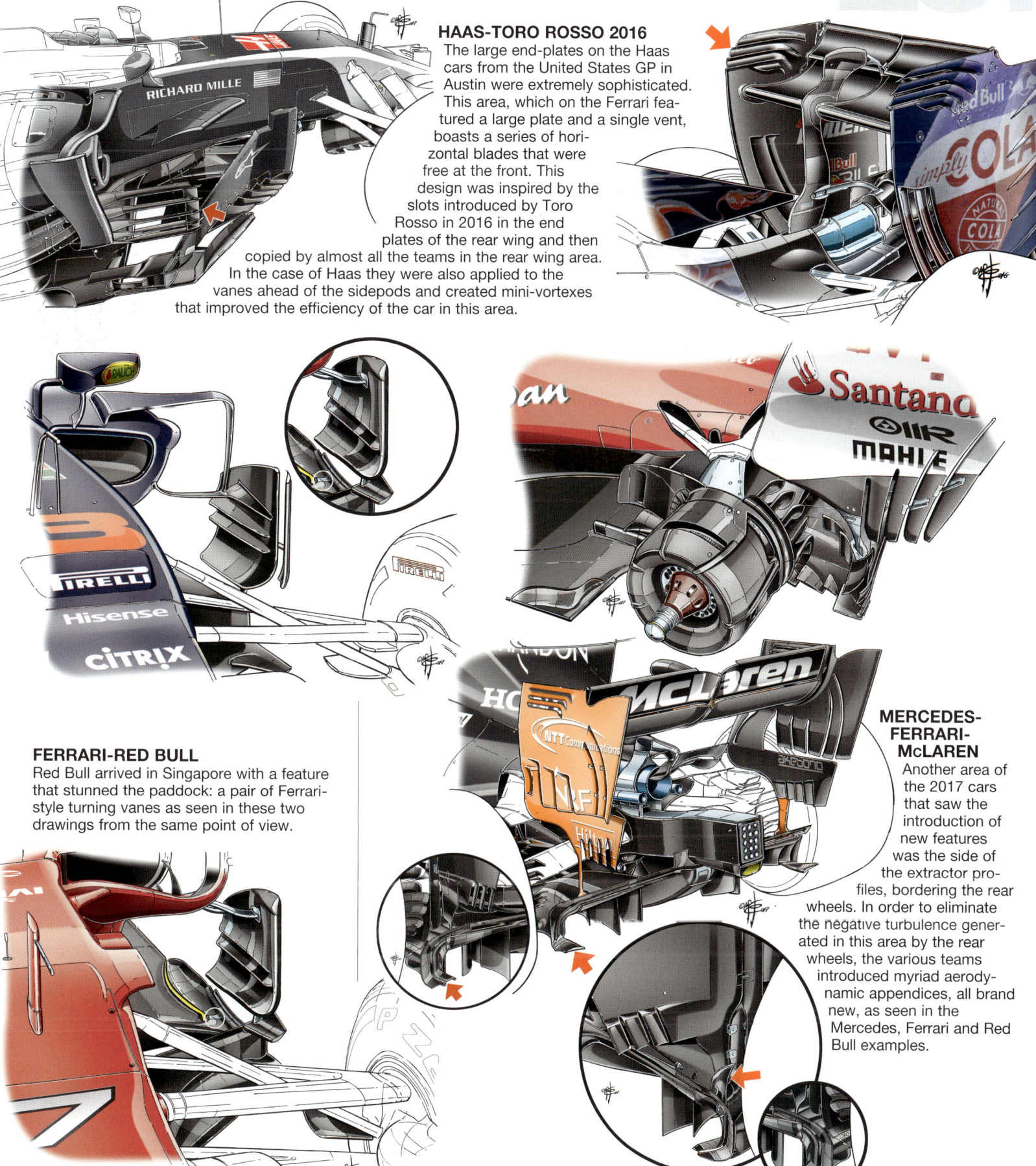

HAAS-TORO ROSSO 2016

The large end-plates on the Haas cars from the United States GP in Austin were extremely sophisticated. This area, which on the Ferrari featured a large plate and a single vent, boasts a series of horizontal blades that were free at the front. This design was inspired by the slots introduced by Toro Rosso in 2016 in the end plates of the rear wing and then copied by almost all the teams in the rear wing area. In the case of Haas they were also applied to the vanes ahead of the sidepods and created mini-vortexes that improved the efficiency of the car in this area.

FERRARI-RED BULL

Red Bull arrived in Singapore with a feature that stunned the paddock: a pair of Ferrari-style turning vanes as seen in these two drawings from the same point of view.

MERCEDES-FERRARI-McLAREN

Another area of the 2017 cars that saw the introduction of new features was the side of the extractor profiles, bordering the rear wheels. In order to eliminate the negative turbulence generated in this area by the rear wheels, the various teams introduced myriad aerodynamic appendices, all brand new, as seen in the Mercedes, Ferrari and Red Bull examples.

FERRARI-McLAREN

With the stepped floor widened by 20 cm (from 140 to 160 cm) it was essential to energize the flow below it. New apertures were therefore cut into the front parts of the sidepods. In this example from Ferrari the cut was long and curving. Early in the year, this slot permitted considerable movement, but was then modified to reduced the flexibility of the area from the Austrian GP. All the teams adopted similar solutions such as this one from McLaren.

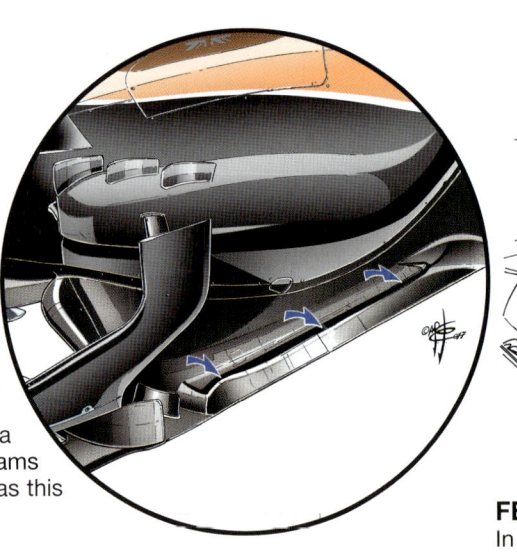

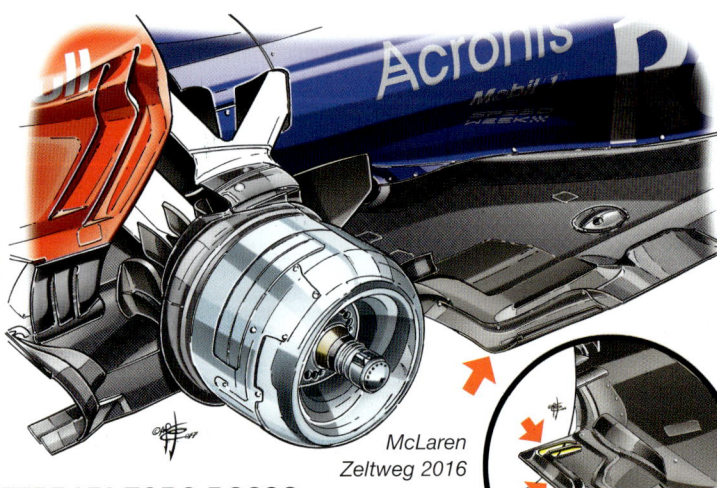

McLaren Zeltweg 2016

FERRARI-TORO ROSSO

In the 2016 season there had already been an increase in the development in the high pressure zone of the floor in front of the rear wheels, in order to reduce the lift effect generated by the tyres. With an increase in the flat area, this requirement was even more urgent with developments introduced for every kind of track. The Toro Rosso solution was interesting, with a longitudinal cut (already seen on the McLarens at Zeltweg 2016) and above all a new upwards curvature to the end part of the floor (red arrow), also highlighted in yellow.

Suzuka

Austin

FERRARI S-DUCT

Ferrari was the last to use the S-DUCT, despite the Maranello team having been responsible for its introduction back in 2008. It did so, however, with a new take on the feature compared with those used on the other cars: the ramp linking the intake in the lower part of the nose with the vent in the upper part of the chassis was twinned rather than being single and central; moreover, in order to obtain a less steep ramp, it had a cross-flow configuration with the air entering the right-hand duct venting in the upper part of the chassis on the left, while that entering on the opposite side naturally vented on the right.

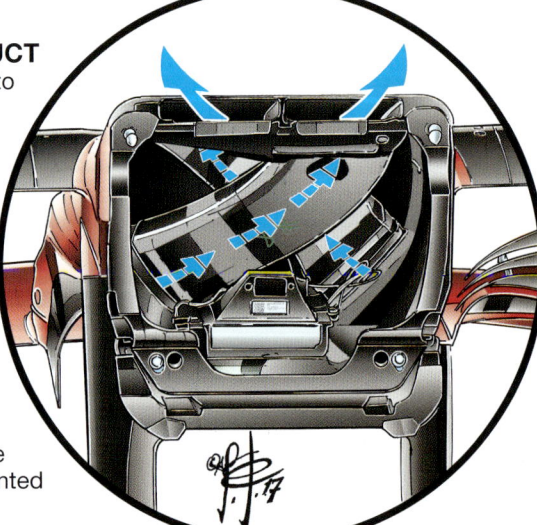

McLAREN

From its debut, the McLaren presented these three conspicuous slots in the long front wing pylons. The solution was designed by Peter Prodomou, the former chief aerodynamicist at Adrian Newey's Red Bull who had been at McLaren since 2016.

RED BULL

The RB13 introduced a steep upwards inclination to the lower part of the front wing end plates to optimize the efficiency of its rake set-up and facilitate maximum flow rate. This configuration avoided grounding while allowing the front end to run as low as possible.

McLAREN

McLaren was the first team to introduce future front wing developments during the season. At Austin the team had this new vent in the main plane close to the neutral central section, highlighted in yellow, which was also repeated in the first flap.

McLAREN

McLaren was also responsible for the most complex rear wing end plate seen in the 2017 season. The most extreme version on the left was introduced in Bahrain and practically divided the end plate into three sections with sophisticated vents to make the flows more efficient in this area of the car.

Talking about **BRAKES 2017**

The technical revolution of the 2017 season brought a new and rather severe technical challenge regarding the braking systems, subjected on the new cars to significantly greater stresses, with an increase in braking torque estimated to be in the order of 25% and decelerations of over 6G. This led the three manufacturers (Brembo, A+P and Akebono) to create slightly stiffer as well as larger and heavier calipers. This was because the FIA had allowed, after 20 years (since 1998 to be precise) an increase in the thickness of the discs from 28 to 32 mm. Consequently, the uprights were also stiffened.

With wider tyres, faster cars capable of generating greater braking torque, the Brembo engineers concentrated on the redesign of the individual elements of the systems, starting with the calipers that had to be adapted to the thicker discs, through to the Brake by Wire components.

The increased size of the carbonfibre discs allowed for the strengthening of the drag area, increasing the braking capacity. Moreover, the greater thickness also provided more space for ventilation holes, determining a further evolution of the cooling of the systems. Customised materials for each team with an average increase of around 200 ventilation holes per disc, reaching a total of 1,400 holes compared to the more than 1,200 of 2016, with a consequent improvement in heat dispersal. It should be remembered that the temperature of an F1 disc may exceed 1,000°C under braking.

Formula 1 requires in-depth personalisation of the braking systems on the basis of the diverse design philosophies of the individual cars. Each of the teams supplied by Brembo adopts a bespoke braking system, closely

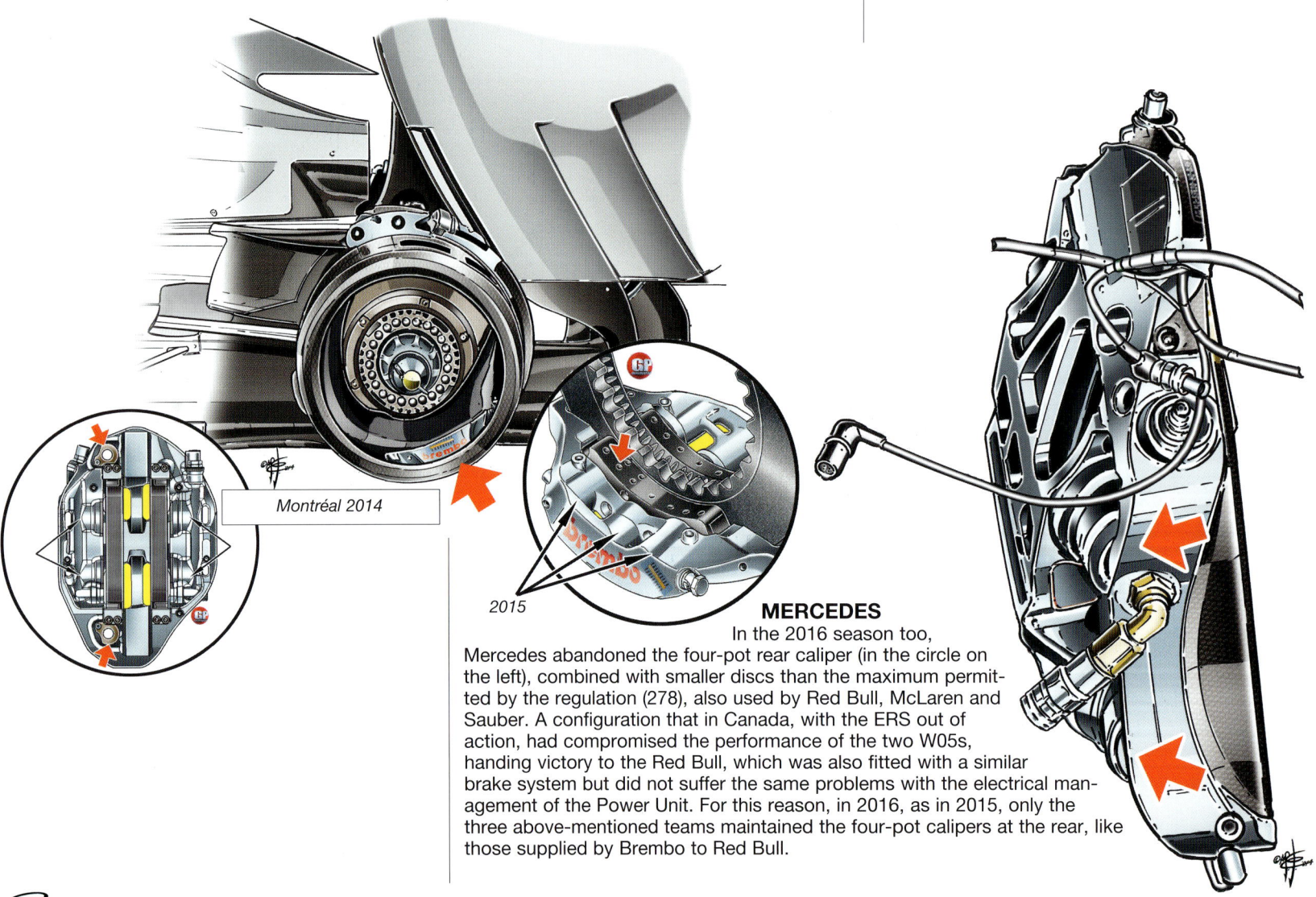

Montréal 2014

2015

MERCEDES
In the 2016 season too, Mercedes abandoned the four-pot rear caliper (in the circle on the left), combined with smaller discs than the maximum permitted by the regulation (278), also used by Red Bull, McLaren and Sauber. A configuration that in Canada, with the ERS out of action, had compromised the performance of the two W05s, handing victory to the Red Bull, which was also fitted with a similar brake system but did not suffer the same problems with the electrical management of the Power Unit. For this reason, in 2016, as in 2015, only the three above-mentioned teams maintained the four-pot calipers at the rear, like those supplied by Brembo to Red Bull.

integrated with the design of the car and subject to continuous development throughout the season.

For this reason too, the brake calipers are completely redesigned for each team supplied, so that they integrate with the aerodynamic configurations of each car, trying to maintain optimum lightness and stiffness values despite the increased size of the disc.

Moreover, development work continued on the individual components of the Brake By Wire system. The engineers continued to extend their skills, designing the BBW components on the basis of the requests of the individual client teams and working on the integration and miniaturization of the various elements.

With regard to the 2016 season, at the rear a number of teams had adopted a radical braking system at the rear, utilising four-pot calipers (Red Bull, Sauber and McLaren) and 260x25 mm discs. Almost like in the pre-parc fermé period when cars would tackle qualification with dedicated calipers and discs that were as light as possible. At the front everyone ran 278x28 mm discs.

The approach for the 2017 season was cautious, avoiding extreme configurations, although obviously at the back smaller diameter discs or with a thickness of 22 and not 32 mm like those at the front were used, in line with the demand fro energy deriving from the recovery system. For some teams, Brembo worked on just a few of the BBW components, while for others their work was

wider-ranging and concerned the entire system: from the simulator on which the rear brake pump acts, to the actuator, governed by the CPU actuating the rear calipers, through to the safety valve system controlling the commutation to traditional braking in the case of a system malfunction.

The introduction of a new material for the discs known as CER, which appeared in 2015 and which evolved over the following years, resulted in considerably reduced wear of the braking system, amplifying its window of use.

All these characteristics allow the driver full control of the braking system. The incredibly low rate of wear and a more efficient thermal conductivity offer reduced warm-up times, a broad range of use, both in terms of pressure and temperature and a very linear friction response with unaltered and repeatable performance from start to finish of the race. However, in the 2017 season, the Brembo brake disc material was reserved for Ferrari and Haas.

The supply of systems was unchanged with respect to 2016, with Brembo calipers on: Ferrari, Mercedes, Red Bull, Toro Rosso,

Haas and Sauber, the British A+P on Williams, Force India and Renault and the Japanese Akebono used exclusively by McLaren.

Over a full season Brembo supplies on average each team composed of two cars with the following material: 10 sets of calipers, between 140 and 240 discs and between 280 and 480 pads.

Given the extreme design performance achieved, the track distance recommended by Brembo for the use of its discs is now 800 km in optimum temperature conditions.

In the diverse use of the braking systems, we have to take into account that the lowering of lap times is due not so much to the increase in straight line maximum speed (paradoxically lower due to the greater contact area of the tyres and a higher Cd value, due in part to greater downforce), but rather to the higher cornering speeds thanks to greater mechanical and aerodynamic grip. At some circuits such as Silverstone, certain braking areas have disappeared because they are now taken flat out with the cars then arriving at the next corner at higher speeds and generating greater decelerations under braking.

WILLIAMS

The prize for the most exotic rear brakes compared with those of the other cars goes to the Williams FW38. Almost all the hot air was expelled to the outside while many other cars channelled it to the inside with the external portion closed. The extreme seal with the wheel rim led to the failure of this last and the explosion of Felipe Massa's tyre in practice on the Friday morning at the Chinese Grand Prix.

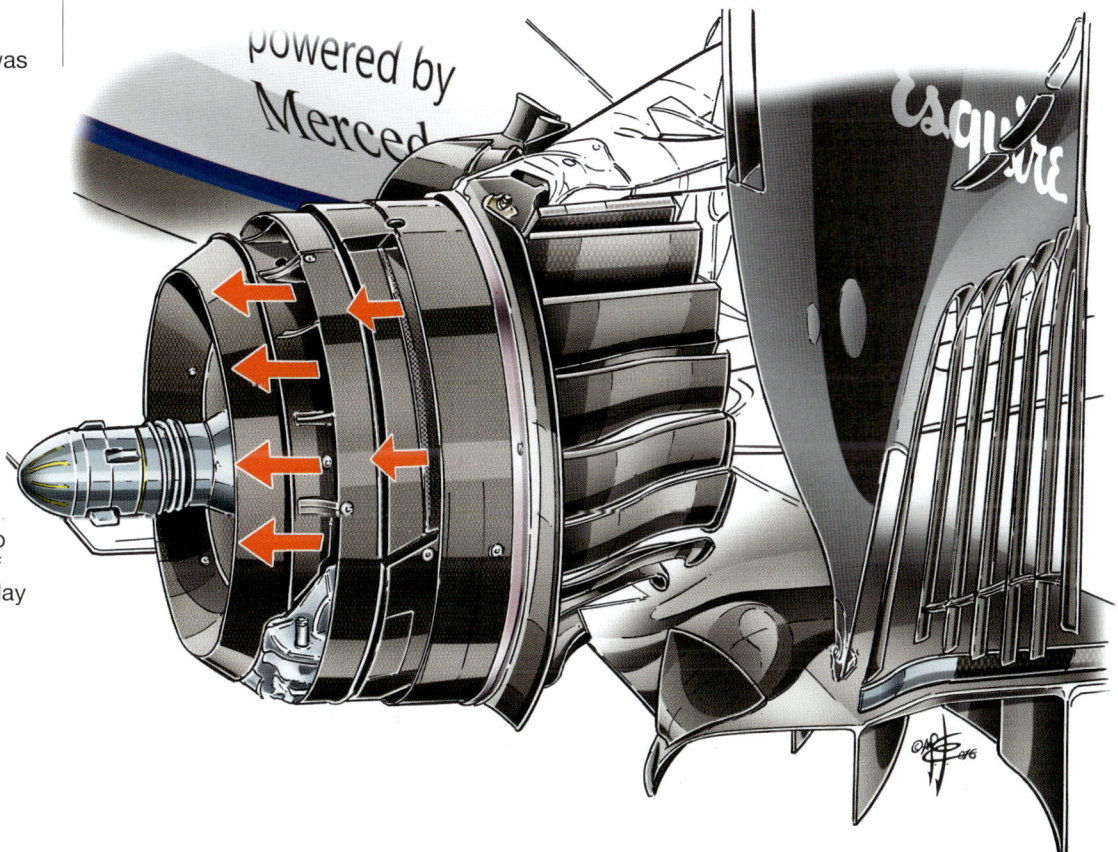

HAAS-FERRARI

Hass also adopted the under-slung front brake caliper location like Ferrari, which adopted this configuration typical of the Red Bulls in the 2015 season; both cars had the possibility of closing the blown hubs, as seen in the drawing, on fast circuits.

MERCEDES BRAKE DISCS

One of the novelties of the 2016 came from Mercedes which asked Carbon Industrie to supply discs with a concave rather than a straight edge (see the insert) to improve the circulation of the air in the closed shrouds. Tested post-race at Barcelona, they were used for the first time in the race at Hockenheim on both cars. In the previous races open shrouds had been used that did not require this feature.

MERCEDES R&D TEST

From the Italian GP, Hamilton experimented with a new front brake system feature. The chief designer at Brackley admitted at the end of the season that this was a R&D department project in view of the 2017 season. The feature required great patience to document as it was concealed inside the carbonfibre shrouds that cover the hub carrier and braking system. In practice it was an additional seal or spacer in carbonfibre either side of the Carbon Industrie disc (1). A pair of metal brackets (2) ensure that that the feature is rigidly fixed so as not to violate the regulation banning the use of more than 6 pistons in the front calipers (3), although the seal of these spaces could be adjusted.

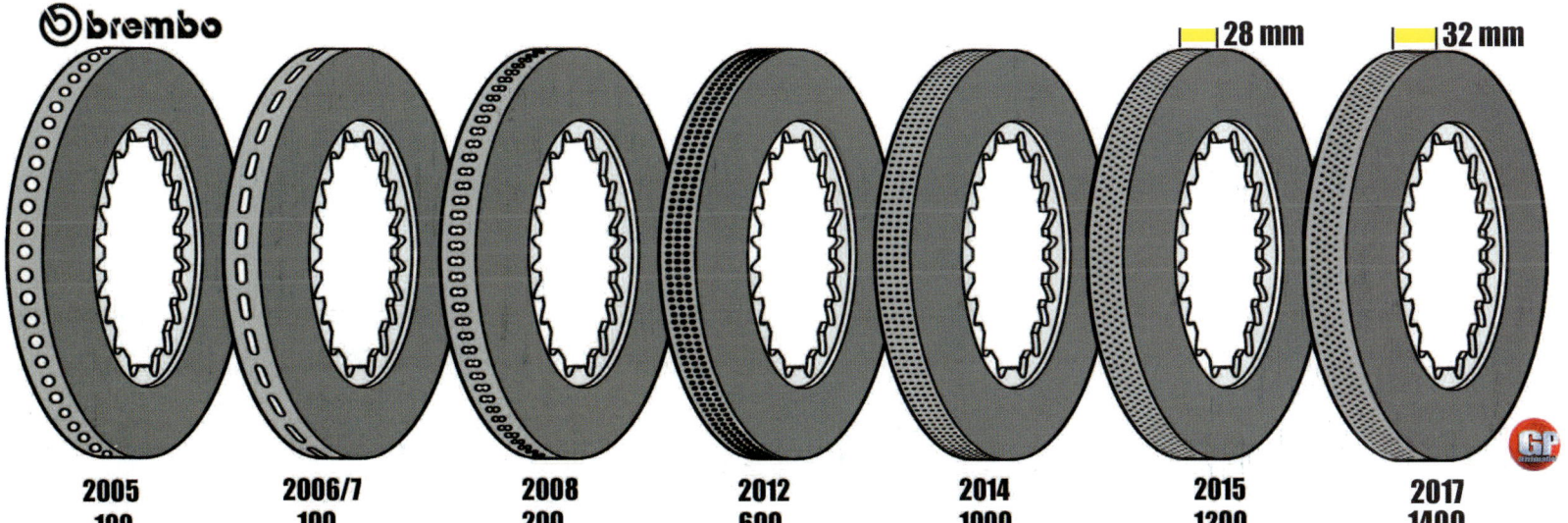

brembo

					28 mm	32 mm
2005	2006/7	2008	2012	2014	2015	2017
100	100	200	600	1000	1200	1400

DISC EVOLUTION

There was a clear turning point for the braking systems: after 20 years the FIA modified the thickness of the discs, increasing it from 28 to 32 mm to adapt the systems to the greater performance of the latest generation cars. The result was that the maximum number of ventilation holes reached 1,400, as shown in this sequence. Previously, the greatest progress made with the braking systems had concerned the material used for the discs. In 2015 Brembo had replaced CCR that guaranteed a maximum wear of 4/5 mm by the end of the race on hard circuits, with CER 100, with wear reduced to 1 mm. Great attention was paid to ventilation, with the number of holes increasing over the years, as documented in this sequence supplied by Brembo and updated ad hoc. The drawing shows the disc with 1,400 holes from 2017 alongside the earlier version with 1,200 holes (5 in oblique rows to allow greater space between one hole and the next).

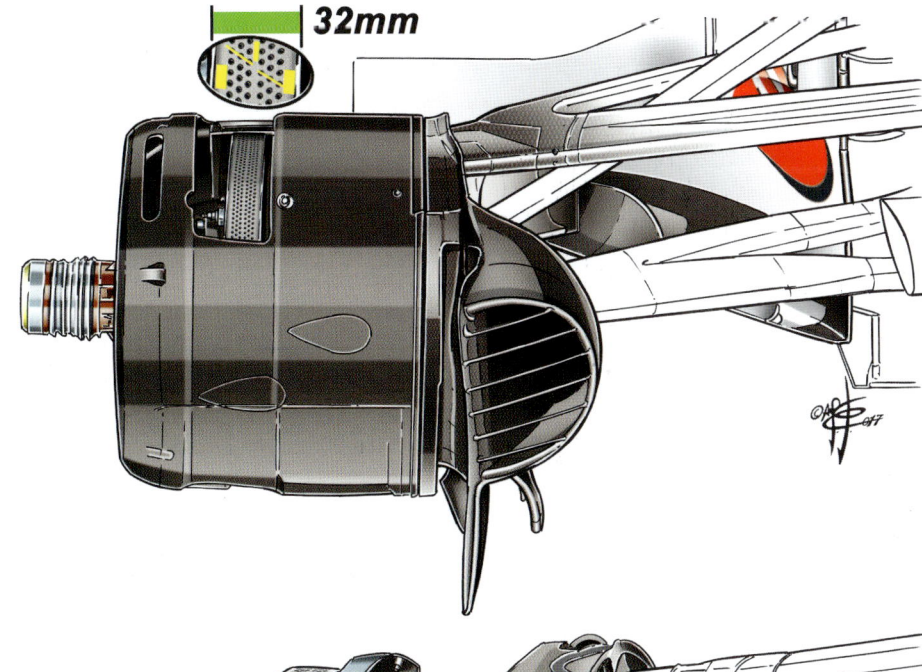

32mm

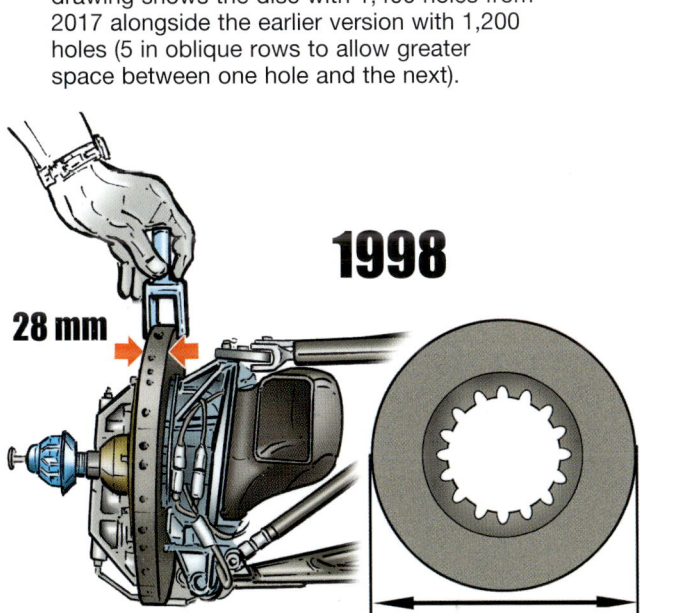

1998

28 mm

278 mm

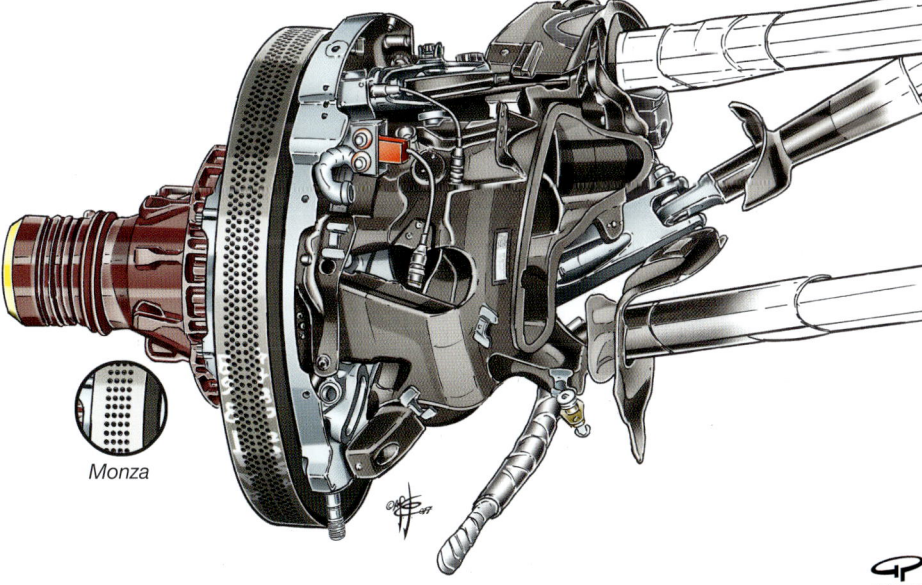

Monza

Talking about **COCKPITS 2017**

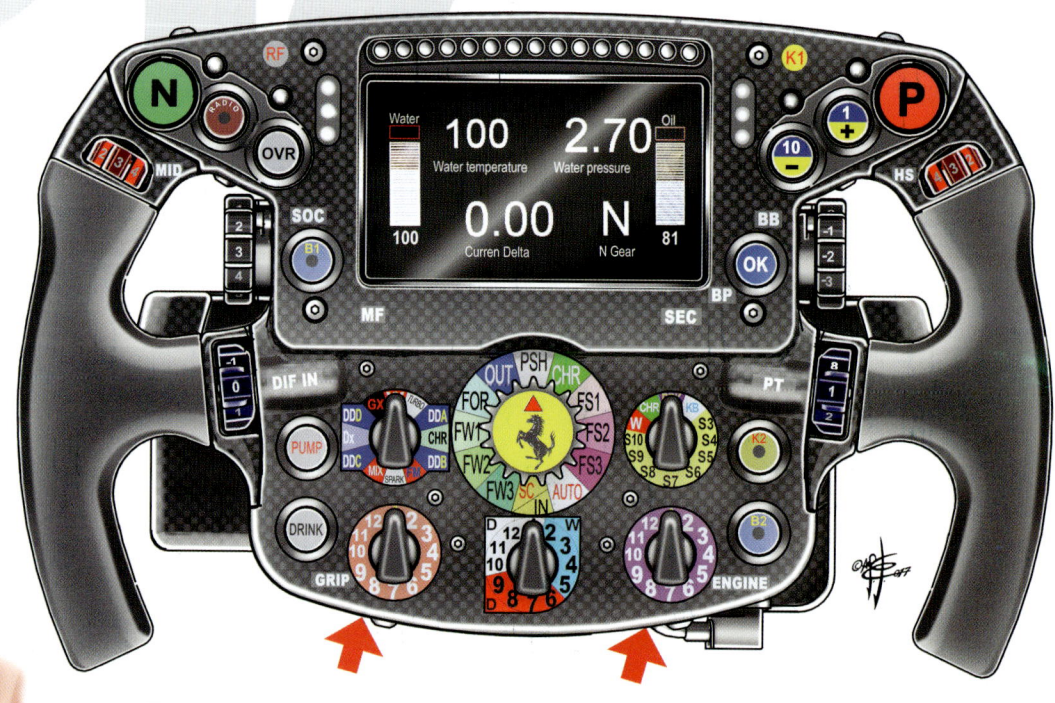

The 2017 season again saw steering wheels subjected to intensive technical development in the wake of further restrictions regarding clutch use.

A limit of 80 mm on clutch paddle travel was introduced to ensure drivers had no electronic assistance at the start.

Mercedes and Ferrari were at the forefront, with the German company surprising everyone in pre-season testing in Barcelona with a steering wheel equipped with two new mouldings for the driver's fingers on the clutch paddles, of which only one of the two was usable at the start. The feature was then copied by Red Bull, McLaren and Williams, with Ferrari eventually adopting it at the Spanish Grand Prix, but only on Vettel's car.

FERRARI COCKPIT-STEERING WHEEL

In order to make the SF70H steering more ergonomic than that of the SF16-H, Ferrari brought into its centre the two knobs that on the 2016 car were located either side of the dashboard, in a vertical position semi-concealed by the steering wheel itself. The Grip knob proved to be particularly important as a tyre management tool.

FERRARI STEERING WHEEL

(1) Multipurpose display with, at the top, the sequence of lights indicating gear change timings.
(2) FIA flag LEDS (blue, green and red).
(3) Warning light for the KERS actuated via the button on the back of the wheel.
(4) Menu navigation function.
(5) Pit limiter.
(6) High speed differential.
(7) Brake bias adjustment.
(8) Control confirm.
(9) Engine braking.
(10) Second KERS control.
(11) ERS strategies.
(12) Spare.
(13) Engine mapping.
(14) Torque.
(15) Multifunction knob.
(16) Traction and tyre strategies.

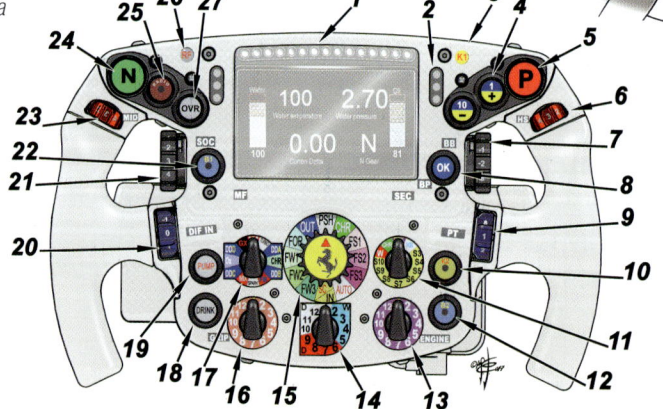

(17) Presets for dry, wet, turbo and torque.
(18) Water.
(19) Oil pump for supplementary lubrication.
(20) Turn-in differential management.
(21) Charging levels management.
(22) Battery recharging.
(23) Centre of corner differential management.
(24) Neutral.
(25) Radio.
(26) Rear flap LED.
(27) DRS.

2017 SEASON

For the 2017 season the FIA imposed a limit of just 80 mm on clutch paddle travel to avoid the creation of a form of manual traction control.

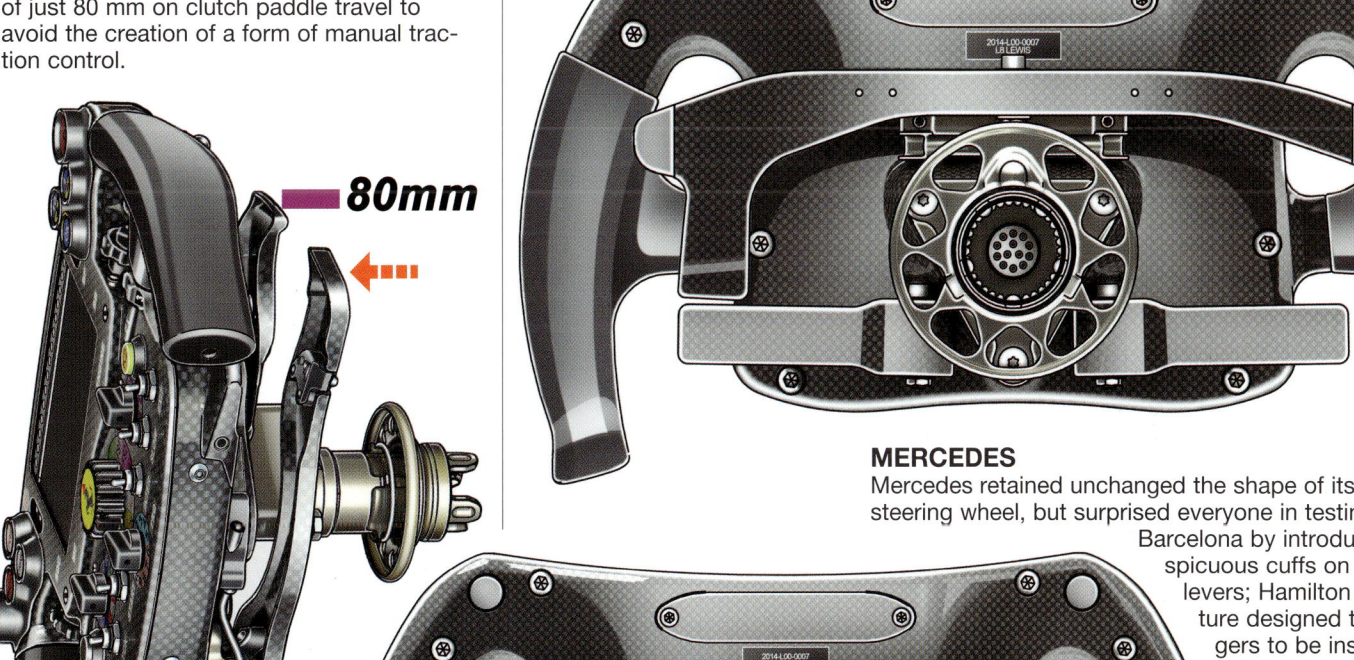

80mm

MERCEDES

Mercedes retained unchanged the shape of its steering wheel, but surprised everyone in testing at Barcelona by introducing two conspicuous cuffs on the clutch levers; Hamilton tested this feature designed to allow the fingers to be inserted to release the clutch very quickly. The cuff system was based on a spring that had a very soft action on the lever, therefore permitting great slow release modularity. The shape of the cuff allowed the driver to return to a neutral position very quickly.

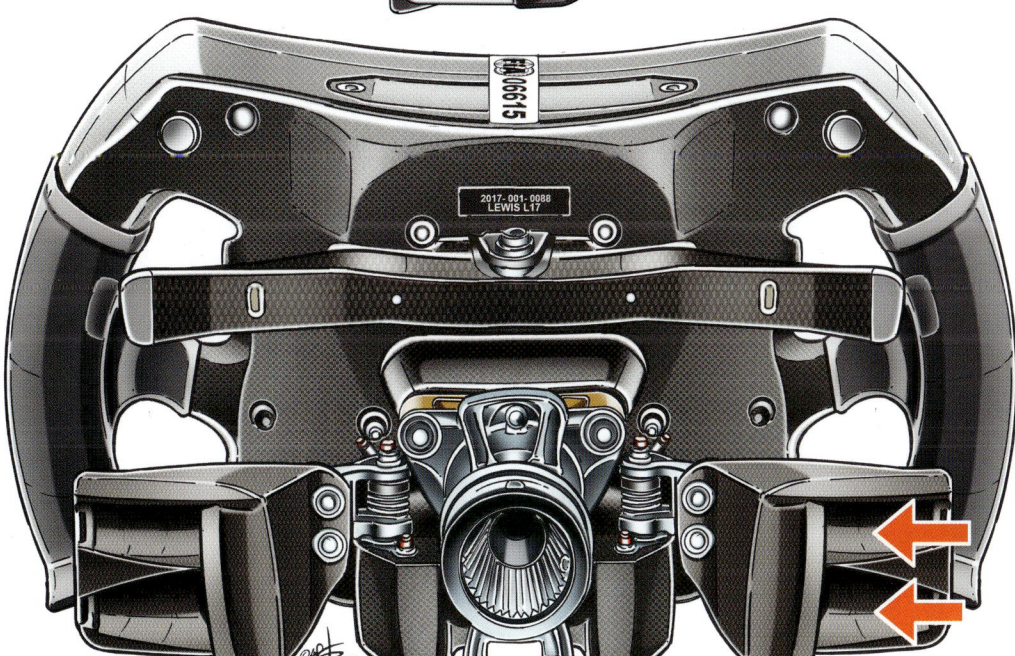

MERCEDES HAMILTON

Only at a later stage was the Mercedes steering wheel redrawn, more detailed than before and executed shortly after the Barcelona tests.

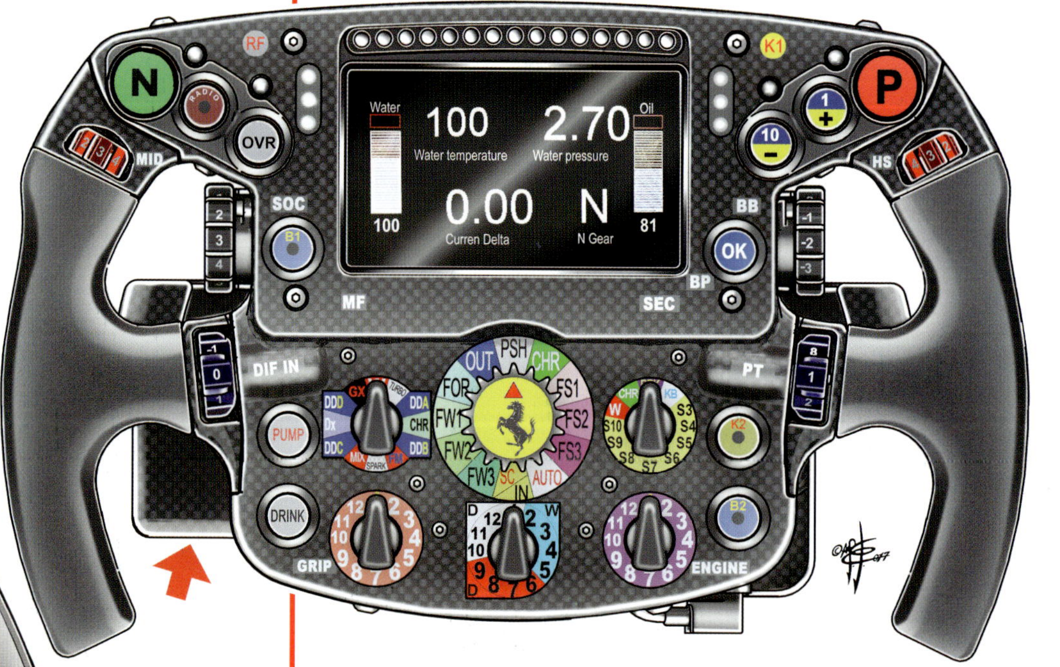

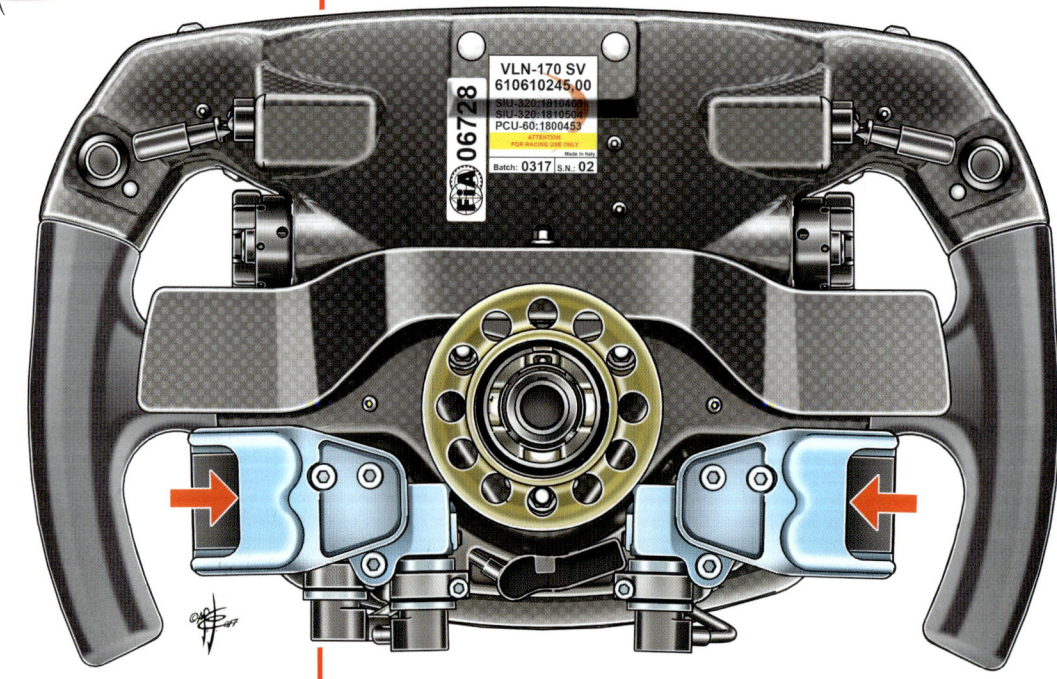

VETTEL AND RÄIKKÖNEN STEERING WHEEL COMPARISON

The three drawings show Vettel's steering wheel front and rear. In the comparison with Räikkönen's wheel we can clearly see the differences in the positioning of the controls between the two Ferrari drivers.

FERRARI VETTEL

Ferrari adopted the Mercedes feature at the Spanish Grand Prix. At scrutineering on the Thursday, Vettel's car had a steering wheel equipped with a single paddle with mouldings in aluminium, as seen in the drawing, but from the Friday morning sessions Vettel used an evolved version with dual finger mouldings.

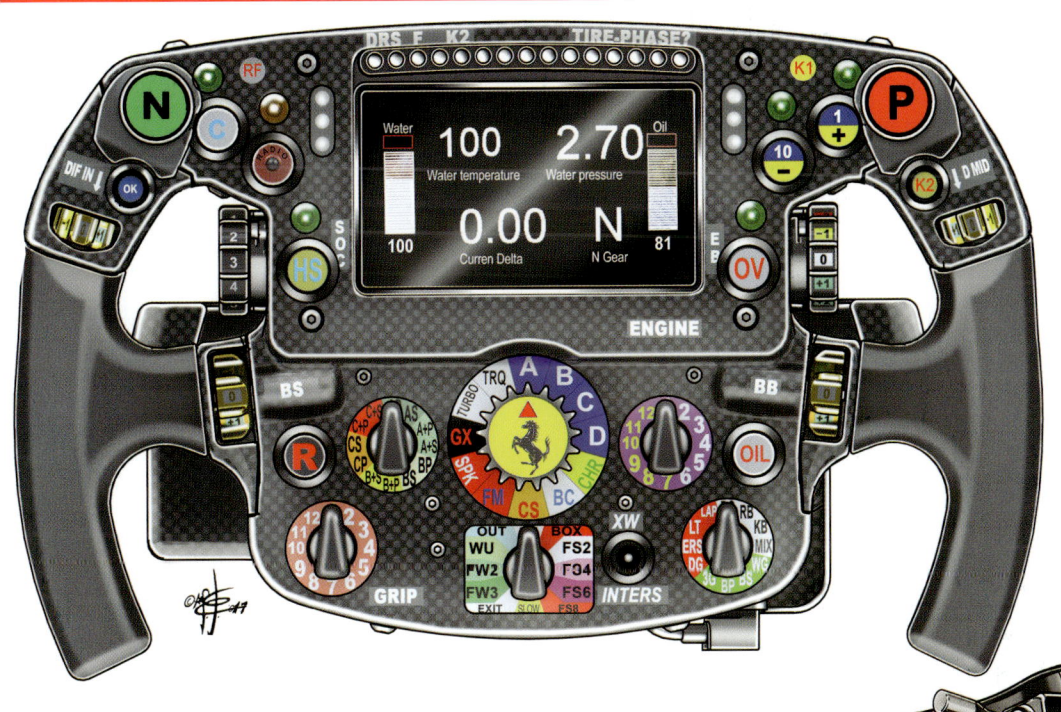

VETTEL

Ferrari later realised the two clutch paddles with more sophisticated shapes and Vettel used this version from the Spanish Grand Prix through to Singapore.

RÄIKKÖNEN

Räikkönen instead never used the version with twin mouldings and his better starts then convinced Ferrari, above all after the incident at the start in Singapore, to use the Finn's version on Vettel's car too.
At Suzuka, the German driver went back to the same clutch paddle configuration for the start, retaining the long rocker through to the end of the season.

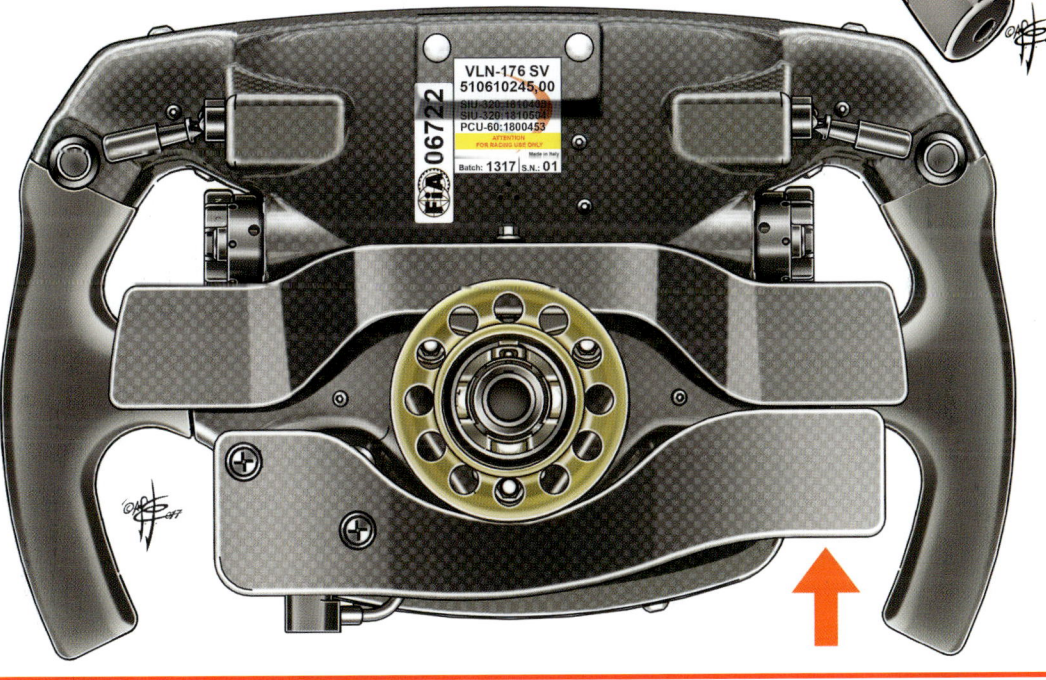

McLAREN

Seen from the front, the McLaren steering wheel was very simple and rational, while at the rear it featured two clutch paddles set some distance from the wheel and had a butterfly-shaped moulding for the driver's fingers, along the lines of the Mercedes moulding.

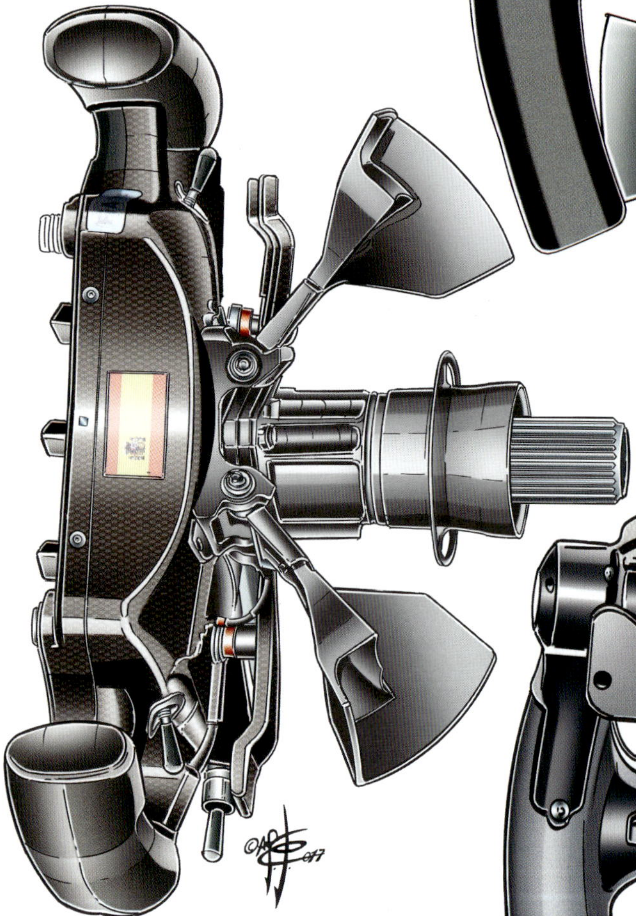

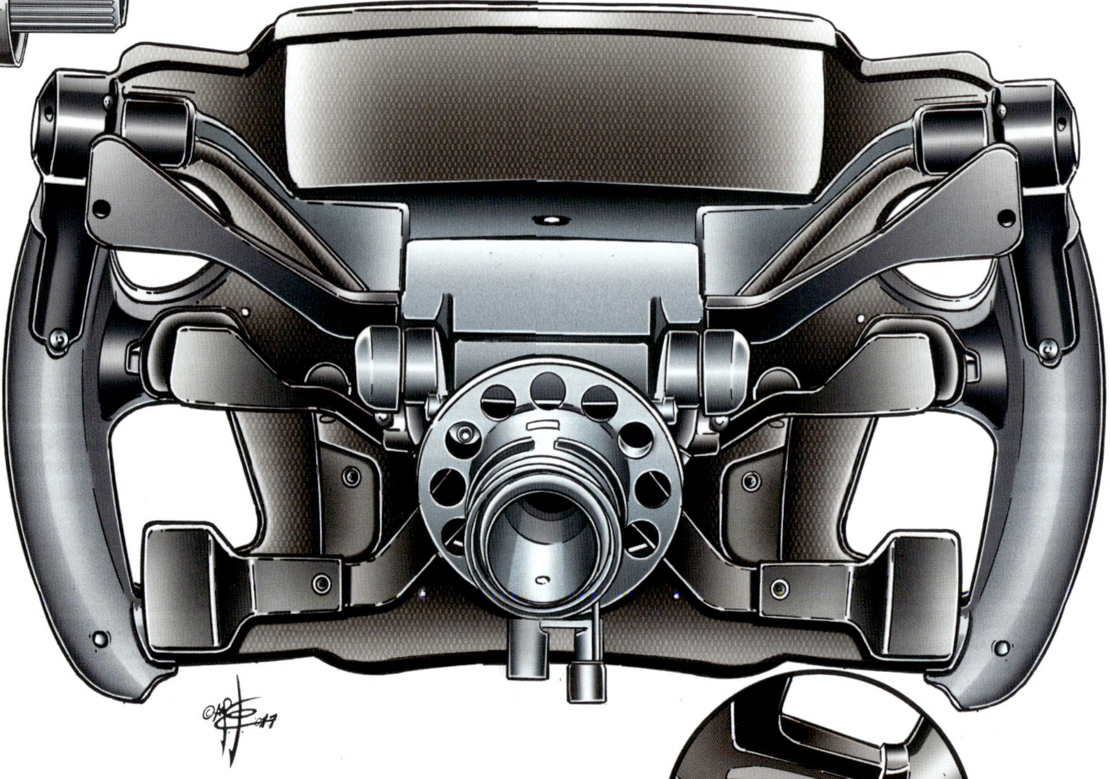

RED BULL

Red Bull simply retained the 2016 steering wheel, updating it with two small finger mouldings added to the clutch paddles.

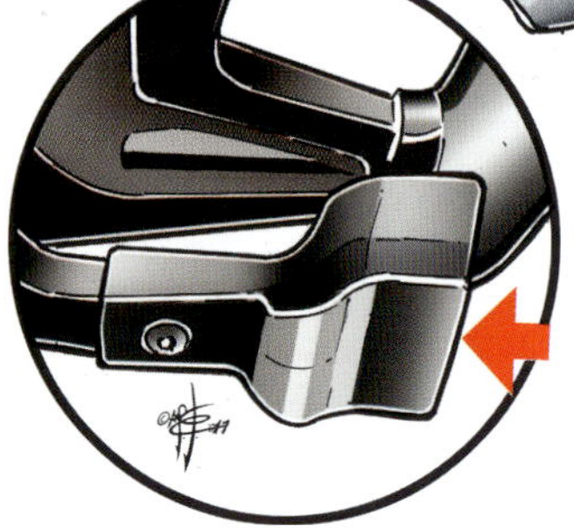

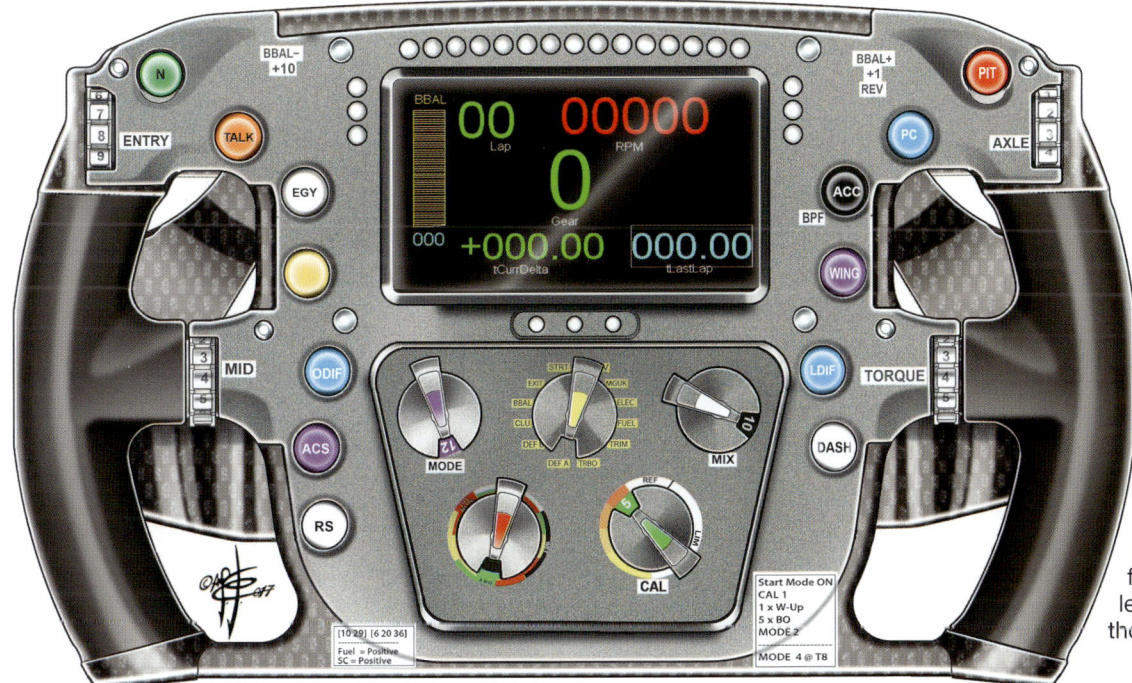

FORCE INDIA

Force India also retained the previous season's steering wheel, with very minor modifications and made with a rapid prototyping technique, as shown by the colour of the material. At the rear, Force India used the finger moulding system, albeit with less extreme shapes compared to those of the McLaren.

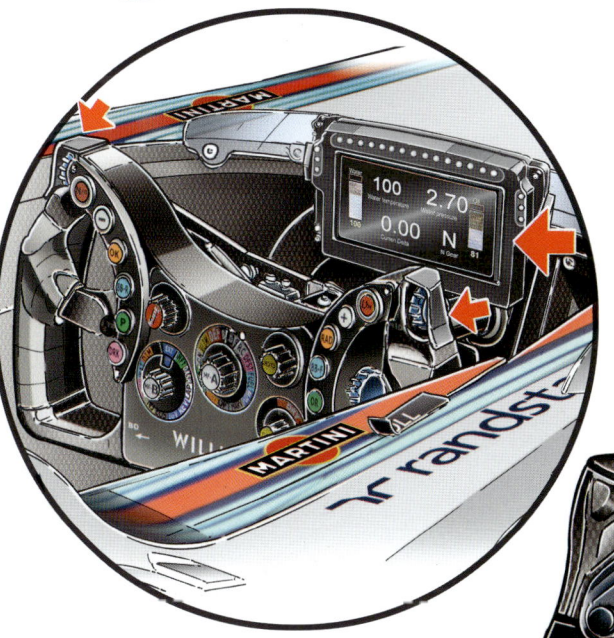

WILLIAMS

Williams was the only team in 2017 to retain the display attached to the dashboard along with a butterfly steering wheel. This was for weight reasons, as was the case in the 2015 and 2016 seasons. The rear view shows that Williams too adopted clutch paddle finger mouldings.

MERCEDES

CONSTRUCTORS' CLASSIFICATION			
	2016	*2017*	
Position	1°	1°	=
Points	765	668	-97 ▼

A fourth consecutive one-two for Mercedes in the 2017 season, with Hamilton winning his fourth World Championship (third with Mercedes) and the W08 topping the Constructors' standings in a season that instead of the total domination of 2016 (19 wins and 20 pole positions in 21 races) saw a close fight with Ferrari, a duel that concluded with the chronic lack of reliability shown by the SF70H in the final races. In 2016 Hamilton had won 10 races and Rosberg (the eventual cham-

pion) 9, leaving Red Bull with the two remaining races, Ricciardo a fortunate winner in Malaysia (a broken engine for Hamilton) and Verstappen successful in Spain thanks to the dramatic first lap collision between the two Mercedes drivers.
In 2017, Hamilton's haul fell to 9 victories, with just 3 going to Bottas and no less than 8 being harvested by rival teams. 5 for Vettel, 2 for Verstappen and 1 for Ricciardo. In the season of the great technical revolution intro-

duced by the FIA, Mercedes maintained the engineering philosophy underpinning the W07, foregoing the rake set-up despite its excellent results on the Force India, a car with much in common with the Silver Arrows given their shared power unit-gearbox assembly. The surprise that was seen from the presentation of the W08 was represented by the wheelbase, no less than 226 mm longer than that of the W07 and significantly longer than all the other cars. A choice dictated by

the absence of positive rake and the possibility of obtaining greater aerodynamic loading it offers. The same result was in any case obtained by distancing the two axles and therefore increasing the surface of the underbody with a longer wheelbase. This configuration brought a number of advantages but an increase in weight that made it more difficult to adapt the car to the characteristics of the various circuits, especially when compared to its Ferrari rival. This handicap was

Mercedes W07
Abu Dhabi

Mercedes W08
Launch

Mercedes W08
Melbourne

eliminated from the Spanish GP onwards with the introduction of an "almost B" version with new aerodynamic features and a considerable reduction in weight. The wheelbase apart, from its presentation the W08 always appeared to be an extreme car with a chassis and a nose that were as narrow as possible to improve penetration and efficiency, while the sidepods were actually reduced in size despite the greater width permitted by the regulations (an extra 10 cm each side). It was no coincidence that the Mercedes was always very fast in a straight line and not just

thanks to its muscular Power Unit. The conspicuous vertical flow vanes in front of the sidepods and that kind of horizontal saw-tooth feature in the lower part (highlighted in the oval) that were introduced at the presentation of the car in 2016 were retained and actually uprated thanks to the liberalisation of the 2017 regulations.
Other features that were introduced on the W07 were also retained, including the division in the front part of the chassis to permit improved front suspension installation, although the configuration did lack some of the sophisticated features used in 2017.
The heave spring controls the vertical forces upon the car. At several races in 2016 Mercedes had run a front heave spring with

asymmetric valves that would allow the front of the car to dive under braking as normal but then rise up only slowly, allowing the leading edge of the floor to remain closer to the ground through the corner. This would partly replicate the effect of a high-rake floor but without the associated penalty in drag.
This system had been under development at Mercedes for several years but was raced for the first time in selected events in 2016. This gave the team the confidence to conceive the 2017 car around the technology and was a key part in the choice of a low-rake/long wheelbase car. However, the technology was banned on the eve of the season (following queries from Ferrari), leaving Mercedes with a less extreme design.

This created a number of problems, with the car being twitchier and with an optimum window of use that was narrower than that of the SF70H which frequently proved quicker in race conditions (Melbourne, Sakhir, Monaco, Budapest, Spa, Singapore, Sepang and Mexico City). Once again Mercedes confirmed its primacy in the exploitation of the Power Unit (see the Engines 2017 chapter), with an important evolution being introduced at Spa.
The last step came in Brazil with engine 5 effectively previewing the 2018 Power Unit and entrusted to Hamilton for qualifying and the race. In climbing from last to fourth place he proved that he could be competitive with a power unit designed to run with a limit on oil consumption of 0.6 litres per 100 km.

Mercedes W08
Barcelona

Mercedes W08
Monza

Mercedes W08
Mexico City

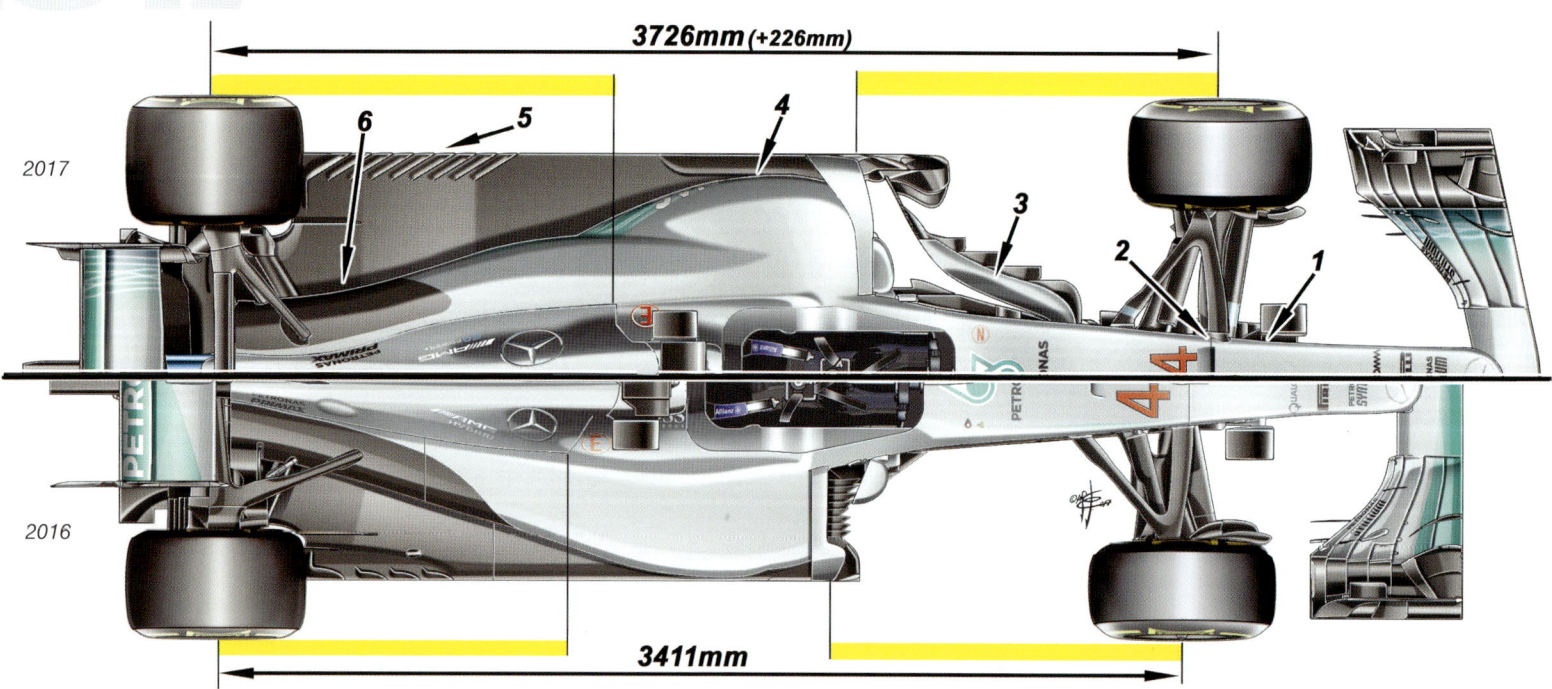

3726mm (+226mm)

2017

6 5 4 3 2 1

2016

3411mm

OVERHEAD COMPARISON

There is a macroscopic difference in the overhead comparison between the W08 and the 2016 car and not just because of the new regulations (wider tyres and greater wing and bodywork-underbody width). The wheelbase was increased by no less than 300 mm, distancing both the front and rear axles from the centre of the car to reduce the toxic effects of front wheel turbulence on cooling and to have a greater floor area ahead of the extractor. 1) Notable narrowing of the front part of the chassis and the nose. 2) S-Duct introduced for the first time.
3) Sophisticated new arrowhead turning vanes in an area liberalised by the new regulations. 4) The sidepods did not exploit the full limit of 160 cm in width permitted by the regulations. Note the longitudinal slot in the floor. 5) The cutaways in the area ahead of the rear wheels were also augmented. 6) The end part of the engine cover was narrower.

FRONT SUSPENSION

Not without a degree of controversy, Mercedes introduced from the presentation of the car (together with Toro Rosso) this feature: a new upper wishbone mount, set very high and protruding beyond the week rim.

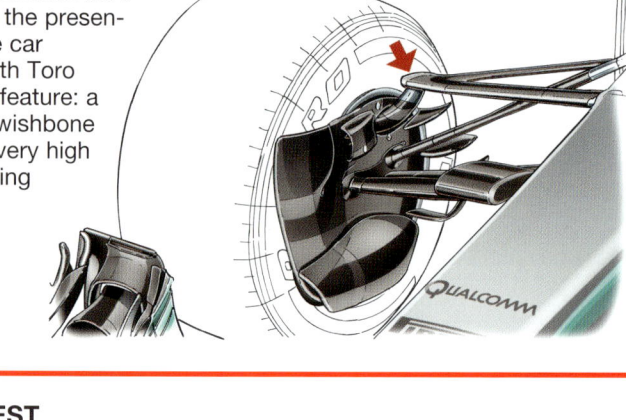

FRONT COMPARISON
W07 (2016) W08 (2017)

The front view clearly shows the technical revolution introduced by the FIA for the 2017 season, with wider tyres, a wider front wing and a lower and wider rear wing. On the body of the car, note the more tapering nose, the introduction of the S Duct and the slightly larger engine air intake.

2016

2017

DEBUT/TEST

From the car's track debut a T wing was used thanks to loophole in the regulations. The three sections of the air box were equipped with pitot tubes to check air speed and pressure and to cool the various components of the power unit. Note the video cameras incorporated in the rear-view mirrors to check the front tyres.

Launch

test

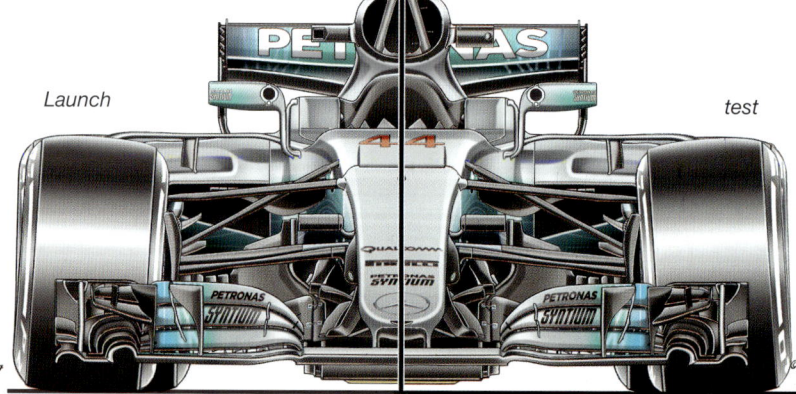

BARGEBOARD 2016

Mercedes had surprised every-one in the 2016 season, intro-ducing an incredible degree of sophistication in the bargeboards ahead of the sidepods with no less than six separate elements, with horizontal "teeth" in the lower part (highlighted in the oval), destined to manage the flow of air towards the lower area and the diffuser. A brand new and efficient feature reprised in 2017 by almost all the teams and of course by Mercedes itself.

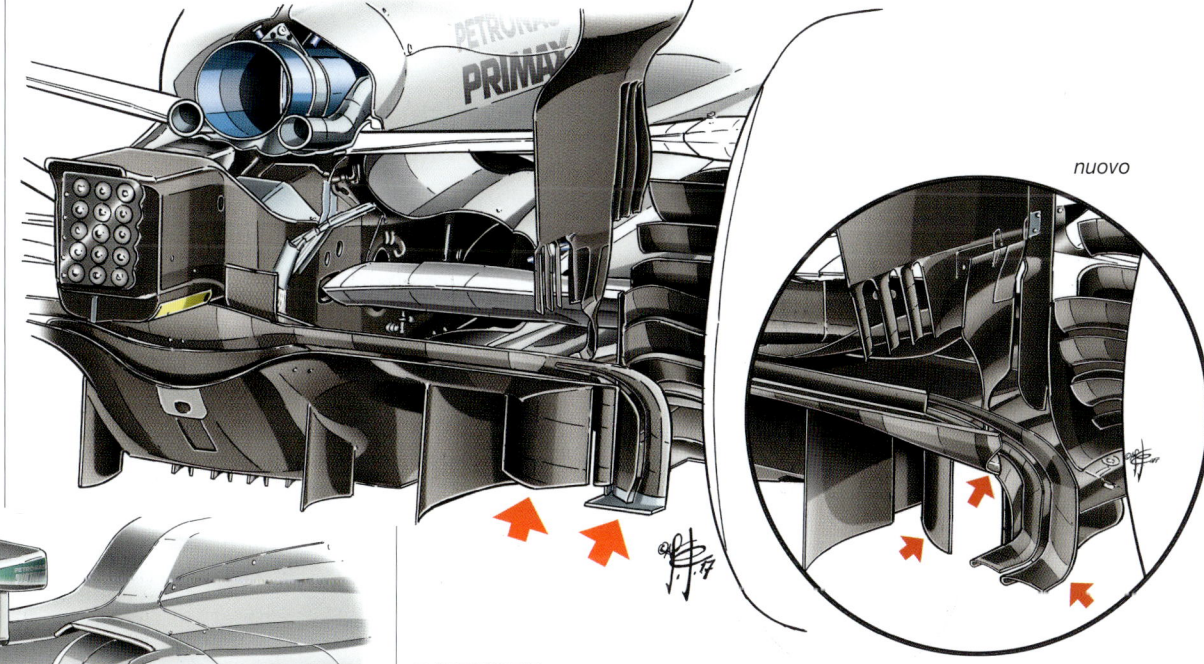

nuovo

DIFFUSER

The diffuser for the 2017 cars was considerably uprated with all-new features such as these horizontal extensions facing backwards and designed to reduce the negative effects of the rear wheels. A compari-son between the testing configuration and the version raced in Melbourne (insert). The differences also extended to the bargeboards and the extractor profile (red arrows).

MELBOURNE

The biplane version of the T wing was used in the opening race of the season.

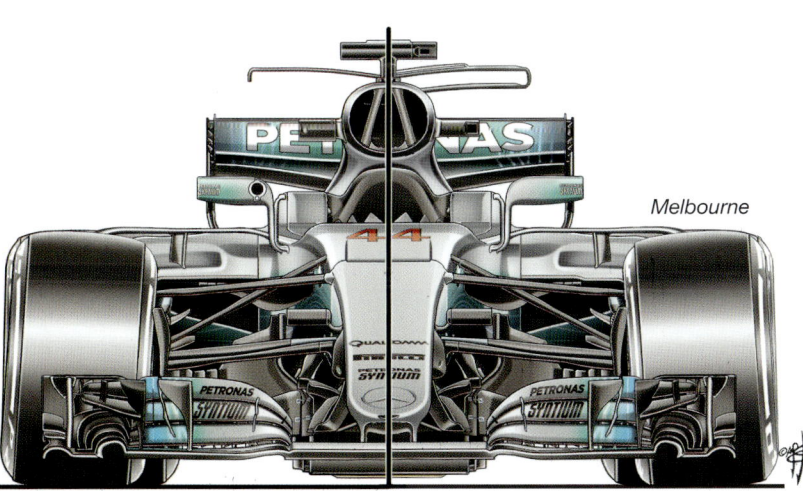

Melbourne

BARCELONA

What was almost a "B" version was raced at Barcelona with a new "cape" feature, a kind of chute in the lower part of the nose, described in detail in the New Features 2017 chapter.

The nose was even more tapered. New turning vanes below the chassis and a reduction in weight of almost 5 kg.

Barcelona

3

1   2

DEBUT MELBOURNE

Mercedes took full advantage of the liberalisation of the area between the front axle and the leading edge of the sidepods. The comparison shows the configuration introduced when the car made its track debut and the one raced in Melbourne. 1) At the base of the turning vanes were no less than four "fingers" facing forwards, while the bargeboards presented (2) vertical slots. 3) The small turning vane in the Bat Wing area (4) also had three distinct elements rather than one.

FRONT SUSPENSION

When Mercedes and Toro Rosso presented their new cars they aroused a degree of surprise with this new upper wishbone mount, set very high and protruding beyond the wheel rim (New Features 2017). The drawing also highlights the vertical slot introduced in the shields inside the front wheels.

FRONT WING

New front aerodynamics for the W08 with a vent (1) in the triangular vane close to the central neutral zone, with a different shape to the cutaway (2) in the large vanes inside the front wheels. The air intake trumpet was equipped with a kind of barred protection (3).

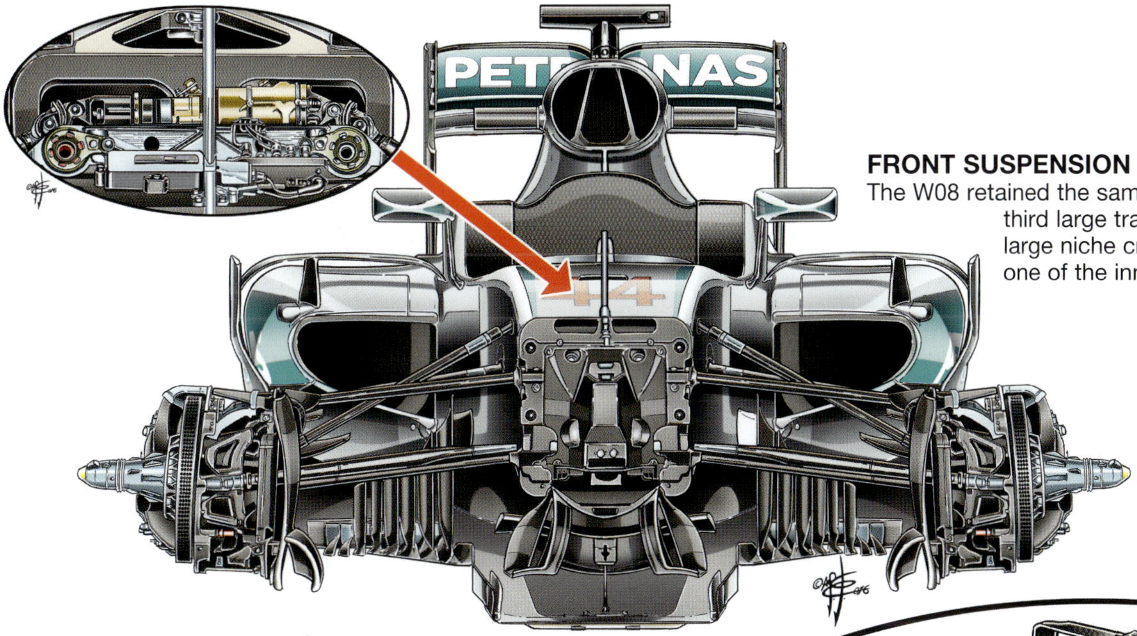

FRONT SUSPENSION

The W08 retained the same front suspension configuration with a third large transverse hydraulic damper housed in the large niche created in the front part of the chassis, one of the innovations introduced on the W07.

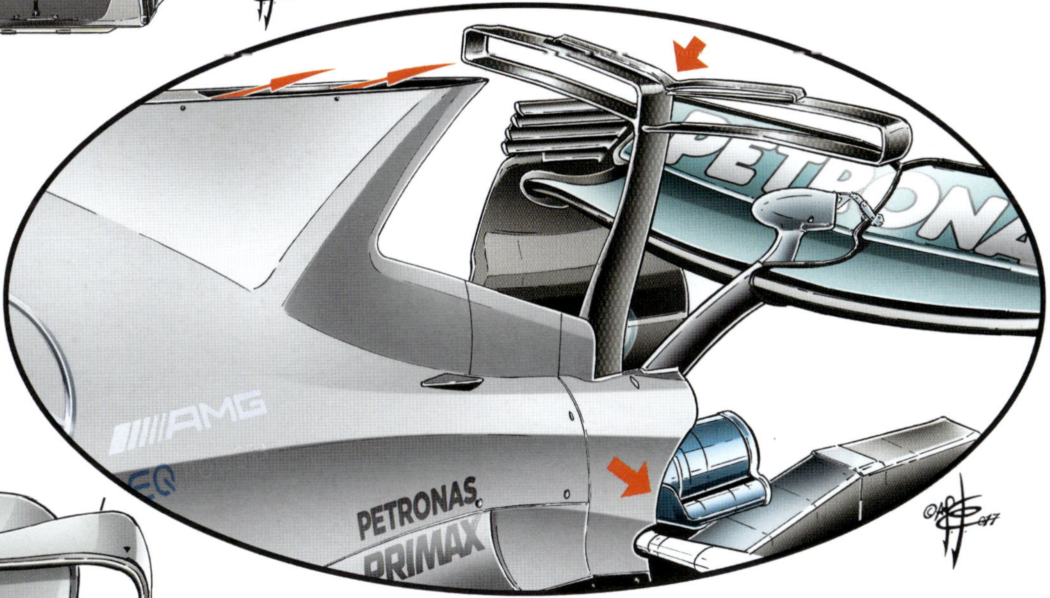

T WING MELBOURNE

A third flap with a gullwing shape, with a small V in the central section (arrow), was added to the central part of the upper profile of the T Wing. In this way another vent was created, while the vertical strap on the edges of the two main elements was eliminated given that the two profiles were linked with a curving element without angular endplates

BAKU

The longitudinal vent that also appeared on the Mercedes to feed the flow of air in the lower part of the underbody is highlighted. The lower part of the sidepods is notably narrower than the floor, the maximum width of which was increased from 140 to 160 cm.

BAKU SPOILER

At Baku a new aerodynamic appendix in the form of a wave was mounted in proximity to the rear arm of the upper wishbone attached to the trailing edge of the sidepods with two metal supports, thus creating a vent that was useful in augmenting the downforce and improving the extraction of hot air.

2016

BARCELONA

There was an aerodynamic revolution for the Spanish GP. The turning vanes (very similar to those of the W07) that had the task of deviating the Y250 vortex were completely redesigned, while the Bat-Wing, the profile anchored below the chassis on the data collection bulb.
In their place a kind of chute was introduced that started high on the sides of the nose and became a form of front diffuser. A new feature, known as the "cape" within the team, proved to be very efficient in directing the flow towards the lower part of the car.

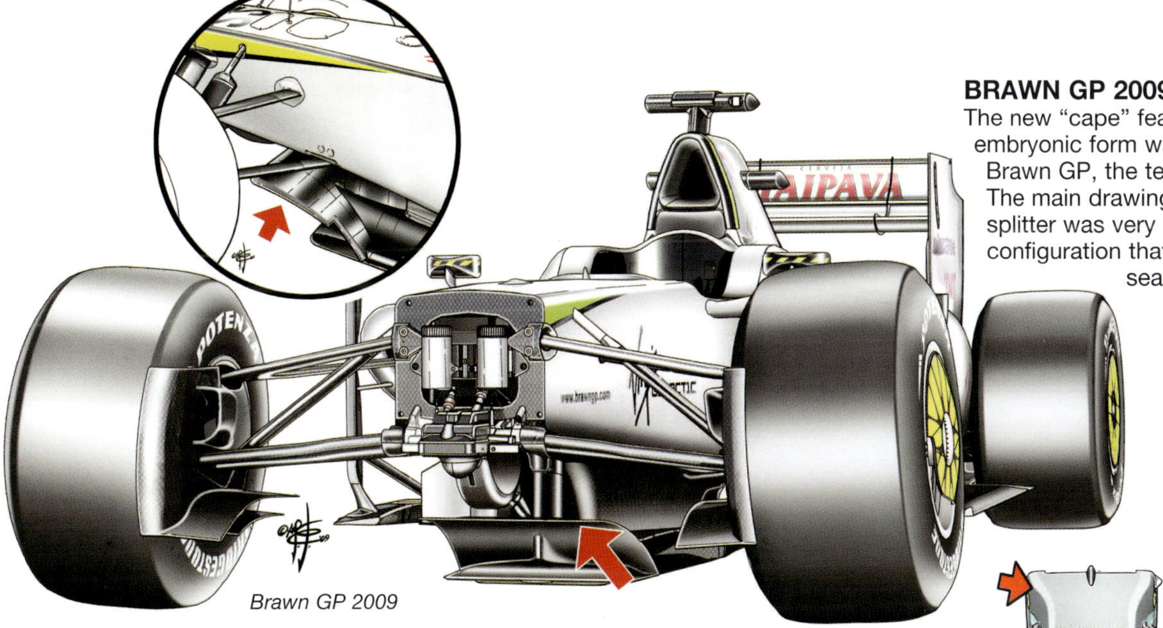

BRAWN GP 2009

The new "cape" feature resembles the splitter that in embryonic form was seen in 2009 on the cars run by Brawn GP, the team that eventually became Mercedes. The main drawing and the detail show how originally the splitter was very small and linked to the car's T-tray, a configuration that was copied by Williams for the 2010 season.

Brawn GP 2009

NARROW NOSE

The narrowing of the nose was clear at the point where it joined the chassis (indicated by the arrow). Naturally the new component had to pass the homologation crash test.

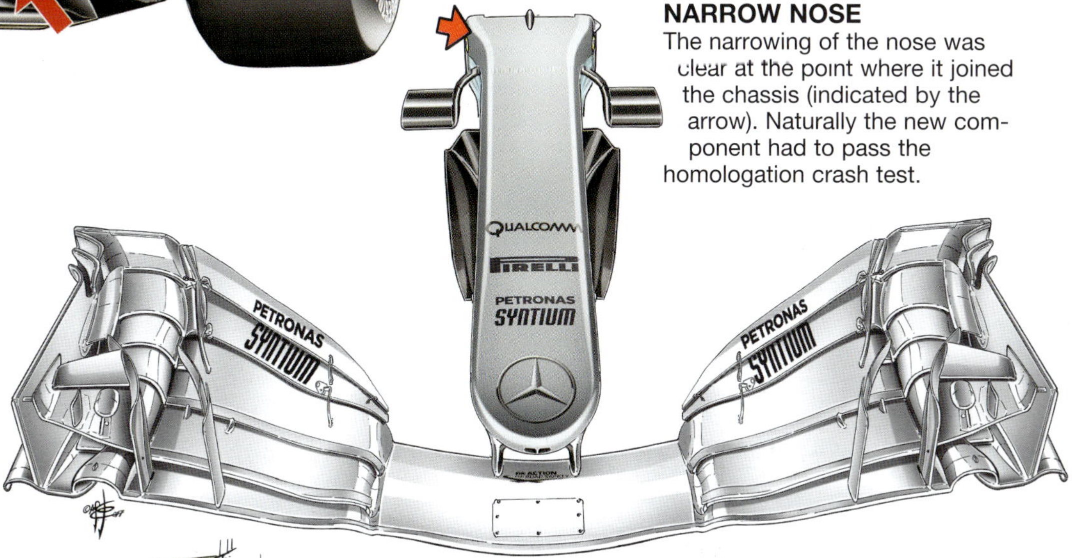

MONACO

The latest detail modification to the brake air intake, with the notch in the upper even more evident. The insert shows the feature as it was at the beginning of the season.

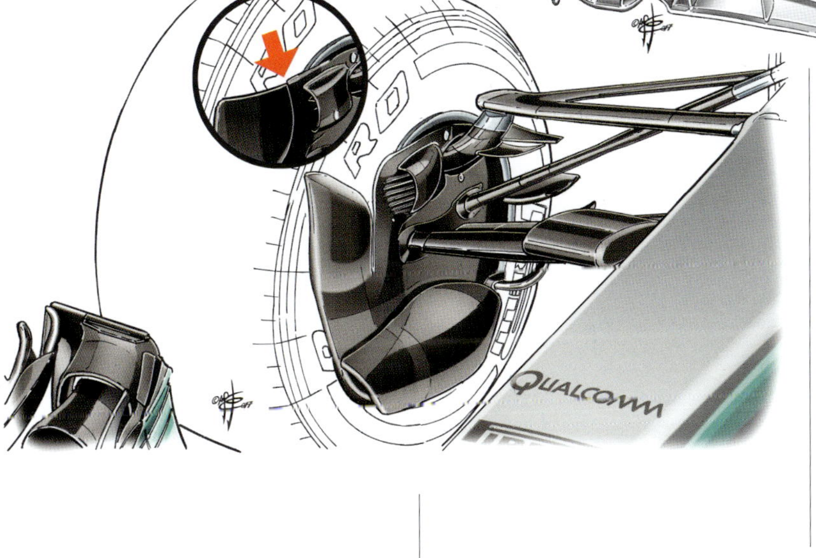

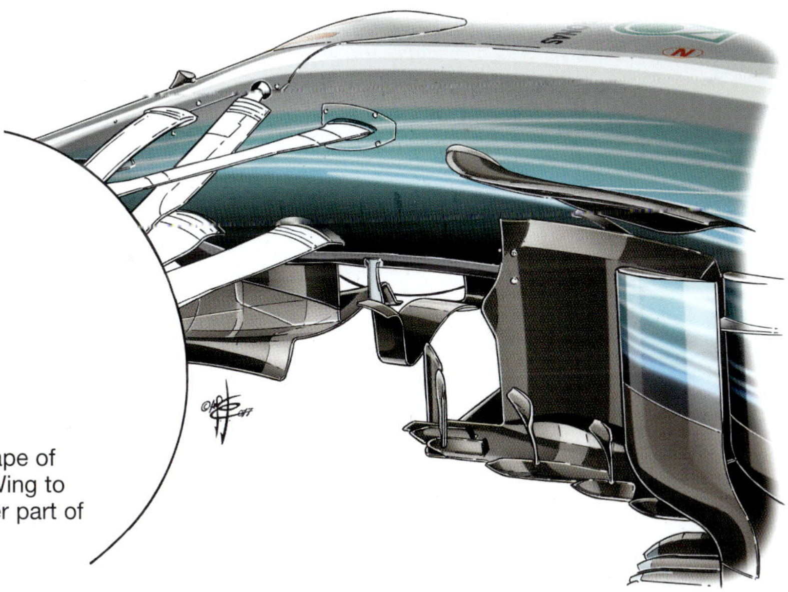

"CAPE"

This view clearly shows how the shape of the "cape" integrated with the Bat Wing to better groom the air flow in the lower part of the car.

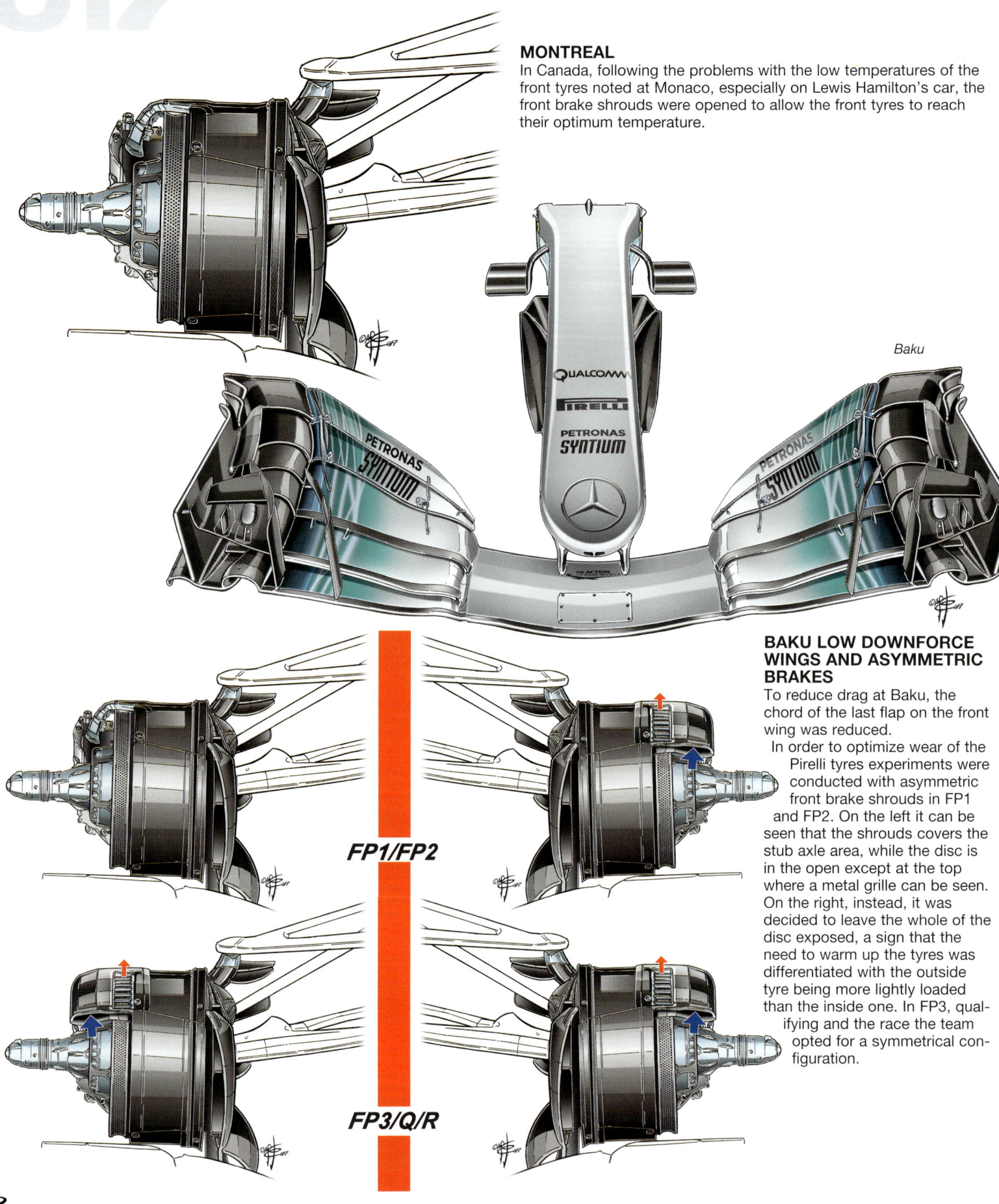

MONTREAL

In Canada, following the problems with the low temperatures of the front tyres noted at Monaco, especially on Lewis Hamilton's car, the front brake shrouds were opened to allow the front tyres to reach their optimum temperature.

Baku

BAKU LOW DOWNFORCE WINGS AND ASYMMETRIC BRAKES

To reduce drag at Baku, the chord of the last flap on the front wing was reduced.

In order to optimize wear of the Pirelli tyres experiments were conducted with asymmetric front brake shrouds in FP1 and FP2. On the left it can be seen that the shrouds covers the stub axle area, while the disc is in the open except at the top where a metal grille can be seen. On the right, instead, it was decided to leave the whole of the disc exposed, a sign that the need to warm up the tyres was differentiated with the outside tyre being more lightly loaded than the inside one. In FP3, qualifying and the race the team opted for a symmetrical configuration.

FP1/FP2

FP3/Q/R

ZELTWEG

In Austria the end section of the "cape" was subjected to a slight modification with a vent used in qualifying and the race by Hamilton alone. The winner Bottas instead raced with the standard configuration.

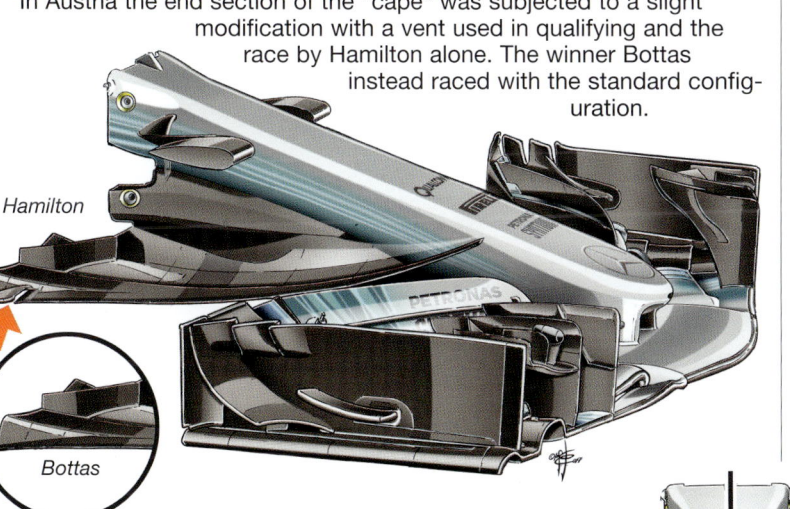

Hamilton

Bottas

SPA

On the fast track at Spa, Mercedes provided both drivers with a "cape" vented in the end part and combined with a low downforce rear wing with the final flap having a reduced chord.

DIFFUSER

A new profile appeared at the top of the diffuser that had a dual function, associating the aerodynamic grooming of the air flow with that of limiting flexing of the extractor at the point where detrimental vibration can be generated.

MONZA

At Monza Mercedes fielded a rear wing with a very short chord and a delta shape, which was then reprised in the design of the mobile flap.

Combined with this miniaturization of the main plane were endplates with three horizontal vents in the upper leading edge (Toro Rosso-style and already seen on the W07 at Silverstone – see the inserts). The Silver Arrows also retained the turning vanes at the end of the keels to groom the wake at high speed.

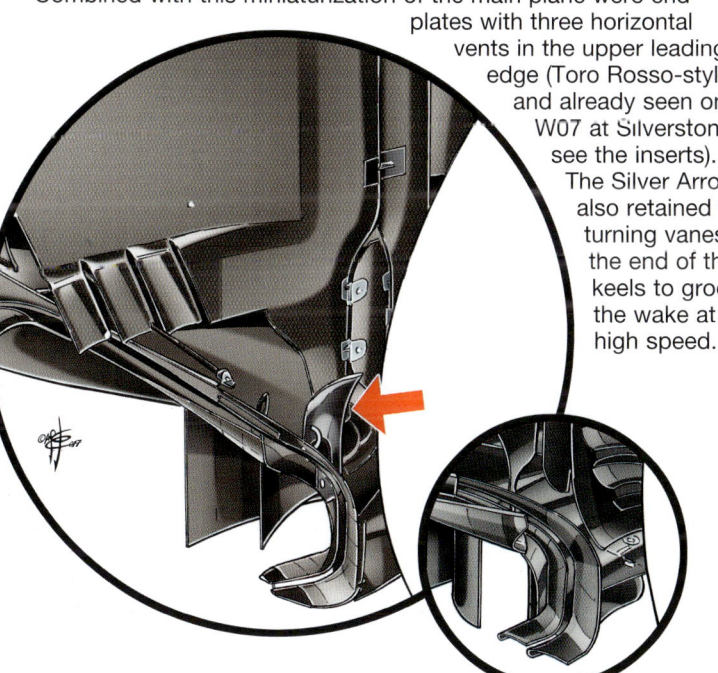

W07 Silverstone 2016

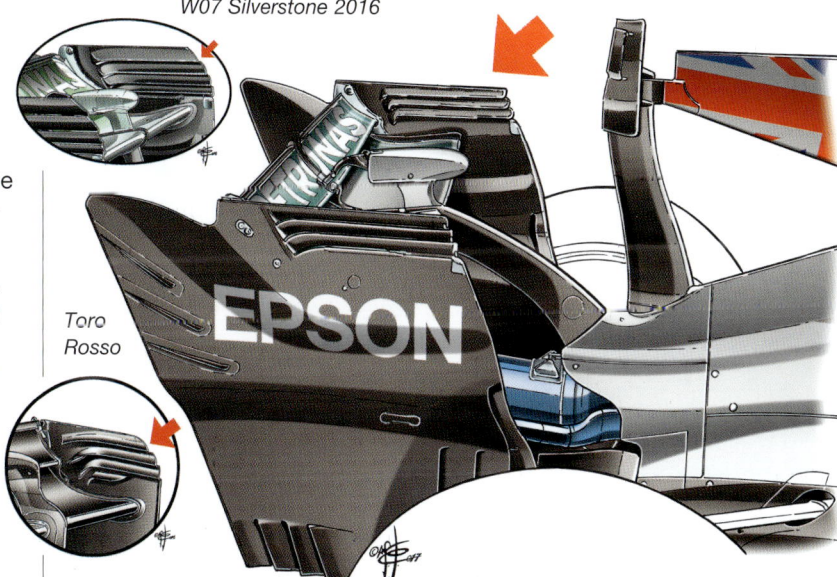

Toro Rosso

DIFFUSER

The latest modification to the diffuser: in detail the area around the rear wheel, with a Williams-style vertical fin (red arrow) and a different shape to the two extensions at the base.

GIORGIO PIOLA

SEPANG

A new aerodynamic package for Malaysia: not only was the width of the leading edge of the "cape" increased, but also the dimensions of the highly complicated barge boards that was affected by the different rates of flow generated by the front aerodynamic element. The three vortex generators set on a floor designed to flex as speed and downforce increased were also orientated differently.

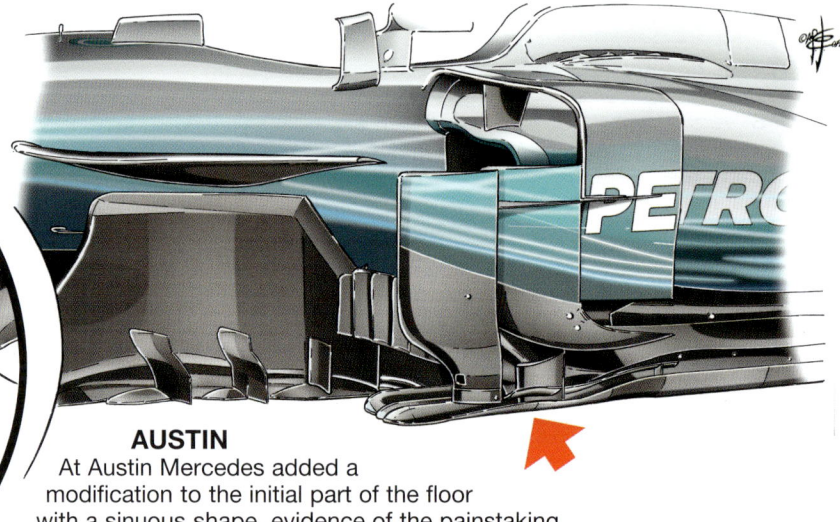

AUSTIN

At Austin Mercedes added a modification to the initial part of the floor with a sinuous shape, evidence of the painstaking attention paid to every minimal detail.

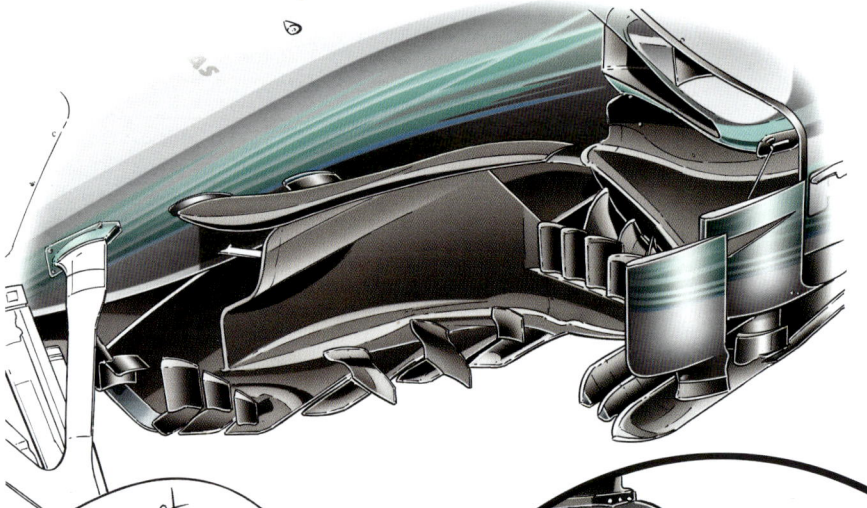

MEXICO CITY

Mercedes opened two gill-like vents to cool the power unit immediately behind the driver's head protection. They were vents that served to evacuate the hot air from the W08's sidepods and were added to the traditional slots either side of the cockpit.

A feature that had already been seen on the Williams at the hottest Grands Prix. The circuit dedicated to the two Rodriguez brothers is situated at an altitude of almost 2,300 metres above sea level with a rarefied atmosphere and for this reason it was more difficult to disperse the heat from the radiators.

INTERLAGOS

In Brazil Mercedes also worked on features with an eye to the 2018 season. Between the FP1 and FP2 both cars were fitted with a modification to the front suspension.

Lewis Hamilton and Valtteri Bottas used the standard configuration in the first session and then between the two the Brackley engineers moved the upper wishbone attachment point to collect data to transfer to the 2018 project.

BARCELONA

What was almost a "B" version was introduced at the Spanish GP: (1) The most innovative feature was the front chute. (2) The lower part of the barge boards was completely revised, as was that of the horizontal zone (3) at the base of the turning vanes. (4) There was also a new kind of "handle" in the area ahead of the rear wheels.

Barcelona

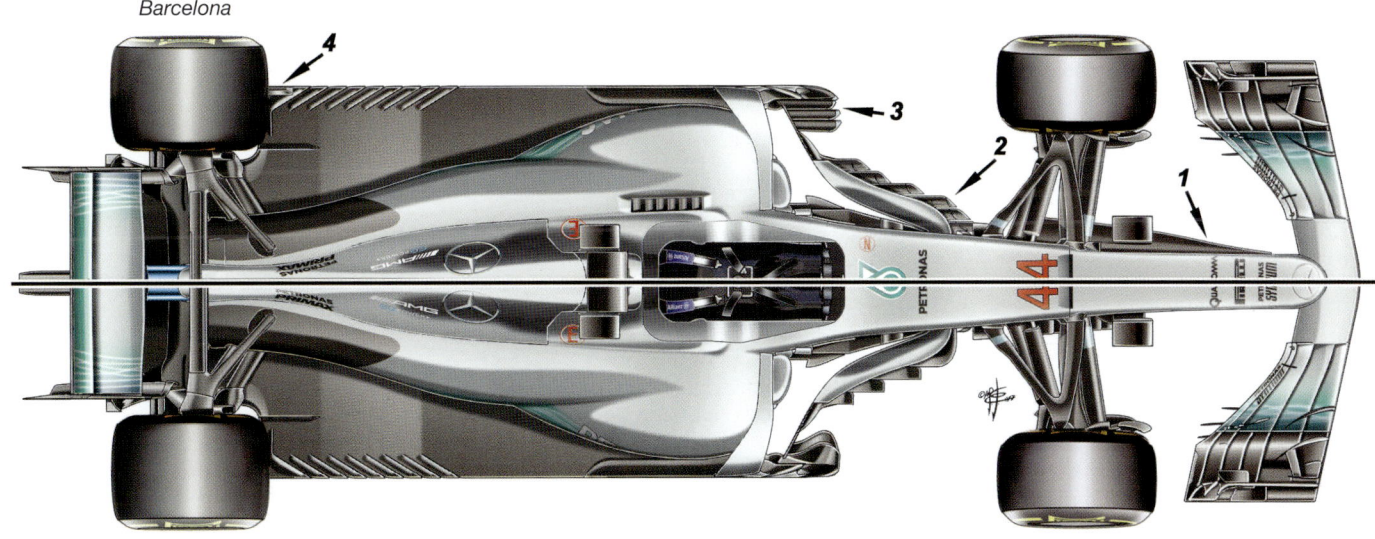

MONZA

The aerodynamic package for the Monza track: (1) Front wing with a reduced chord; (2) the slots either side of the cockpit were retained but the end part of the sidepods was lower and closed. (3) The rear wing had profiles with a significantly reduced chord and a delta shape.

Monza

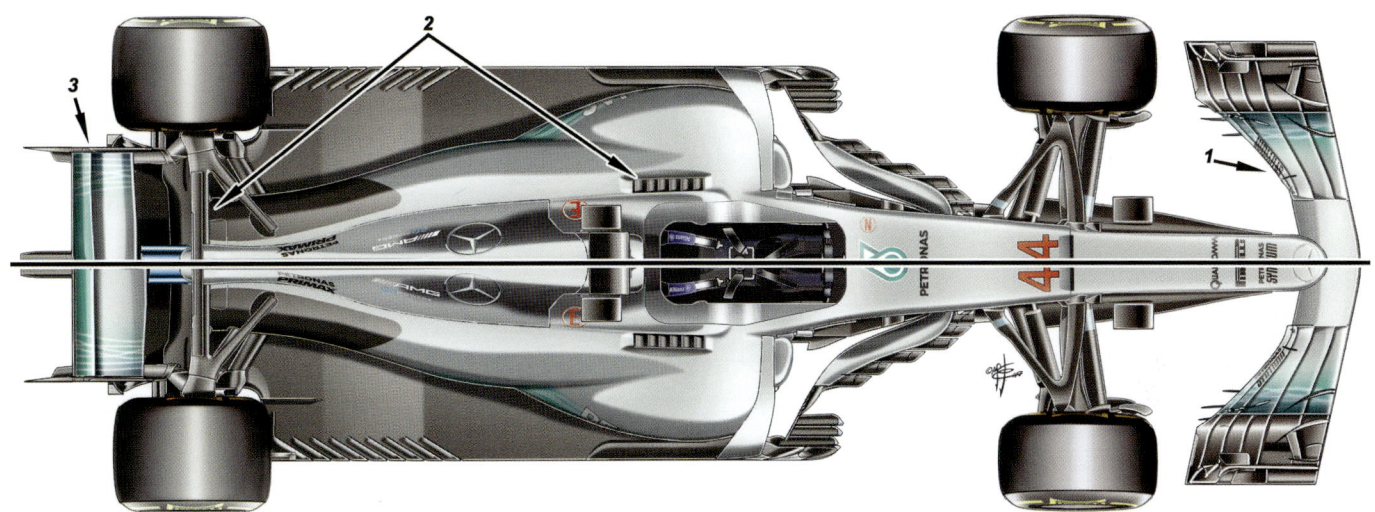

MEXICO CITY

The last major aerodynamic package was seen in Mexico with the "cape" (1) with a larger leading edge; this conspicuous slot (2) was opened in the engine cover fin to improve heat dispersal. (3) Larger rear vents and (4) a maximum downforce rear wing.

Mexico City

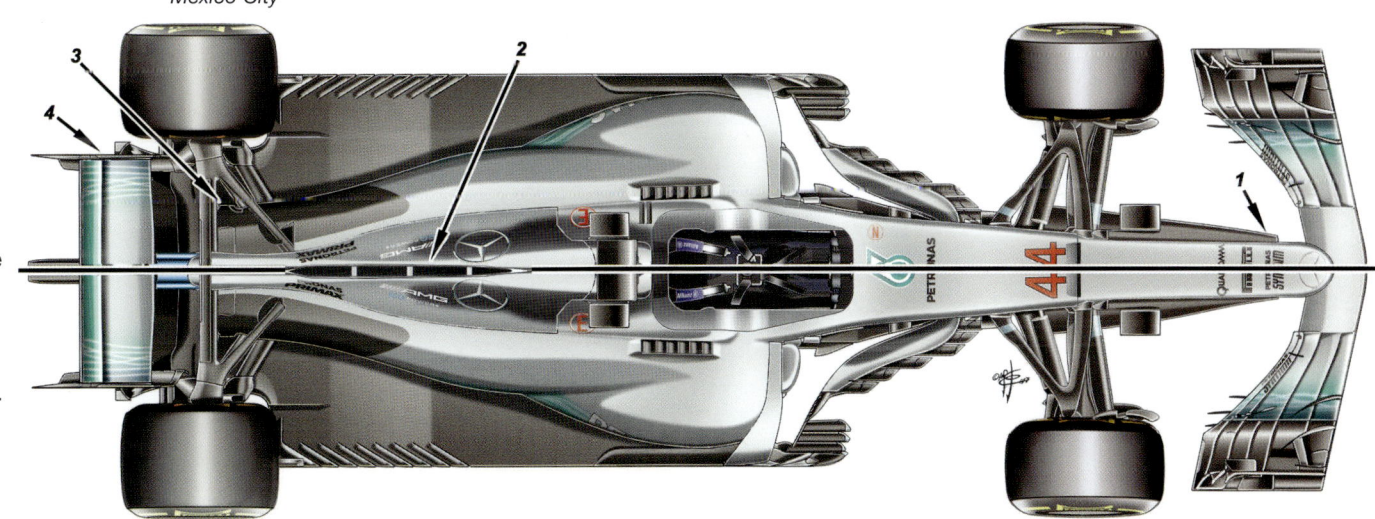

FERRARI

CONSTRUCTORS' CLASSIFICATION

	2016	2017		
Position	3°	2°	+1	▲
Points	398	522	+124	▲

With the SF70H Ferrari wiped out the disappointment of the 2016 SF16H (which served as the starting point for certain features such as the return to push-rod front suspension.) Above all it returned to playing a leading role, fielding a car that was the most innovative of the pack, with new features that would be widely copied. The greatest surprise came from the interpretation of the regulation relating to the shape of the sidepod openings and the positioning of the protection structures, with a location and shape rigidly imposed by the regulations. On the SF70H they were in fact preceded by a wing profile/duct that actually contained the protection structure (generally located within sidepods), exploiting a loophole that talked of bodywork and not the sidepods in particular. In this way, the sidepod openings were set further back from the front axle without the need to lengthen the wheelbase. The Ferrari's wheelbase was 175 mm shorter than the Mercedes,' which permitted more efficient use of ballast.

The 75° arrow-shaped inclination of the front of the sidepods required by the regulations was thus respected by the large aerodynamic profile, while the true sidepods were effectively set further back at an optimum 90° to the chassis. This configuration provided a great advantage in that it distanced the turbulence generated by the front wheels. Moreover, this aerodynamic duct led to a further innovative feature: a flush intake that draws air from the upper part of the car thanks to the vacuum effect generated by the new forward-set profile.

In a season in which it started from scratch, for the first time Ferrari fielded the most innovative car on the grid by exploiting

Ferrari SF16-H
Launch

Ferrari SF70H
Launch

Ferrari SF70H
Melbourne

loopholes in the regulations, where certain aspects were not described and therefore not forbidden. A different design philosophy that saw a greater contribution from the talented Rory Byrne who worked alongside Simone Resta during the development of the SF70H project, evolving the aerodynamic concepts produced by David Sanchez.
Not since 2008 had Ferrari built an all-new car that sailed close to the win in terms of the regula-

tions and it was no coincidence that there were many similarities with the F2008.
An example of this was the S-Duct, introduced as a great innovation on the F2008, banned by the FIA at the end of the season and reprised in 2012 in reduced form by Sauber and then adopted by many other teams but not by Ferrari, and continuing with the aerodynamic assembly between the front axle and the start of the sidepods.
It should be pointed out that these similarities were permitted by the freeing-up of the 2017 regulations in an area previously subjected to strict restrictions. This was particularly the case with the arrow-shaped profiles in the upper part of the chassis introduced to train the flow the

designers wanted to direct towards the upper sidepod aperture. The turning vanes also resembled those of the F2008. They were integrated with the sophisticated bargeboards, which had involved some extreme design research.
The candelabra with a dual horizontal vent was attached to the sidepod at the top and the raised lip of the floor at the bottom: the assembly was supported by the end section of the bargeboard. The extent to which all these aerodynamic appendages acted in synergy is evident.
Moreover, these appendages were in part copied by Red Bull in the aerodynamic package they introduced in Singapore. The generous cutaways in the floor were also new, both at the front,

with a design that earned the nickname "scimitar" in the area in front of the rear wheels, the excessive flexing of which was the object of clarification by the FIA, as was the variable blowing front hubs, which were eventually outlawed from the Austrian GP. The engineers led by Mattia Binotto (who had taken over as technical director following the departure of James Allison and Dirk de Beer in May 2016) took a path that was very different to that of Mercedes, which had instead stretched the wheelbase in search of the greatest surface area to generate downforce. The nose configuration was also different to that of the Mercedes, with the small nose of the SF16-H being retained. The sidepod openings were very different,

Ferrari SF70H
Zeltweg

Ferrari SF70H
Budapest

Ferrari SF70H
Abu Dabhi

located almost horizontally and very high up, suggesting that the aerodynamicists, led by Enrico Cardile, focused on the creation of small intakes and a larger hot air vent at the tail that would help disperse the heat from the radiator packs. The result was a greater flow of air to channel to the rear diffuser where it would create downforce.

The turning vanes were new, inspired by those of the 2016 Mercedes, while the size of the brake intakes was surprising and indisputably the largest of all the cars. Ferrari retained its blown hubs, with a variable function that was outlawed ahead of the Austrian GP.

The front suspension retained the push-rod layout of the SF16-H: the upper wishbone was aligned with the steering arm while the lower wishbone had a tuning fork shape. The novelty lay in the deep trench created in the cockpit survival cell (along the lines of the 2016 Mercedes,) which allowed a notably large third hydraulic damper to be housed. The rear suspension had a pull-rod layout like the Mercedes, with a very short tie-rod and a configuration similar to that of the SF16-H, combined with a stiffened gearbox to avoid the flexing that caused frequent reliability problems in the 2016 season. In contrast with its rivals, the SF70H used a dual rear wing pylon and a conservative wing design with flared end plates as per the regulations.

There was also notable technical development despite the two consecutive episodes of unreliability in the Malaysian and Japanese GPs that weighed heavily on the outcome of the season for SF70H, that on paper undoubtedly deserved better. There was not only continuous development in the aerodynamic sector but also in terms of the suspension, firstly at the rear and then at the front. From the Belgian GP, Ferrari repeatedly tested during the Friday morning sessions, especially with Räikkönen, a new front suspension configuration with a bracket of a different length for anchoring the strut. The feature was always rejected by Vettel but then tested by Red Bull too, with the steering

angle also contributing to the management of ride height. However, this system, never used in the race by either team, was declared to be illegal by the FIA.

Intensive work in the engine department over the winter break had allowed Ferrari to close the gap to Mercedes in terms of power output. This result was obtained on the basis of the 2015 engine rather than the 2016 unit, augmenting the turbo boost pressure and therefore the ability of the ERS to recover energy. Both Ferrari and Mercedes exploited lubricant to "enrich" the fuel in the combustion chamber, burning oil that oozed from the segments and "boosted" the petrol for the few qualifying laps

that served to record the fastest times in Q2 and Q3 and in those moments of the race when the drivers were trying to overtake. There was talk of an increase in power of more than 70 hp with respect to the 2016 Power Unit. In this sector too Ferrari exploited the full potential of the grey areas in the regulations.

On the SF70H, in fact, there was not only the usual oil reservoir for the engine lubricant, but also a second container for the oil that was heated in the crankcase before being drawn into the plenum chamber of the six-cylinder engine and then the turbocharger as the recirculation circuit had to be a closed system to avoid dispersing oil on the track.

Following a request for clarification from Red Bull, the FIA specified with two successive declarations (after the Canadian GP and just before the race in Hungary) that from the Italian GP onwards, the permitted consumption of oil would be reduced from 1.2 litres per 100 km to 0.9 l/100 km, anticipating the 2018 ruling that was to introduce a limit of 0.6 litres per 100 km.

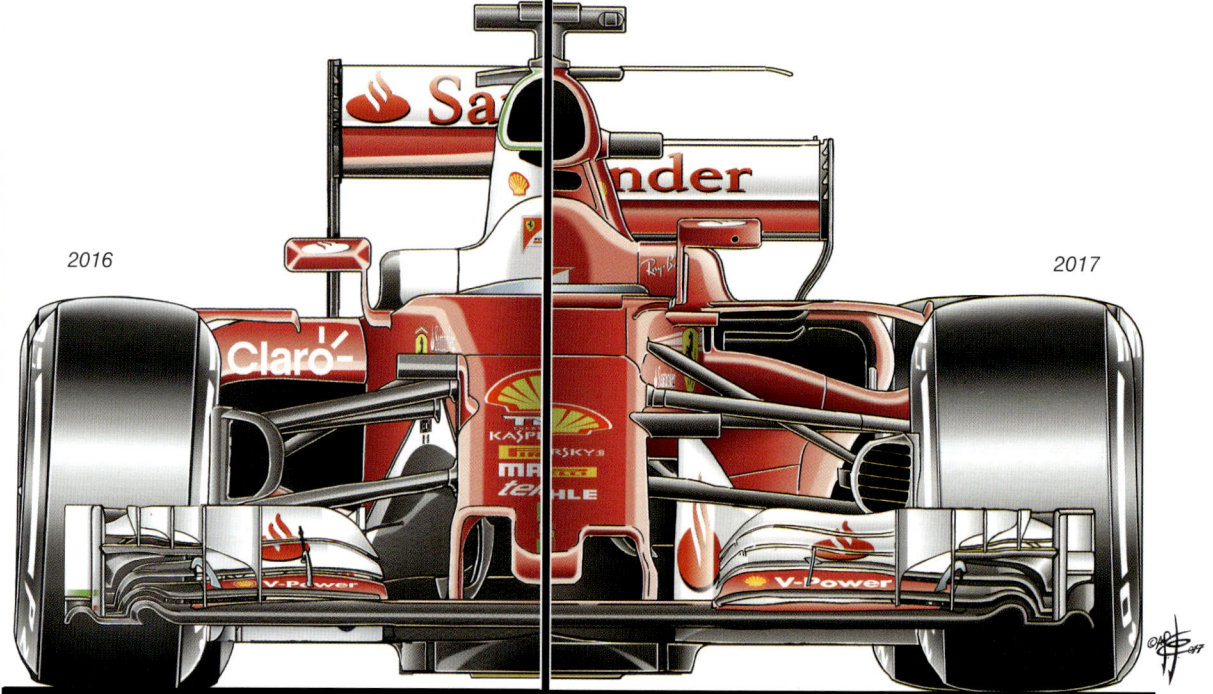

2016

2017

REVOLUTIONARY SIDEPODS

The front view highlights the macroscopic differences associated with the introduction of the technical revolution imposed by the FIA (wide tyres and new wing dimensions).

While the shape of the nose was almost unchanged, in this view we can see the major innovation introduced by Ferrari with the conspicuously different shape and location of the sidepod mouth. Ferrari set the sidepod

intake mouth further back thanks to an intelligent interpretation of the regulations.

The designers in fact located the side-impact protection outside the sidepod (in the detail) within a flow vane that among other features respected the 75° angle imposed by the regulations.

In this way not only were the sidepods distanced from the turbulence of the front wheels, but the turning vane could act to

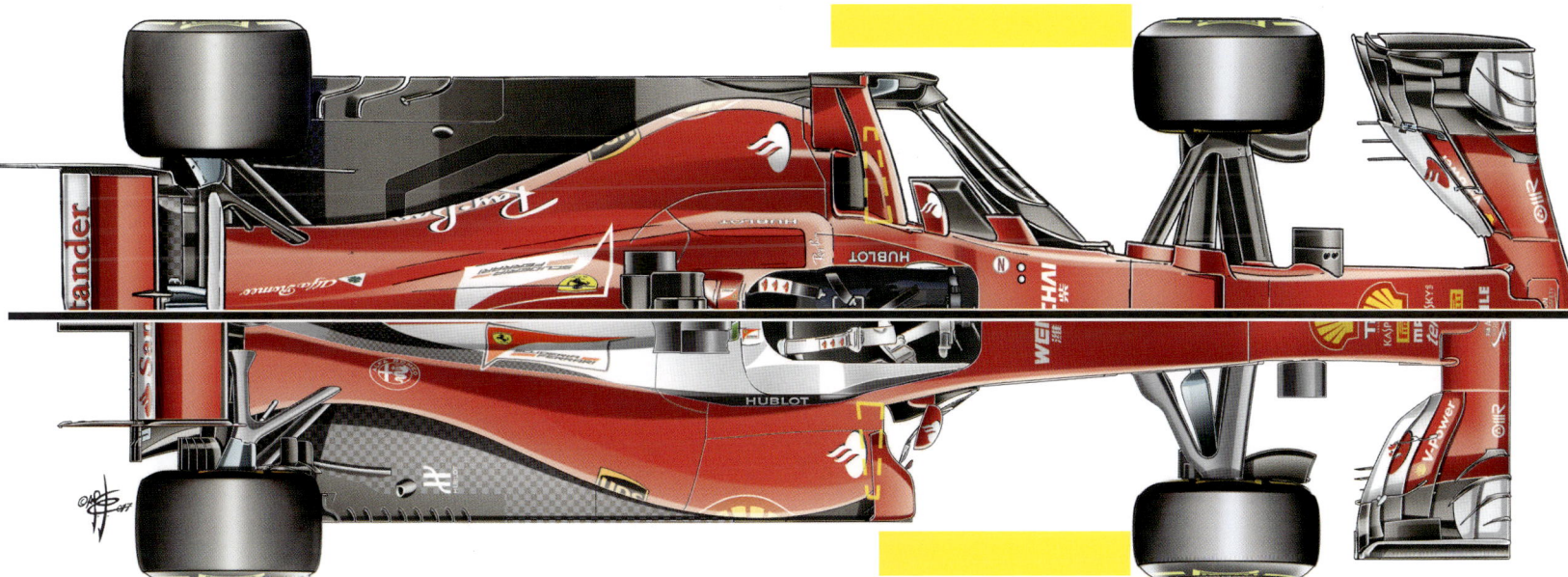

separate the air destined for the cooling system (which was drawn from the upper part of the sidepod) from that channelled to

the rear end and the diffuser to increase downforce.
In the detail drawing we can see how the cone that protrudes

from the front part of the turning vane has been set lower, while still respecting the regulatory dimensions in full.

Note the F2008-type arrowhead profile in the upper part of the chassis and the greater freedom allowed in the bargeboard area.

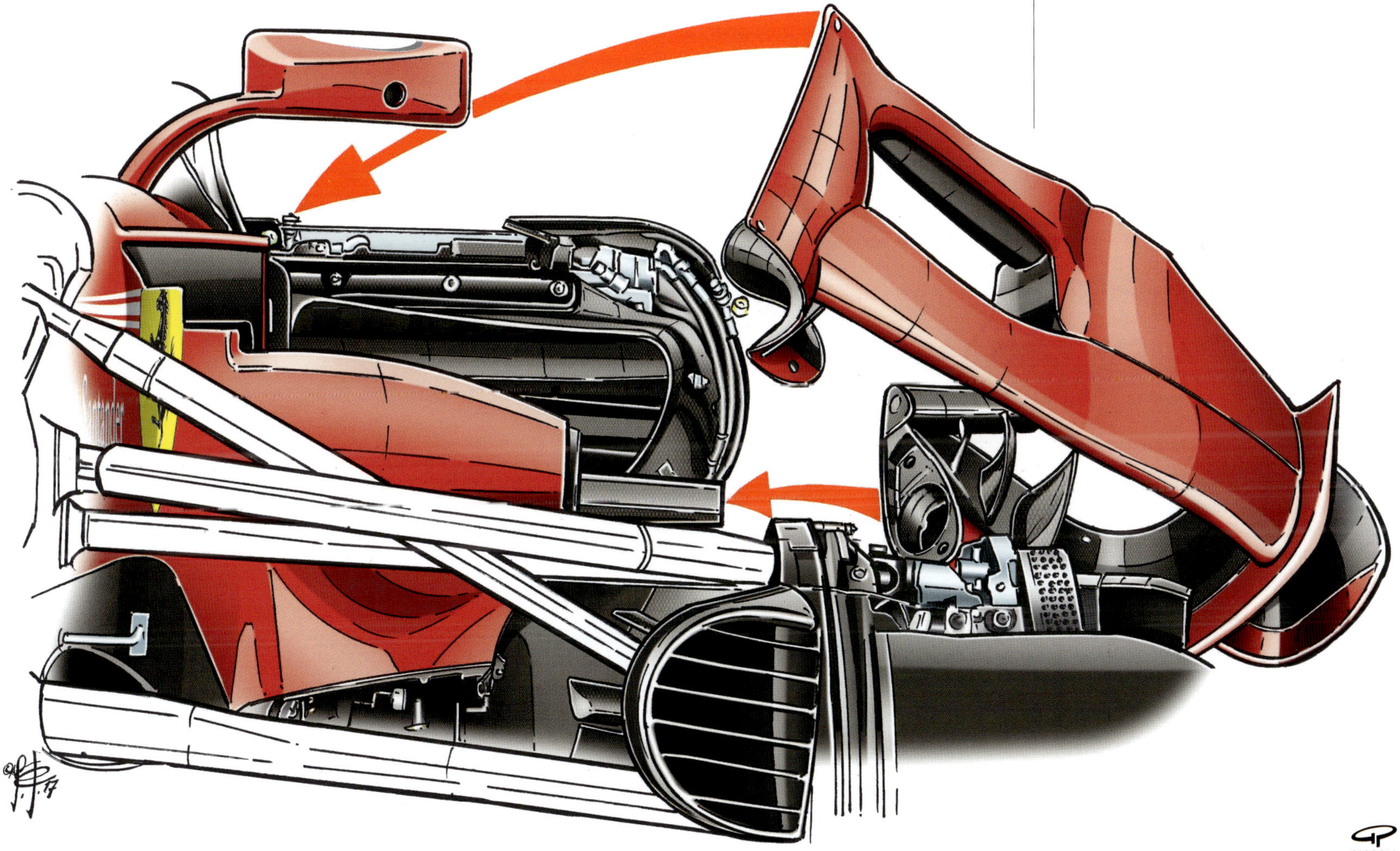

F2008 LEGACY

The greater regulatory freedom had the effect of rendering the area between the front wheel and the start of the sidepods very complex and rich in aerodynamic devices as on the 2008 cars; this was the case with the arrowhead fins on the upper part of the chassis (reprised by Mercedes and Williams) and the large barge-boards (see the New Features chapter) highlighted in the SF70H drawing.

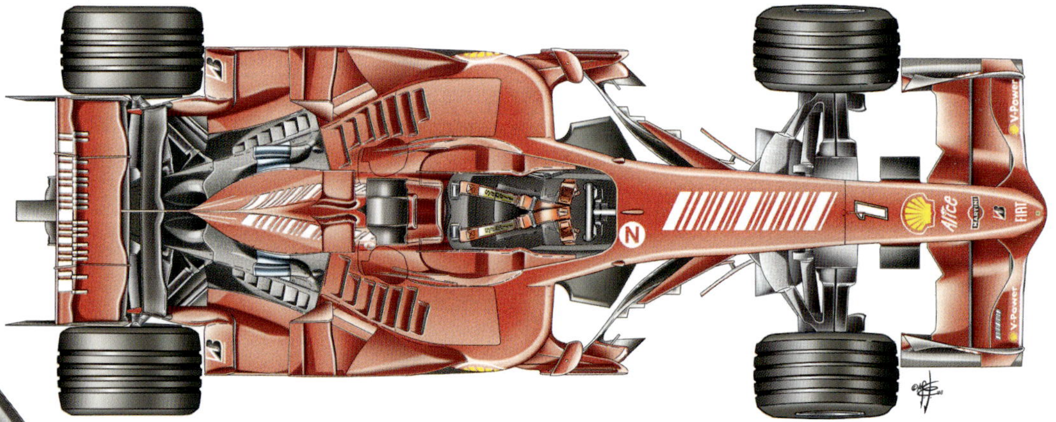

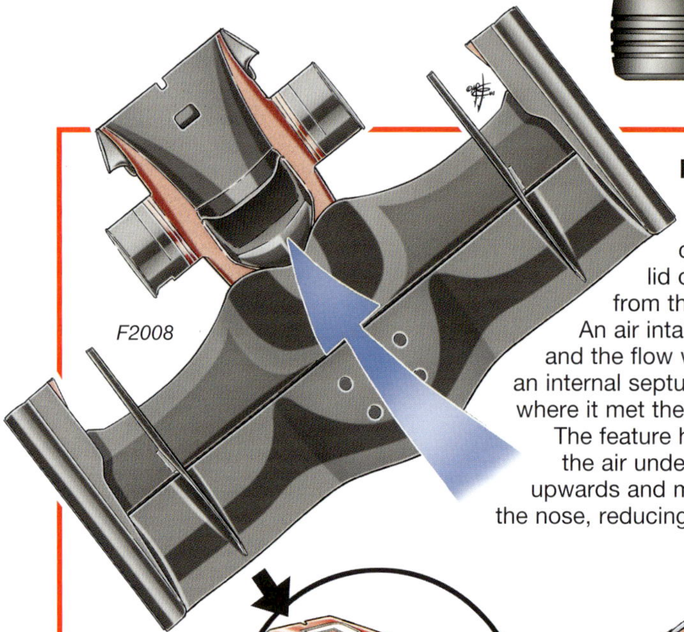

F2008 S-DUCT

It was the F2008 that surprised everyone that season with the introduction of the S-Duct, concealed at the presentation under a kind of lid on the upper part of the nose and only revealed from the Spanish GP.

An air intake had been let into the lower part of the nose and the flow was channelled via a carefully shaped duct with an internal septum towards the upper part of the nose cone where it met the flows from the front of the car.

The feature had the dual effects of increasing the speed of the air under the nose which was partially "aspirated" upwards and met up with the flows over the upper part of the nose, reducing the quantity of air directed to the lower sec-

tion and therefore below the car's underbody. In this way, all the air that passed under the bottom of the car could be extracted correctly and generate greater downforce. For this reason it was used on all the high downforce tracks. At the end of the season the FIA imposed such severe restrictions as to outlaw this feature.

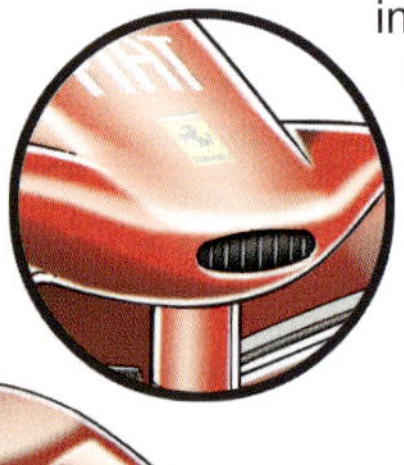

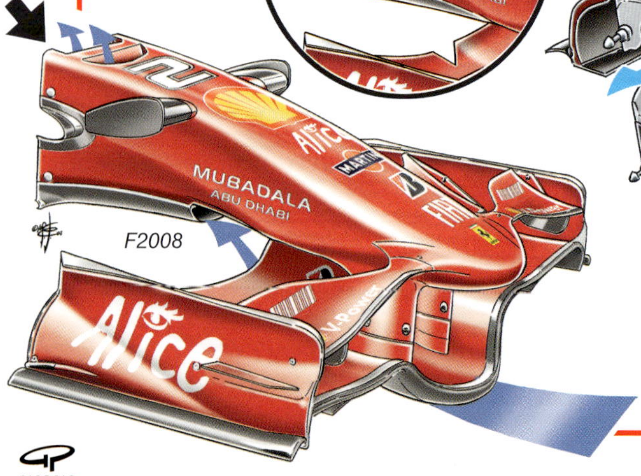

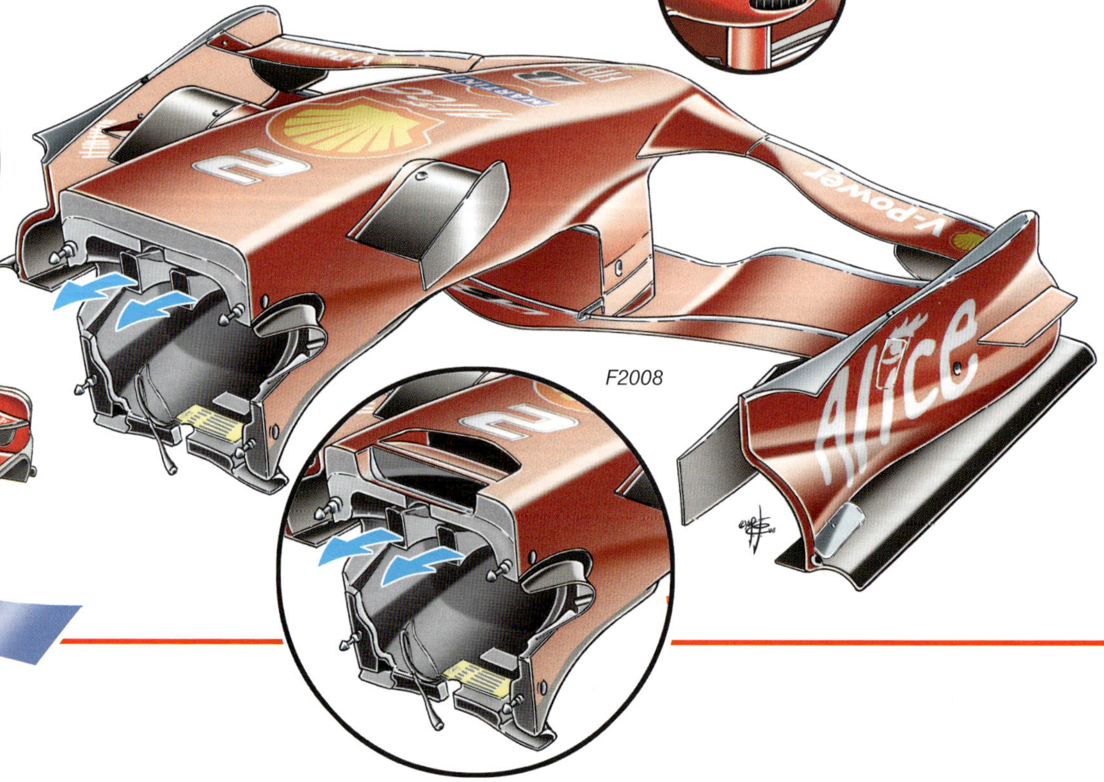

F2008

F2008

F2008

F2008

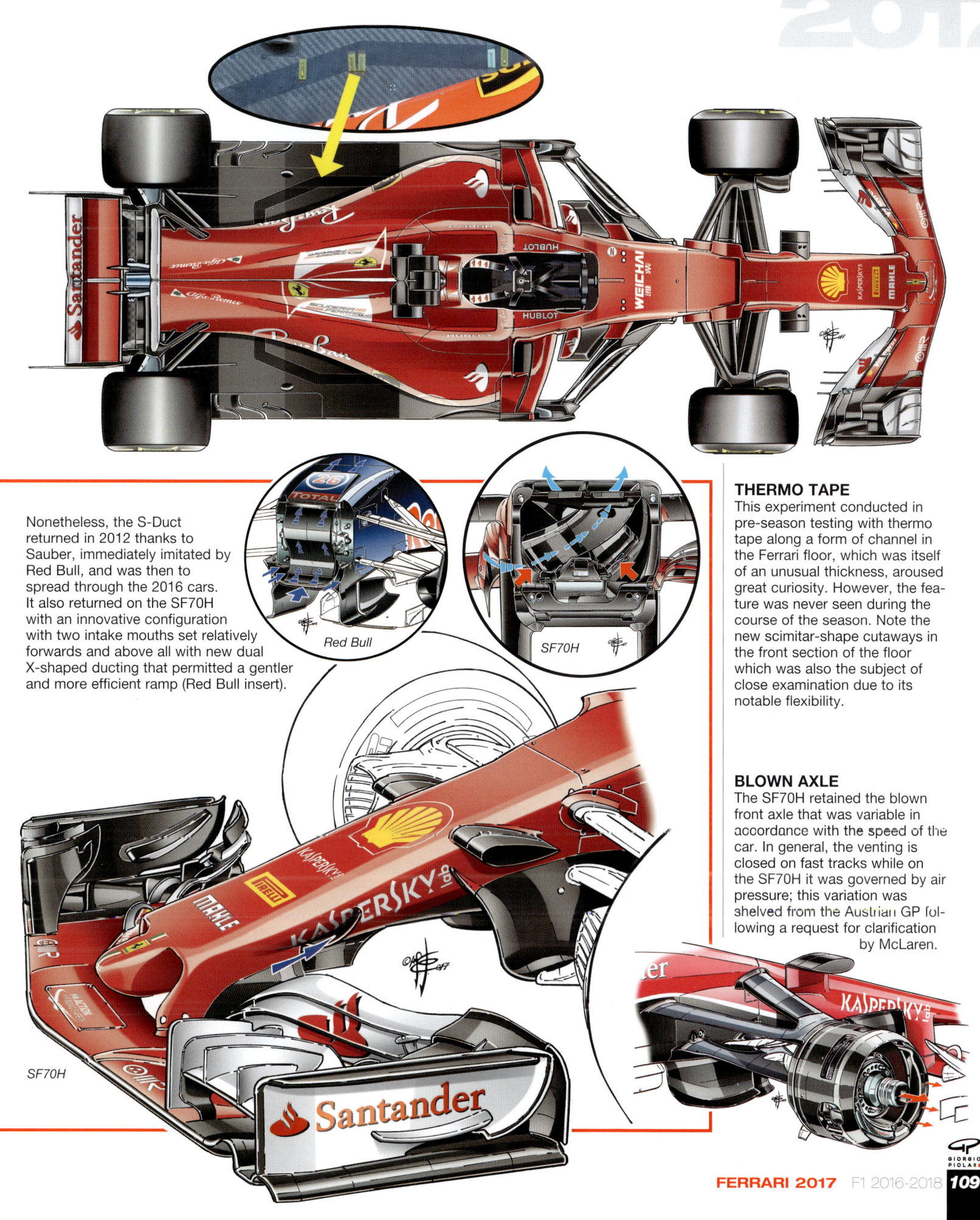

Nonetheless, the S-Duct returned in 2012 thanks to Sauber, immediately imitated by Red Bull, and was then to spread through the 2016 cars. It also returned on the SF70H with an innovative configuration with two intake mouths set relatively forwards and above all with new dual X-shaped ducting that permitted a gentler and more efficient ramp (Red Bull insert).

Red Bull

SF70H

THERMO TAPE

This experiment conducted in pre-season testing with thermo tape along a form of channel in the Ferrari floor, which was itself of an unusual thickness, aroused great curiosity. However, the feature was never seen during the course of the season. Note the new scimitar-shape cutaways in the front section of the floor which was also the subject of close examination due to its notable flexibility.

BLOWN AXLE

The SF70H retained the blown front axle that was variable in accordance with the speed of the car. In general, the venting is closed on fast tracks while on the SF70H it was governed by air pressure; this variation was shelved from the Austrian GP following a request for clarification by McLaren.

SF70H

FLANGE

The SF70H had a kind of second skin that made the mechanical organs very clean, to the benefit of cooling and internal fluid dynamics: this is the flange between engine and gearbox.

ENGINE

Ferrari abandoned the 2016 engine project for the 062, going back to the genealogy of the 059/4. Note the shifting of the heat exchanger from its original location in the V to the front part, permitting the use of variable height intake trumpets.

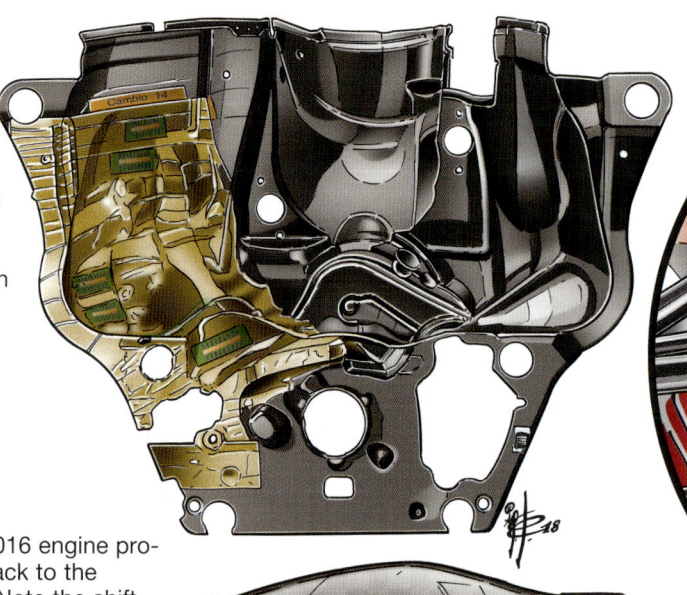

FRONT SUSPENSION

Ferrari had the merit of having created the simplest front suspension installation to modify, with all the various components adjustable from the outside located in a fully open niche in the upper part of the chassis. 1) Anti-roll bar, previously internal. 2) Torsion bar. 3) Potentiometers. 4) Third horizontal damper. 5) Steering column at the same height as the upper wishbone.

GEARBOX

A good deal of work went into the SF70H gearbox casing to avoid the transmission flexing that had caused frequent reliability issues with the SF16-H. Naturally, the second casing in carbonfibre over the titanium box was retained. This feature had been introduced by Ferrari in 2004 to stiffen the structure and it became a second shell in 2013, to which the Mercedes suspension was fixed. The rear suspension reprised the layout of the 2016 car, with a very short tie-rod and the kinematics between gearbox and differential like on the Mercedes.

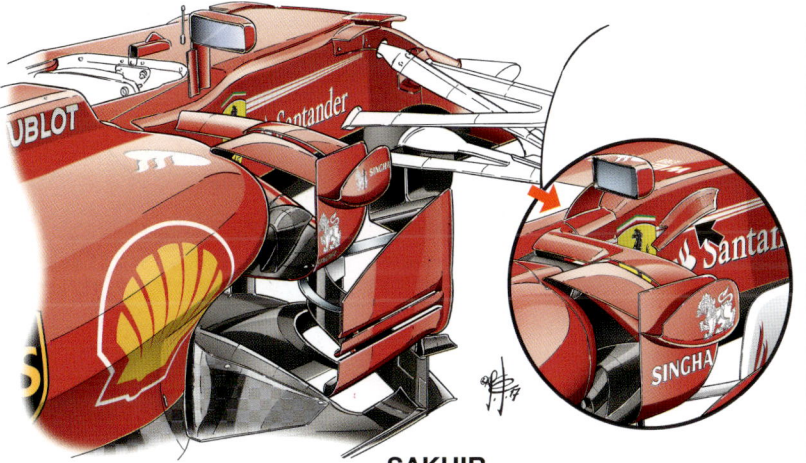

SAKHIR

Among the various adaptations for this track, Ferrari also modified the rear-view mirror mount, revealing painstaking attention to every detail in order to manage the flows to the rear-central part of the car to best effect.

SOCHI

At Sochi, the innovative scimitar-like opening at the front of the floor suspected of permitting excessive flexing at full speed was stiffened (the titanium insert is highlighted with a circle).

DUAL MONKEY SEAT

At Sochi reprised the dual Monkey Seat already seen in free practice at the Bahrain GP but subsequently rejected for qualifying and the race on the Al Sakhir circuit. The triplane winglets above the deformable structure of the gearbox were joined by a dual profile anchored to the twin pylons supporting the rear wing and equipped with a vent.

F2004

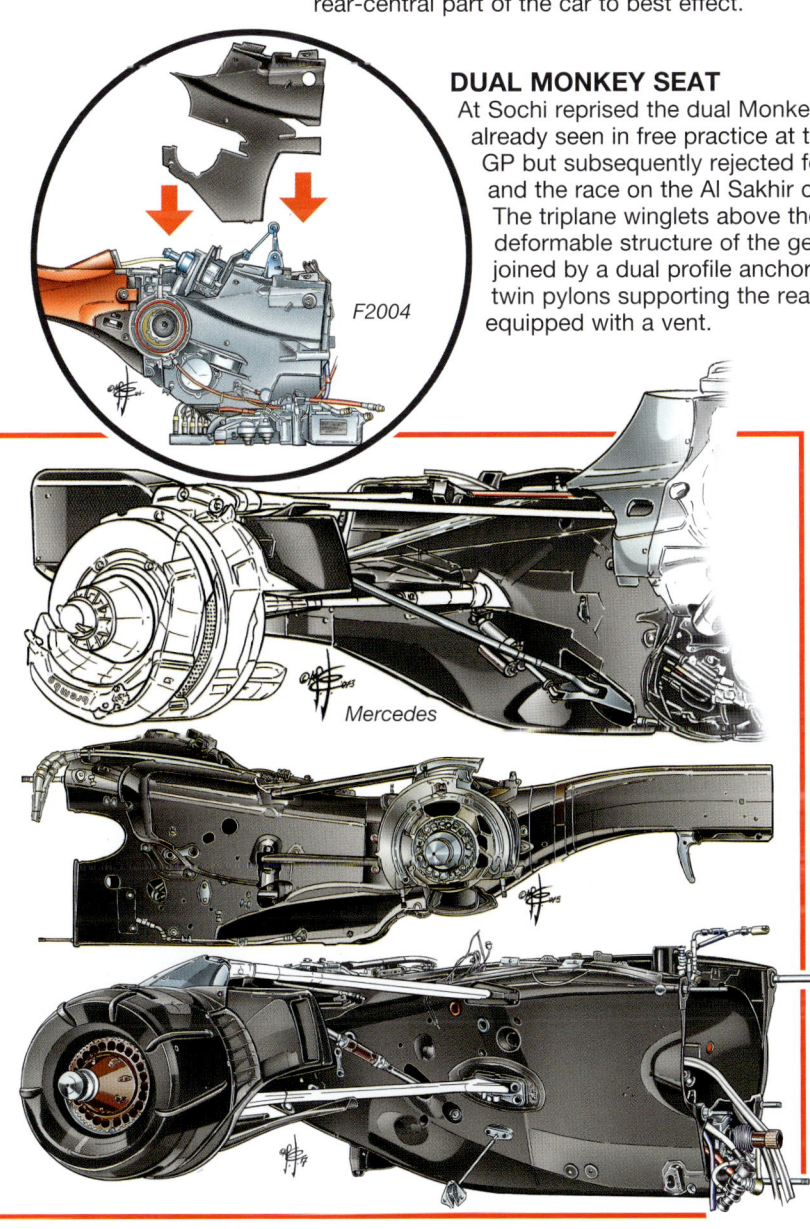

Mercedes

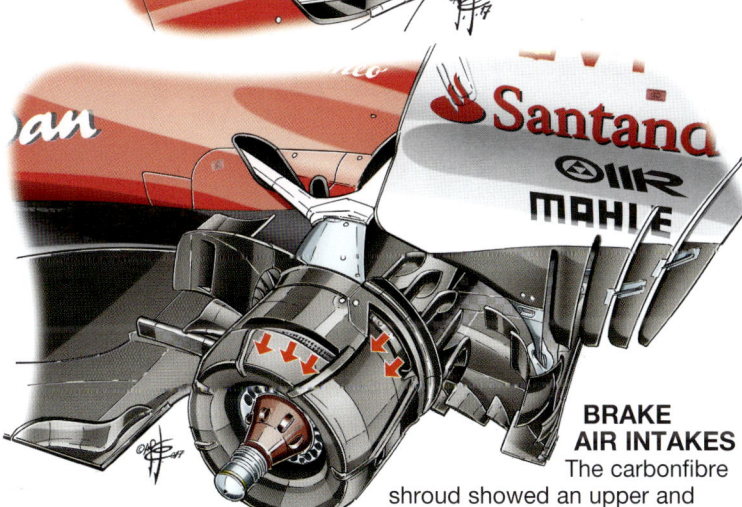

BRAKE AIR INTAKES

The carbonfibre shroud showed an upper and lower aperture in proximity to the carbonfibre disc (smaller in diameter with respect to that permitted by the regulations) designed to irradiate the heat from the brakes to the wheel rims and ensure the tyres got up to temperature earlier.
Note the metal bracket that shows how the upper rear suspension wishbone was set outside the hub carrier, facilitating camber recovery. The floor instead was of the same configuration seen in Australia, with a wealth of aerodynamic devices cleaning the flow of the turbulence from the rear wheels.

BARCELONA

At the opening race of the European season in Barcelona, Ferrari sported a new bargeboard that was initially fitted to Kimi Räikkönen's car, while Vettel's retained the more traditional version.
The modification two vertical vents in the upper part that reached almost mid-way up the aerodynamic device mounted either side of the chassis, in proximity to the front splitter.

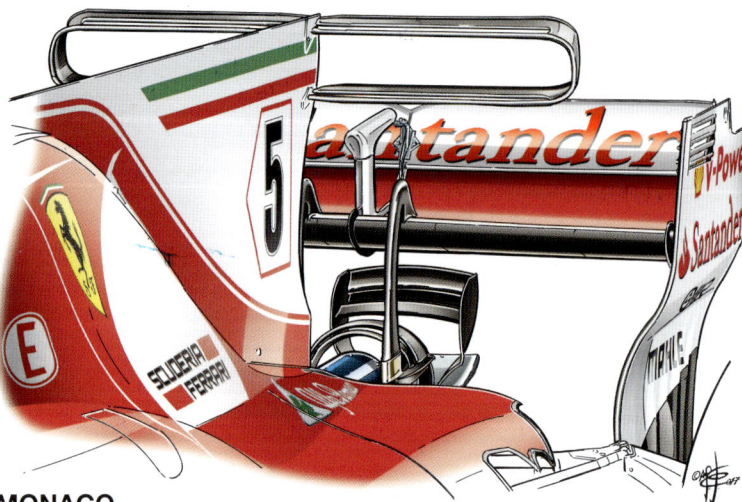

MONACO

A biplane T Wing similar to that on the McLaren, combined with the dual Monkey Seat seen in Barcelona, in search of maximum downforce for the Monaco street circuit.

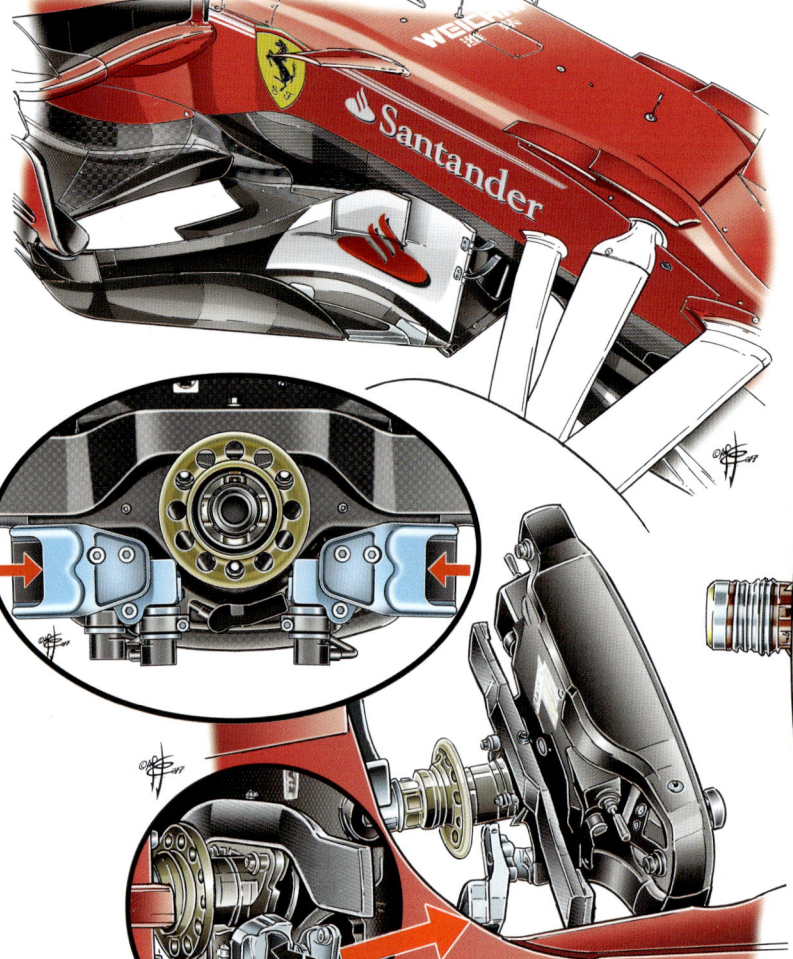

MONACO BRAKE SHROUDS

In order to get the front tyres up to temperature immediately, at Monaco along with the upper window in proximity to the brake disc, two vertical slots were let into the outside edge to allow the heat to irradiate towards the centre, that is to say, the central part of the tread. Note in the oval insert the Brembo brake discs with the most extensive ventilation possible (oblique rows of six holes) and a larger amount of material on the outside.

VETTEL STEERING WHEEL

Only Vettel had a steering wheel without the long offset clutch control rocker. In its place was a single rocker seen at scrutineering on the Thursday, fabricated in titanium and inspired by the one Mercedes had introduced in the second winter test session and equipped new shells on the lever for the finger, offering greater sensitivity in the search for the clutch biting point while accelerating the release phase. The feature was then used by the German alone.
By the Friday the rockers with shells for the fingers had become two, as on the Mercedes (also in titanium) but were replaced for the following races by carbonfibre versions.

BARGEBOARD

In Montreal, the bottom part of the turning vane outside the sidepod where the transverse slot can be seen was revised: the concept had already been noted in Spain, but the shapes were slightly modified to improve the area in which the floor had been designed to favour programmed flexing that would help channel the air from the upper to the lower part and on towards the rear diffuser.

Montréal

MONTRÉAL BRAKES

In Montreal, Ferrari opened three droplet apertures in the brakes shrouds to increase heat dispersal given the severe stresses to which the braking system was subjected: on the circuit dedicated to Gilles Villeneuve there are four major braking point, one of which particularly intense at corner 13. The high average

MONTRÉAL REAR WING

Ferrari took two rear wings for Räikkönen, the one with the straight main plane, and one for Vettel, a dished wing tested in free practice for the Spanish GP in Barcelona. The mobile flap was characterised by two V's in proximity to the metal supports, while the endplates had three horizontal slots opened in the leading edge in the style of the wing introduced by Toro Rosso the previous year and then adopted on most of the cars in the field.

The wing support pylons were also modified and were now separate and pivoted under the main plane, while through to the 2017 Monaco GP they had an upper arch that allowed them to be attached to the main plane from above.

Venerdì   *Q/R*

speed permits good brake cooling, but discs with the greatest number of holes (more than 1,400) are required to ensure optimum ventilation. Note that in 2017 the maximum disc thickness was increased to 32 mm from the 28 mm imposed through to 2016.

BAKU

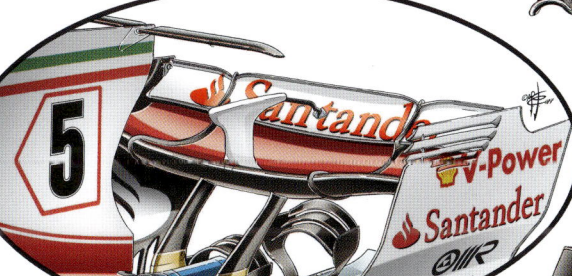

On the Friday Ferrari conducted tests ahead of the Austrian GP with a higher downforce wing and straight profiles (left), before returning to the correct aerodynamic configuration for the Baku circuit, with a dished wing as in Montreal. Note in the large drawing, the appearance of two additional planes cantilevered with respect to the sidepods to increase the heat extraction effect.

FRONT WING

A new front wing made its debut at Zeltweg, with a revised main plane shape, no longer straight (1) in proximity to the central neutral zone, but now slightly arched to favour the creation of the Y250 vortex. The main plane had a long vent (2) while the "kink" (3) that creates the "tunnel" carrying air away from the front wheel to avoid an aerodynamic stall was larger. There was also a new horizontal appendix (4) equipped with a vent inside the endplate (highlighted in the insert drawing).

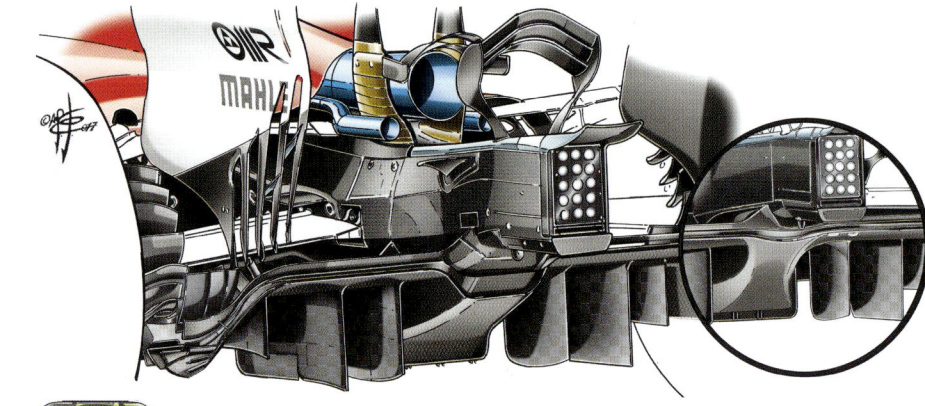

ZELTWEG

At Zeltweg, Ferrari had had to block part of the floor immediately behind the front wheels, equipped with a conspicuous "scimitar" cutaway to avoid excessive downward flexing as the speed increased, significantly increasing downforce. This feature, from its first appearance, provoked reactions from the other teams and the FIA intervened in two phases: firstly requiring a metal bracket to be fixed in proximity to the vertical turning vane and then demanding that the tip of the "scimitar" be secured.

DIFFUSER AND FLOOR

In the drawing a part of the Budapest aerodynamic package: a new floor in both the diffuser section with a radically modified central section (in the insert, the preceding configuration) to create greater downforce, but also in the area ahead of the rear wheels. Modifications were made to the six "s"-shaped slots that helped reduce tyre squirt, the vortex that is generated by the deformations of the shoulder of the tyre when it is subjected to lateral loads or when it passes over kerbs.

BUDAPEST BARGEBOARD

Ferrari introduced an important aerodynamic package at Budapest. The most important modification concerned the bargeboards: on the floor, between the two vertical elements there were three turning vanes. On the front splitter, two vortex generators directed the air towards the "scimitar" which now had a horizontal slot capable of channelling the air laterally and no longer to the back of the carbonfibre blade after the FIA had banned excessive flexing.

SPA FRONT ENDPLATE

At the low downforce track of Spa we saw a new aero package with light downforce wings, derived from the Montreal configuration.
There was a new front endplate that no longer featured the long horizontal flap that acted as a "lid", but just a small profile visible just beyond the leading edge.

GIORGIO PIOLA

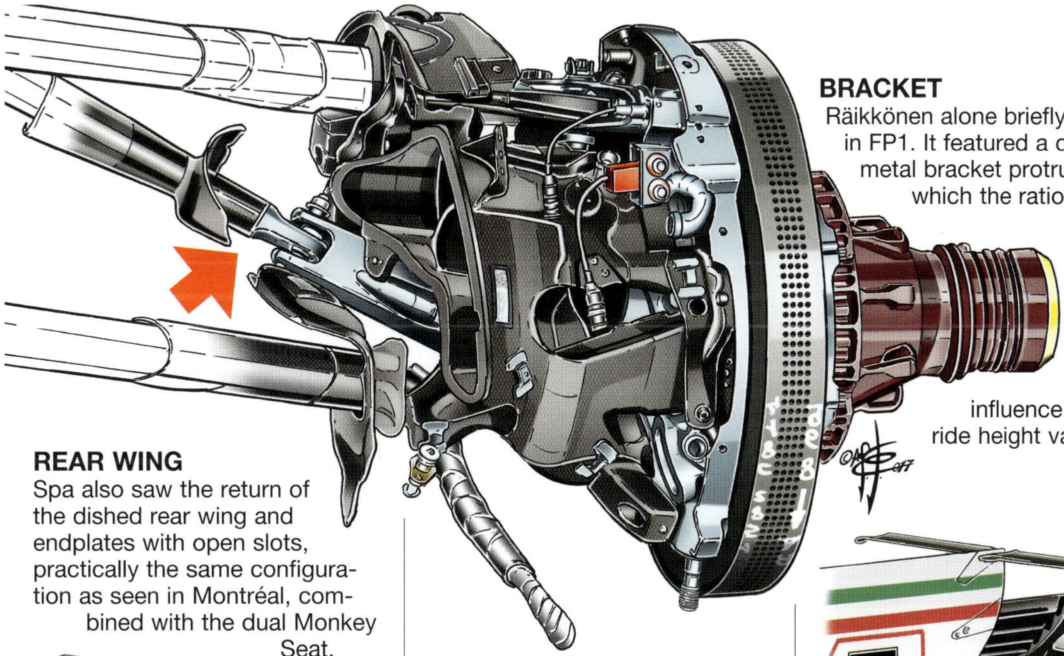

BRACKET

Räikkönen alone briefly tested a new front suspension configuration in FP1. It featured a different strut mount on the hub carrier with a metal bracket protruding from the box section structure, thanks to which the ratio of the lever with respect to the work of the dampers could be changed. The experiment was continued at all the following races and copied by Red Bull, but the FIA then declared it illegal due to its influence on steering and ride height variations.

Monza

Q/R

REAR WING

Spa also saw the return of the dished rear wing and endplates with open slots, practically the same configuration as seen in Montréal, combined with the dual Monkey Seat.

Spa Francorchamps

MONZA REAR WING

Two different rear wing configurations were tested at Monza, with a specific design for the ultra-fast circuit with straight profiles and a significantly reduced chord (in qualifying and the race) and a Monkey Seat. Note the leading edge curving upwards that contrasts sharply with the dished rear wing that was seen at Spa.

FLOOR

In reality, this detail, noted only at Monza thanks to observation from above, was part of the aerodynamic package introduced at Budapest. We can see the vent on the floor of the SF70H (indicated by the black arrow): this was a reinforcement of the front part that takes on an aerodynamic function. What should have been only a support in reality permits part of the flow to be directed towards the hole characterising the vertical flow vanes mounted to the sides of the radiator mouths.

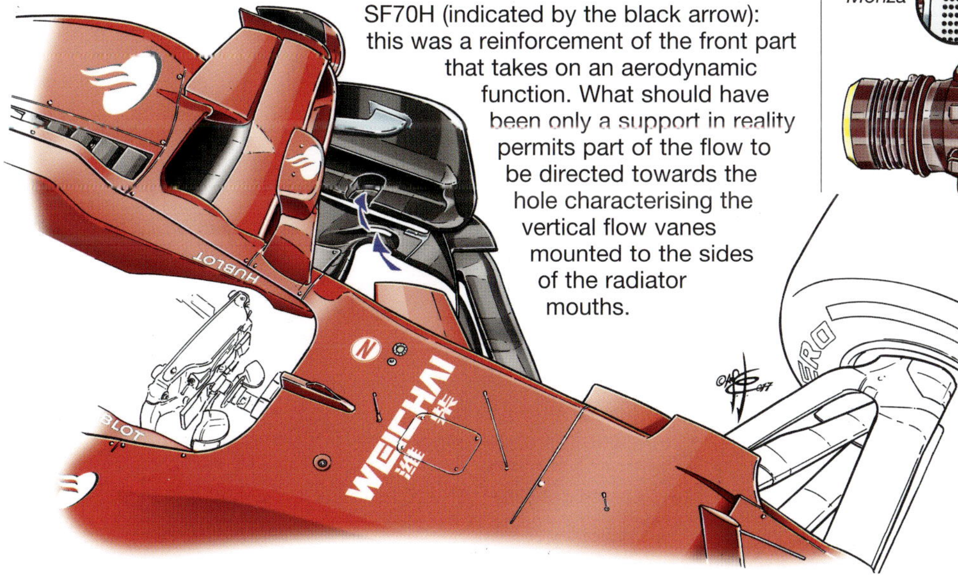

Monza

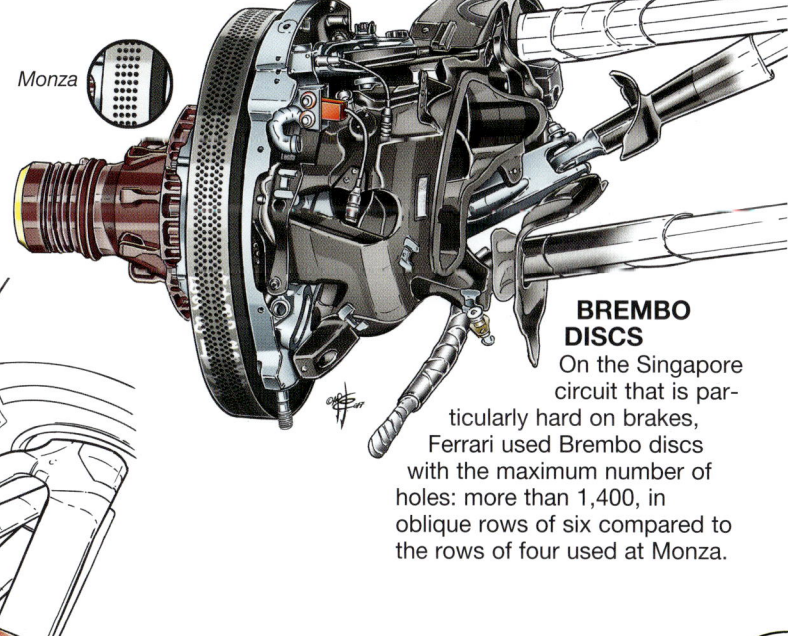

BREMBO DISCS

On the Singapore circuit that is particularly hard on brakes, Ferrari used Brembo discs with the maximum number of holes: more than 1,400, in oblique rows of six compared to the rows of four used at Monza.

SEPANG

For Malaysia, Ferrari made significant modifications to the SF70H including new bargeboards: a vertical element was added to the leading edge with the new aerodynamic element completing a system that combined three vents in the upper part. At the base of the new bargeboard were five mini-flaps curving downwards that served to direct the flow below the floor of the SF70H.

SUZUKA-AUSTIN

The new front wing endplate introduced at Suzuka was modified for Austin with the addition of a small T wing at the trailing edge that was curved towards the outside (red arrow). Objective: to try to direct (with the white upper flaps) the flows towards the outside of the front wheel and feed the vortexes useful in grooming the very turbulent wake of the front wheel, improving the quality of the air towards the rear end.

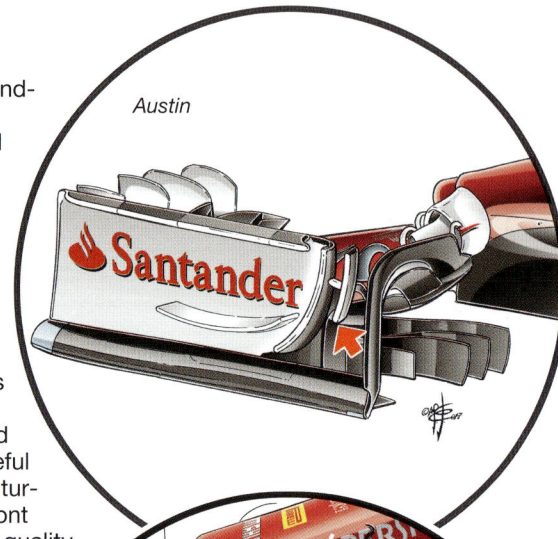

Austin

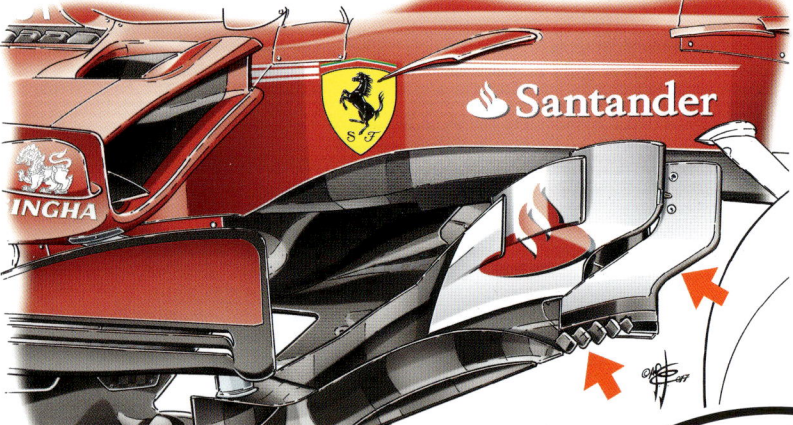

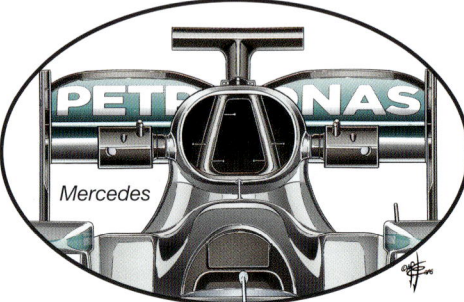

Mercedes

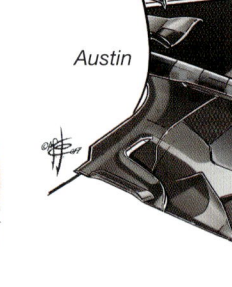

Suzuka

AIRBOX

At Sepang, Ferrari modified the airbox, adding two Mercedes-style ears to the air intake to improve cooling of the Power Unit, in view of what was expected to be a very hot and humid weekend; this modification was then retained for the rest of the season. The two intakes did not, however, improve the aerodynamics given that drag increased and the rear wing was partially obscured, thus reducing downforce a little. In the drawing of the dismantled component we can clearly seen how the flows were handled separating from the main intake duct.

Suzuka

Austin

AUSTIN FLOOR

A further modification at Austin: in the area of the floor ahead of the rear wheels, Ferrari reduced the slots that cut diagonally the lateral edge from five to four, but with a deeper cut, while the fifth element with an L-shape remained, outside which was the Mercedes-style arch.

Vettel

DIFFUSER

At Austin Ferrari introduced the rear diffuser that had been seen briefly in post-race testing at Budapest. This was a revised and corrected version with a minor adjustment to the central section that was now straight rather than curved. At the sides of the gearbox protection structure there was a further small turning vane that worked in synergy with the winglet that had always been present slightly higher.

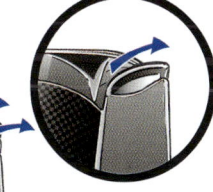

FRONT WING

This drawing shows two details of the SF70H's new front wing with a different arched channel, outside the endplate and a the new small flow separator (red arrow) with a very extensive flat footplate area.

Räikkönen

OLD DIFFUSER

The old diffuser returned at Abu Dhabi so as to adapt the SF70H to the particular requirements of the Emirates track that demanded a high downforce configuration. The most recent extractor had been designed to improve efficiency on fast circuits where drag was the factor to be privileged, while at Abu Dhabi there was more emphasis on downforce. The vents previously eliminated returned in the central section.

NEW DIFFUSER

Räikkönen tested a new diffuser inspired by the concepts seen on the Red Bull RB13 and the Mercedes W08: the outside edges of the diffuser that on Vettel's SF70H were characterised on the trailing edge by no less than six mini-flaps fitted shutter-style. These elements were themselves precedes by three vertical turning vanes arching upwards. On Kimi's SF70H we could observe a cleaner configuration with the two Gurney flaps creating two vents in the outermost part.

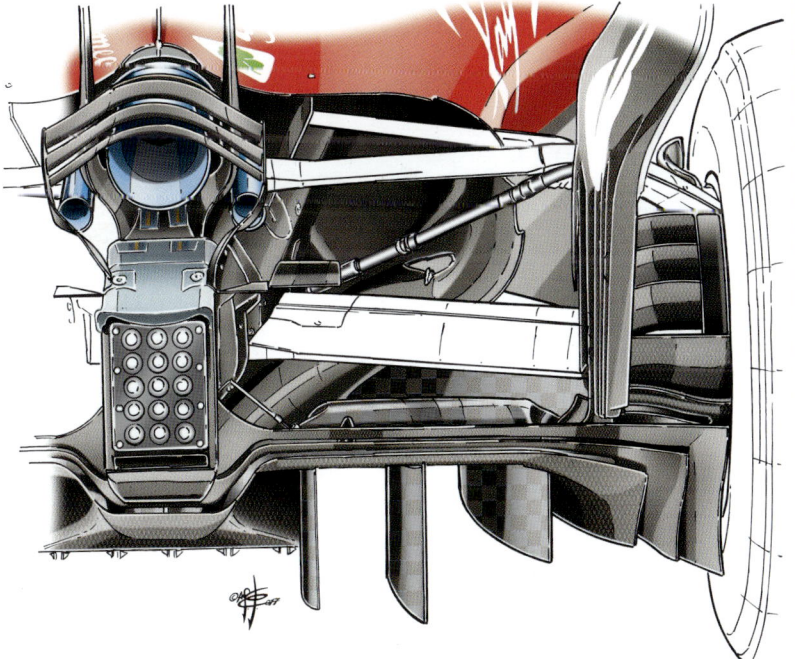

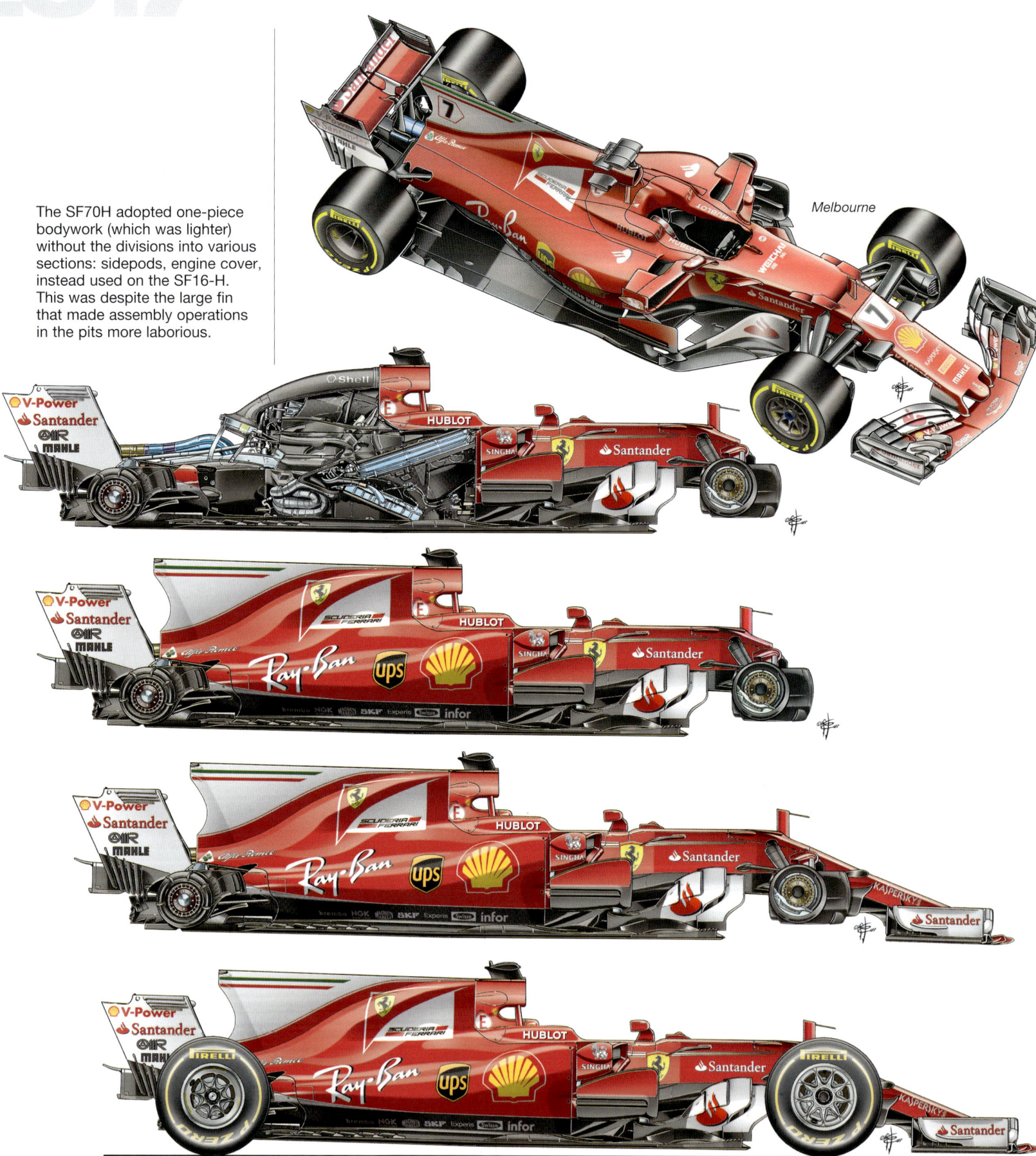

The SF70H adopted one-piece bodywork (which was lighter) without the divisions into various sections: sidepods, engine cover, instead used on the SF16-H. This was despite the large fin that made assembly operations in the pits more laborious.

Melbourne

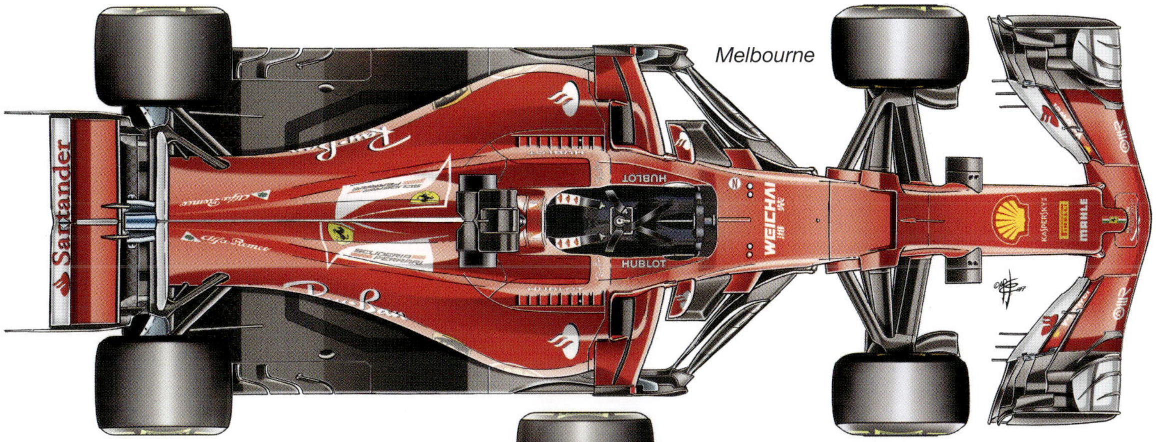

Melbourne

Zeltweg

Budapest

Development was continuous throughout the season, with two major steps following the debut in Melbourne: Zeltweg and Budapest. In particular, at the Hungarian circuit, following the disappointment of Silverstone, the SF70s turned up with modifications to all sectors of its aerodynamic package: front wing, barge boards, floor and diffuser to mention just those elements not associated with the adaptation to a high downforce configuration for Budapest such as the rear wing and the double T Wing.

Budapest

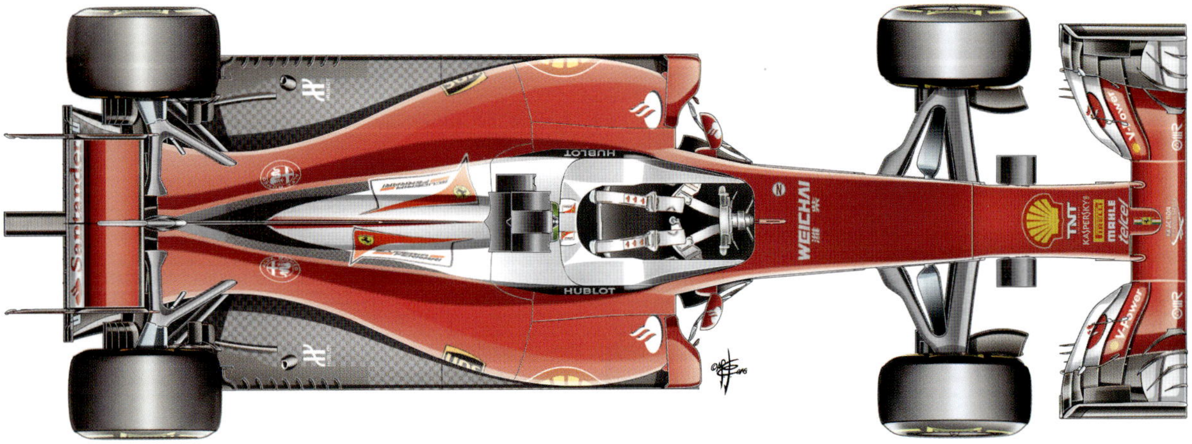
HUBLOT
WEICHAI
HUBLOT
Santander

SF16-H

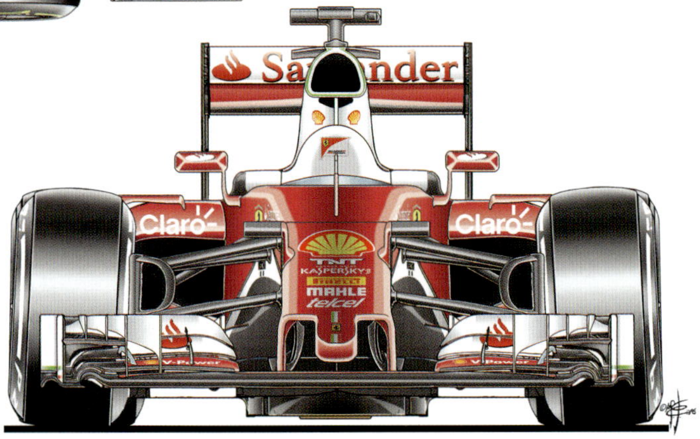
Santander
Claro
Claro
TNT
KASPERSKY
MAHLE
telcel

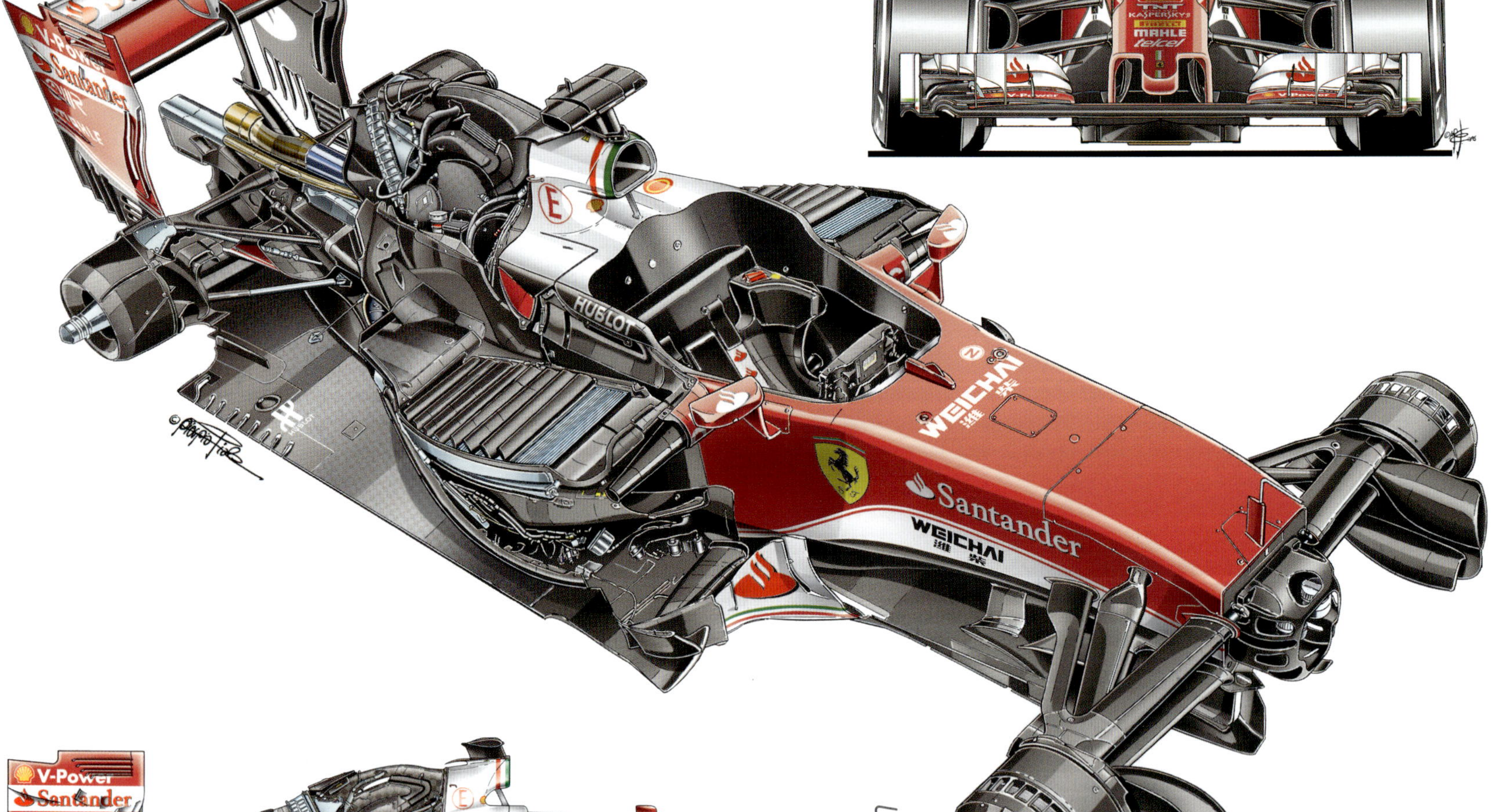
Santander
V-Power
Santander
HUBLOT
WEICHAI
Santander
WEICHAI

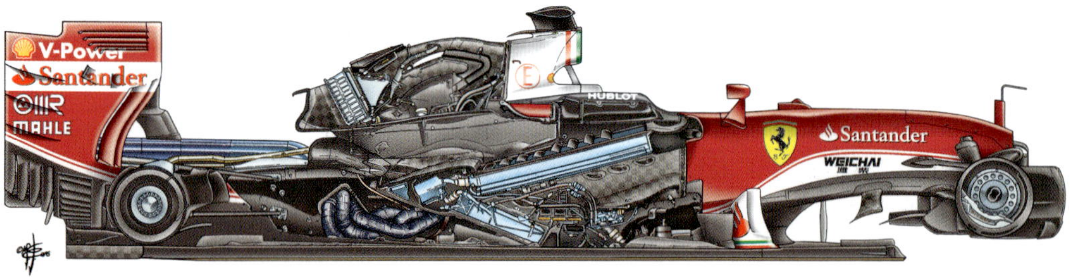
V-Power
Santander
MAHLE
HUBLOT
Santander
WEICHAI

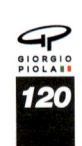

SF70H

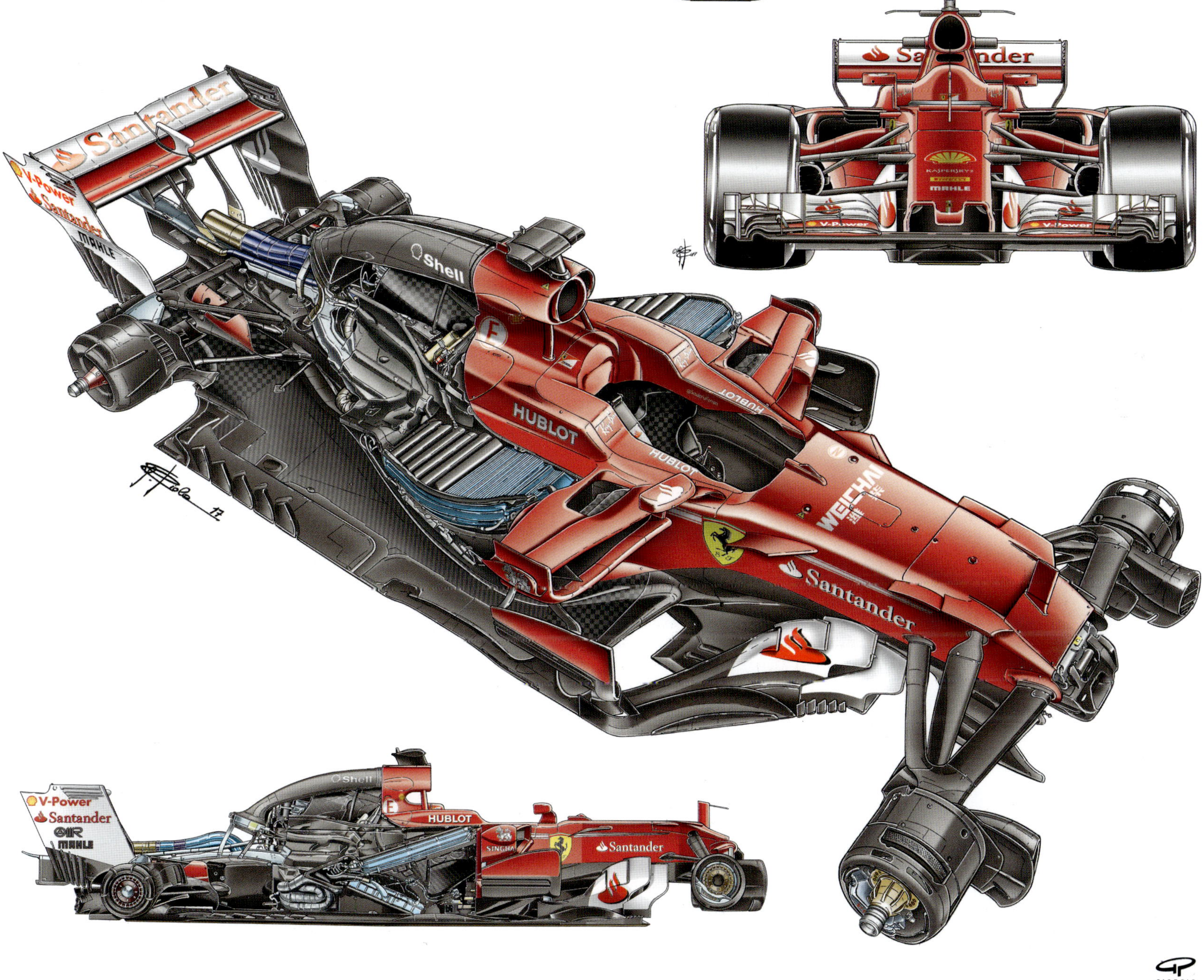

RED BULL

CONSTRUCTORS' CLASSIFICATION				
	2016	*2017*		
Position	2°	3°	-1	▼
Points	468	368	-100	▼

Major surprises were expected from Red Bull, as had generally been the case with every technical revolution introduced by Adrian Newey, frequently capable of stunning F1 observers as he did, for example, in 2009 with the RB5. Instead the RB13 was possibly the most "banal" of the cars he has designed, perhaps due to the upheaval in the technical regulations introduced by the FIA for 2017. It's also been claimed that Newey was not that interested and it is also the case that there was a less than perfect correlation between the data from the wind tunnel and that seen at the track. The result of this situation was unfortunately evident right from the car's debut at pre-season testing in Barcelona. In fact, when the car was unveiled, many thought that the team must have been playing it safe before bring-

ing updates for a competitive debut in Australia.

However, it was soon clear this was not the case and that the RB13 was a simple evolution of the 2016 car, designed on the basis of two fundamental concepts: low drag and the positive rake set-up, with a wheelbase that proved to be the shortest in the field.

In practice, Red Bull had made little effort to exploit the freedoms granted by the regulations in the area between the front wheels and the sidepods, where its rivals, in particular Mercedes and Ferrari, had introduced sophisticated and extreme features.

Despite the slight progress made with the Renault Power Unit, the RB13s suffered from a considerable power gap compared to the Mercedes, a shortfall it was unable to make up quickly

enough. Only at the eighth race of the season on the fast track at Baku did an aerodynamic package arrive capable of making the Red Bulls competitive, fast on the straights and comfortable through the corners, despite a very light aerodynamic load. Ricciardo took the win, helped by the infamous in-race clash between Hamilton and Vettel. The second major step came at a circuit with opposite characteristics to Baku, the slow track at Budapest, where new sidepods appeared, new end plates, a new diffuser and new bargeboards that notably improved the car's handling in slow corners. Again at Budapest, Red Bull Racing was invited by the FIA to put a bracket on the front wing to prevent the external arched step from flexing, as had clearly been the case in the British Grand Prix. The most important development

and the most visually obvious came in Singapore, where the Red Bull arrived with turning vanes in front of the sidepod openings that might have been made in Maranello, so similar were they to those of the SF70H. These features considerably improved the car's handling, making for a far more competitive season finale. At the US GP at Austin, Verstappen used the Friday morning practice session to test a new front suspension configuration with a bracket in the mount between strut and lower wishbone, a layout used by Ferrari from the Belgian GP. Its influence on the car's steering characteristics was highlighted by the need to change the steering column and links when the Dutch driver tested the feature. However, as had been the case at Ferrari, this modification to the suspension was never used in qualifying or the race and at the end of the season the FIA ruled that it should be outlawed.

Red Bull RB12
Abu Dhabi

Red Bull RB13
Launch

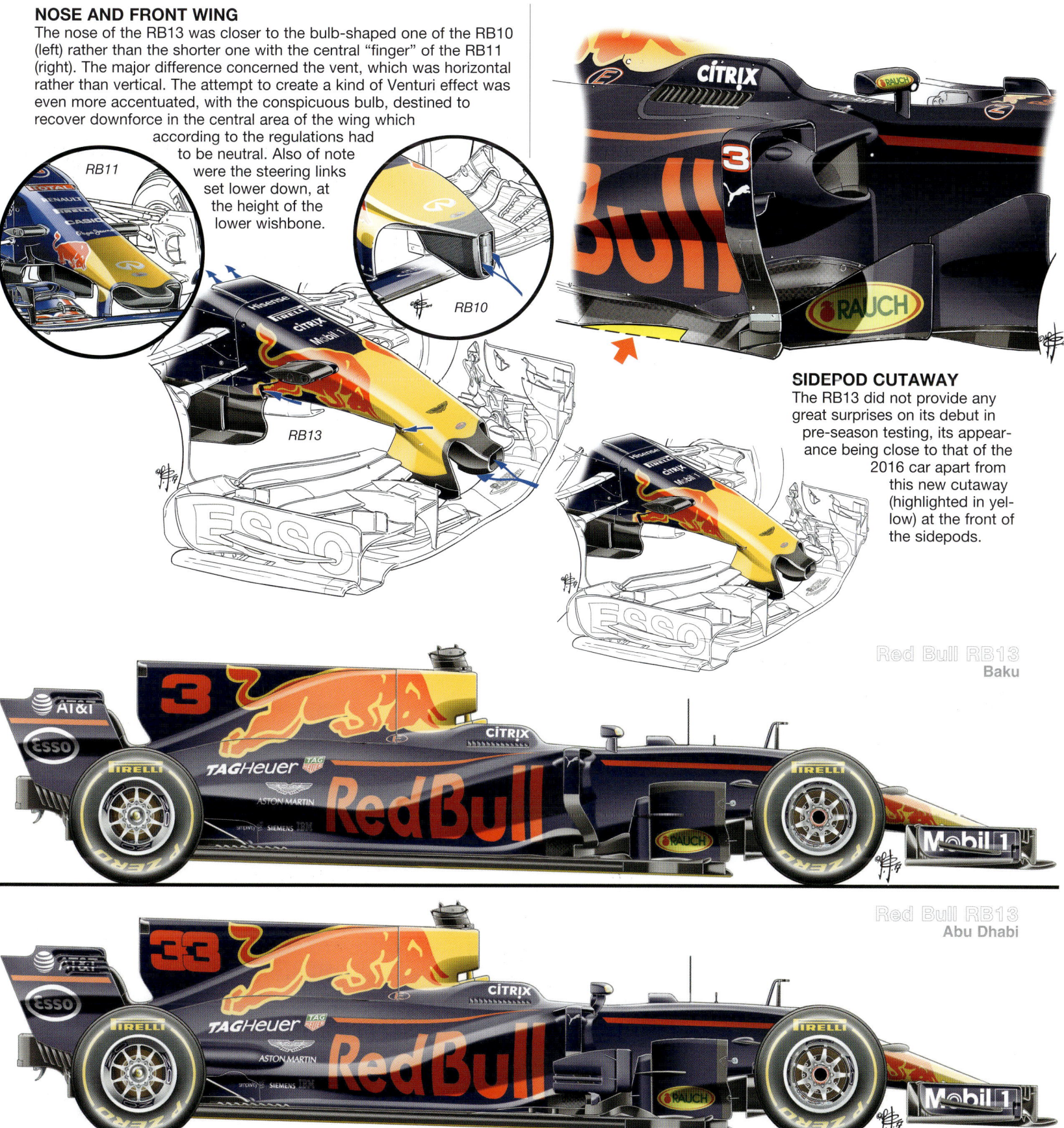

NOSE AND FRONT WING

The nose of the RB13 was closer to the bulb-shaped one of the RB10 (left) rather than the shorter one with the central "finger" of the RB11 (right). The major difference concerned the vent, which was horizontal rather than vertical. The attempt to create a kind of Venturi effect was even more accentuated, with the conspicuous bulb, destined to recover downforce in the central area of the wing which according to the regulations had to be neutral. Also of note were the steering links set lower down, at the height of the lower wishbone.

RB11

RB10

RB13

SIDEPOD CUTAWAY

The RB13 did not provide any great surprises on its debut in pre-season testing, its appearance being close to that of the 2016 car apart from this new cutaway (highlighted in yellow) at the front of the sidepods.

Red Bull RB13
Baku

Red Bull RB13
Abu Dhabi

MELBOURNE

Adrian Newey began introducing aerodynamic updates from the first race of the season. The bridge-like turning vane in the upper part of the radiator inlets, displays a triangular section that extends towards the rear. The bargeboards were also completely revised with respect to those used in Barcelona: the main element was stepped in the upper part from mid-way along its length, while three-quarters of the area of the second were removed to improve the flow around the central part of the car and destined for the cooling of the Renault Power Unit. The two-part step with a vent permitting the air flow to be channelled below the bargeboard was also revised.

FRONT SUSPENSION

A great deal of work went into the front suspension to ensure the positive rake configuration was exploited to best effect. The transverse third front suspension element controlling ride height and roll was easily accessible and retained the Belville spring discs as on the RB12.

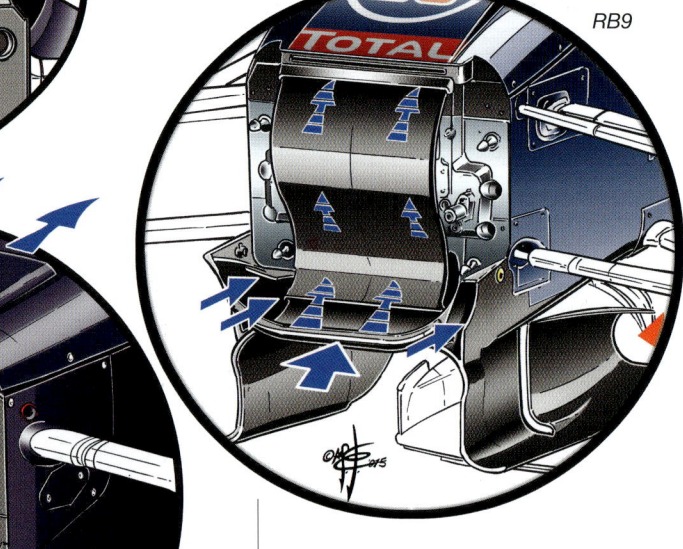

RB9

RB11

SHANGHAI

A new turning vane was introduced in China with the separation of the three elements that in Melbourne, below, had been attached to the same step, so as to create two additional slots on the footplate, in an attempt to increase downforce.

S-DUCT

The RB13 revived the S-Duct employed by Sauber in the 2012 season and immediately fitted on all the Milton Keynes cars. Compared with 2016 (see the insert, right), the ramp taking air from the lower to the upper parts was narrower and less sharp. The slots taking air to the electronic CPUs in the lower lateral part of the chassis were retained.

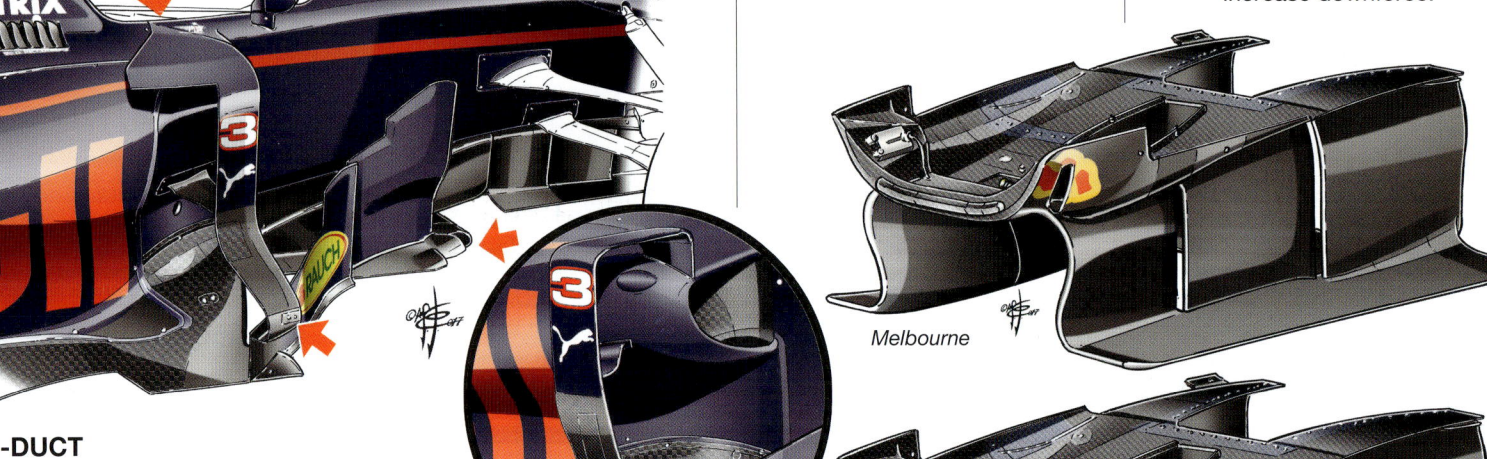

Melbourne

Shanghai

BARCELONA

At the Spanish GP, Red Bull modified the bargeboards with three vertical elements and a horizontal flap curving slightly downwards that linked the first two profiles and served to reattach the flows to the floor, but the main aspect was the extension of the step towards the front: there were no serrations as on the Mercedes, but rather slightly raised horizontal flaps at the leading edge designed to channel the air below the floor and increase downforce.

MONTRÉAL

A new aerodynamic package appeared in Canada: a new front wing characterised by a new vertical vent in the end plate, just in front of the sponsor's number 1, new bargeboards with the second vertical element raised and a new floor in the area ahead of the rear wheels. The diagonal vents opened on the outside leading edge of this delicate area were increased.

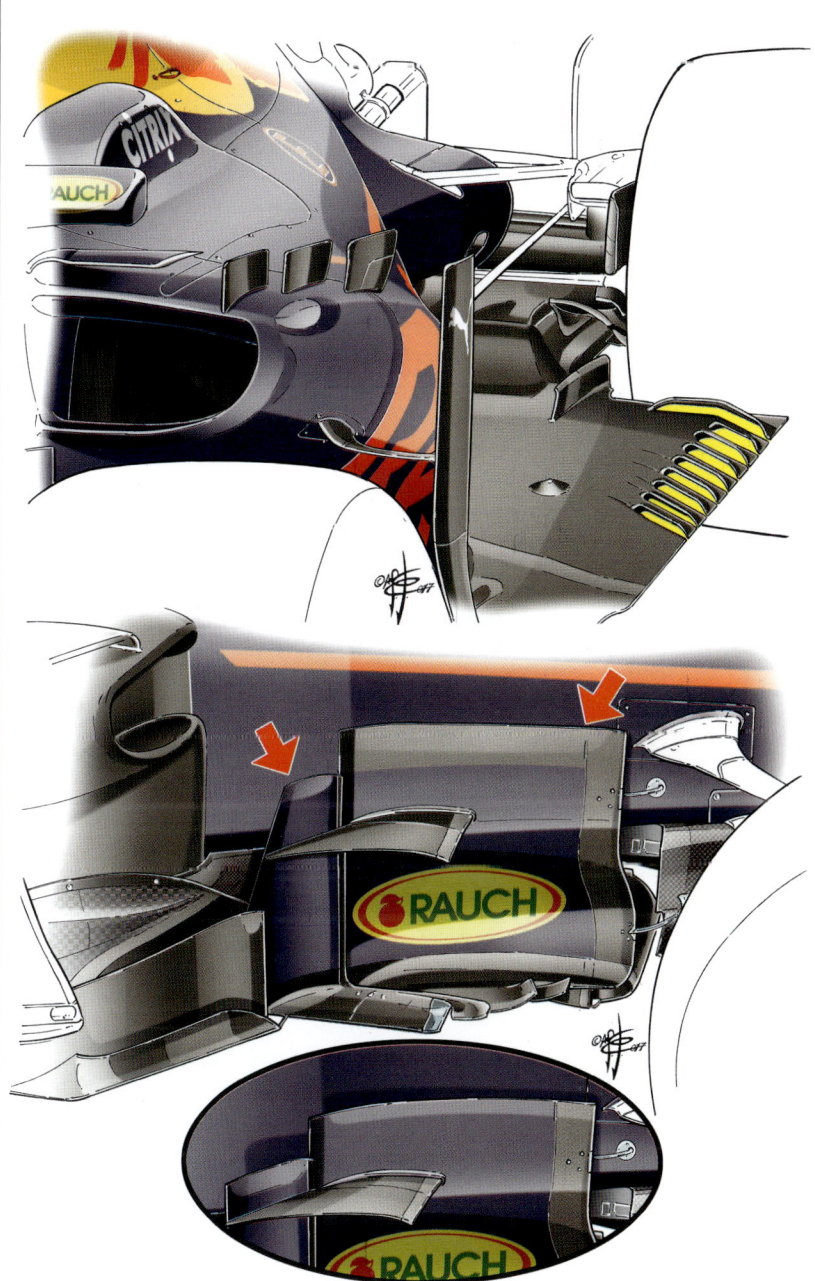

BAKU

A new rear wing was taken to Baku, with less downforce than the medium-load version seen in Canada. The main plane was almost flat, with a raised leading edge in the central section, while the mobile flap had a reduced chord but featured a small Gurney flap and a small V in the centre. The end plates were also revised with just two horizontal vents at the top with a front aperture.

SIVERSTONE FRONT WING

A new front wing debuted at Silverstone composed of the usual six elements that in the outside part became no less than eight to direct as much of the flow as possible away from the front wheels.

Behind the upper flaps were no less than six vents as well as that of the main plane. The Red Bull presented a modest flair to the leading edge, because it was wider than the footplate separating this section from the end plate. Rather than generating a tunnel along the lines of the Mercedes, on the RB13 they preferred to shorten the flaps, creating a large open channel that directed a great quantity of air towards the outside of the rear wheels. In the three-piece upper flaps the vertical strap in the middle of the "bridge" was eliminated.

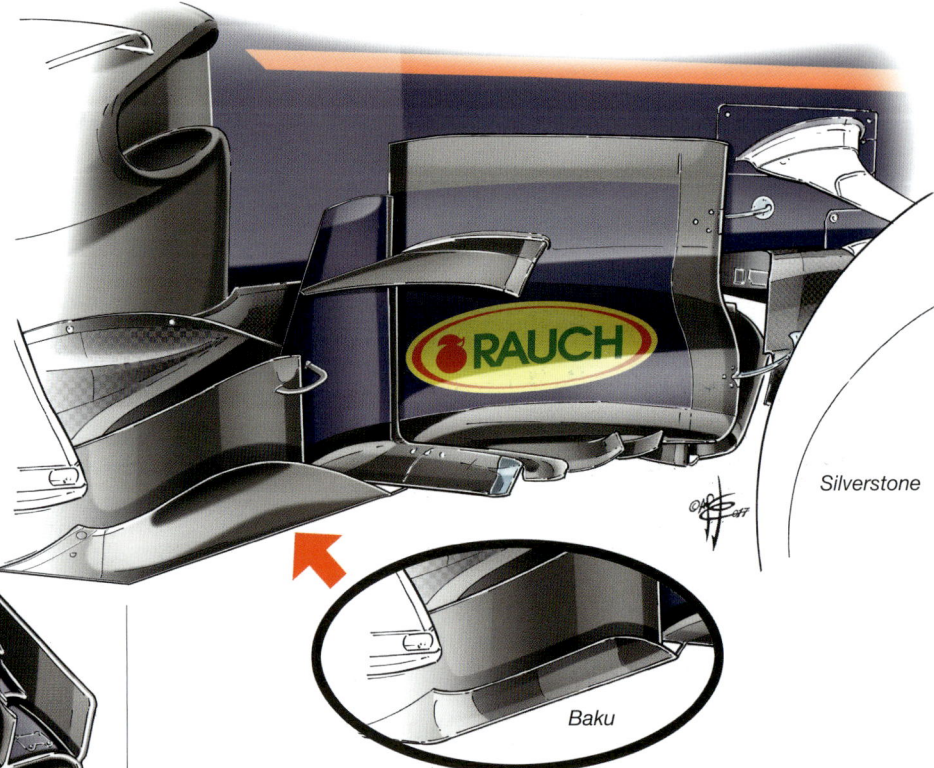

Silverstone

Baku

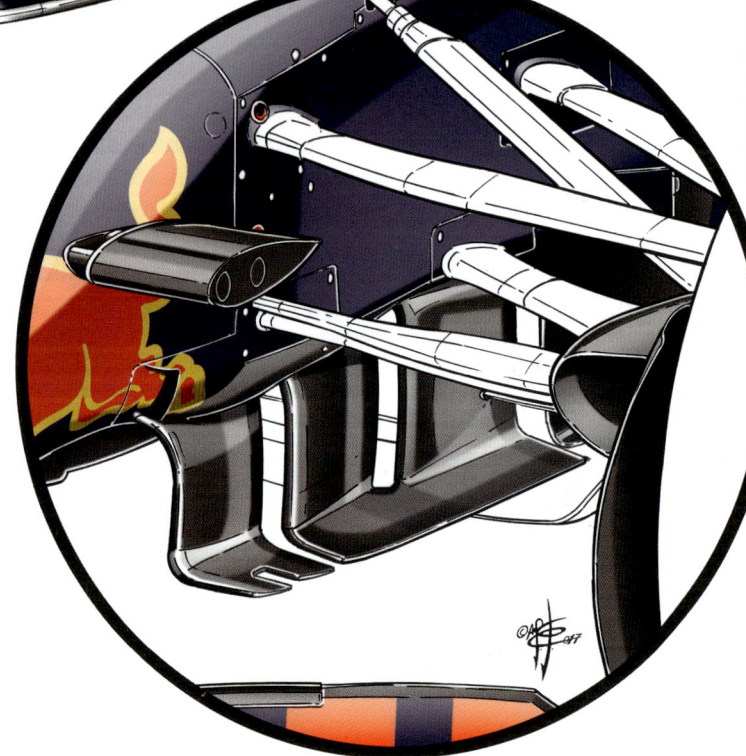

SILVERSTONE BARGEBOARD

The turning vanes and bargeboards were also modified. The turning vanes, which were now characterised by three elements with two vents, had a larger external footplate that was no longer straight (see the red arrow), but appropriately flexed and raised on the trailing edge the better to "harmonize" the flows with the bargeboard. The design of this last was unchanged from Austria apart from the addition of a small lanceolate turning vane anchored to the floor and seen ahead of the bargeboard itself.

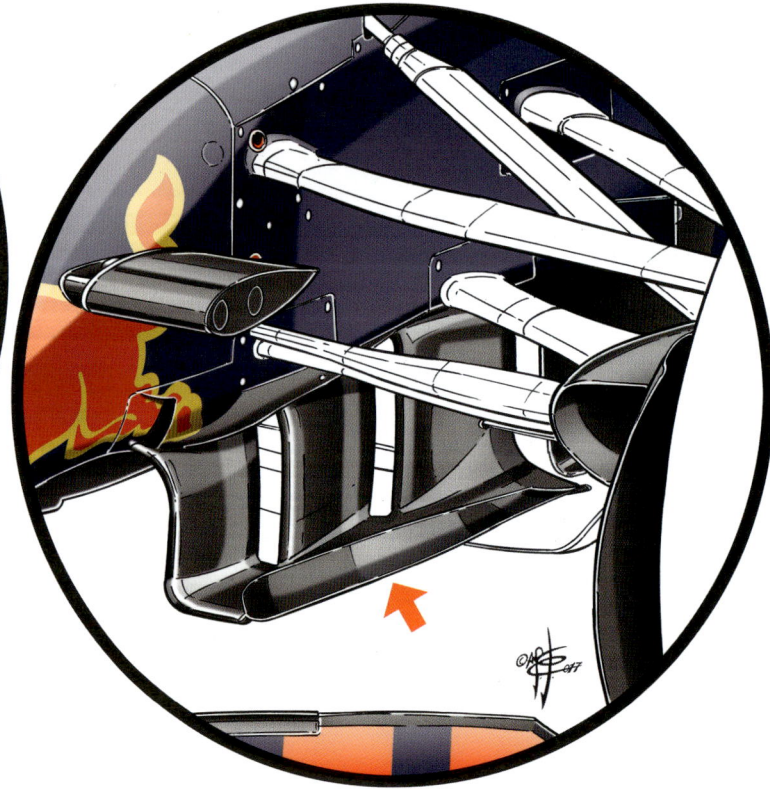

GIORGIO PIOLA

BUDAPEST BARGEBOARD

Highlighted in this drawing are the teeth applied to the lower part of the leading edge of the floor, at the start of the sidepods and designed to better direct the flow towards the lower rear section.

Ricciardo

SPA FRONT WING

The two Red Bull drivers also tested alternative front wings characterised by the final flap with different cutaways, Ricciardo adopting a version with less downforce. In the race both used the version tested by Verstappen. Note the strap (arrow) on the end plate imposed by the FISA following the Hungarian GP.

Verstappen

SPA REAR WING

At Spa Re Bull tested two very low downforce rear wings: Ricciardo with the lightest configuration, equipped with end plates with no slots at the top, while Verstappen had a rear wing with less of a rise on the main plane, a slightly more visible flap and end plates with two horizontal vents. In the race, both used the version with more downforce.

MONZA

There was a further reduction in downforce for Monza with almost horizontal flaps and end plates with no vents to counteract the lack of power from the Renault engine and obtain a good top speed.

SINGAPORE

The new aerodynamic package introduced in Singapore aroused interest above all because the RB13s presented an aero feature already seen on the Ferrari...
The car used by Max Versteppen in free practice had a turning vane at the sides of the radiator mouths, identical to that of the SF70H (insert). This feature was then retained for the rest of the season.

SUZUKA

While this was not a new feature for the Japanese GP, it was noticed when the RB13s were taken to scrutineering: the end part of the upper face of the diffuser was characterised by the presence of two... droplets on either side. These were vortex generators that served to groom the flows as required by the Milton Keynes aerodynamicists. The feature had already been seen in testing at Barcelona, but never examined in detail.

AUSTIN DIFFUSER

The extractor profile in the area close to the rear wheels (arrow) was modified to work in synergy with small fins attached to the rear brake intakes and reduce the turbulence created by the rear wheels.

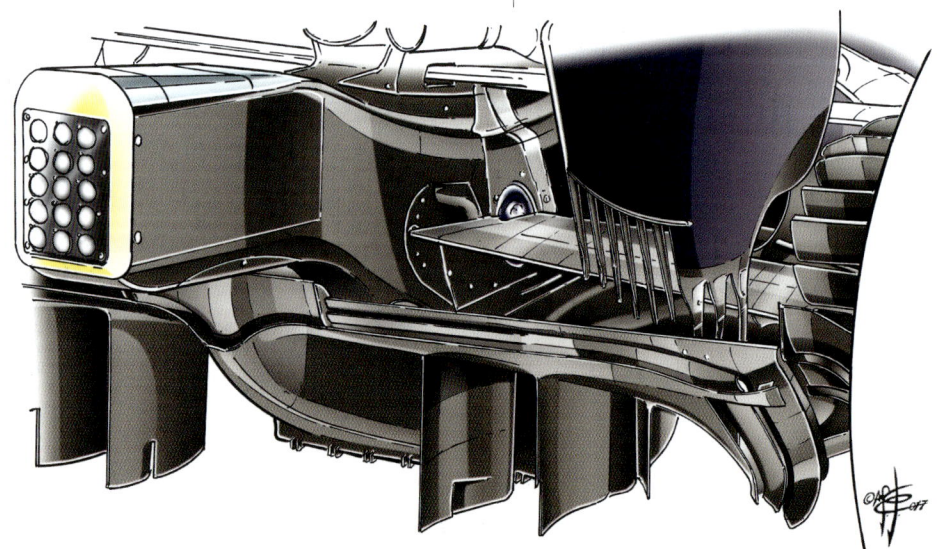

AUSTIN BRACKET

From the United States GP onwards Red Bull reprised a concept developed by Ferrari on the SF70H from the Belgian GP when the new front suspension was introduced. A bracket provided a different strut mounting point, a modification that provoked a strong reaction from Mercedes. As with the Ferrari, this feature was used only In practice on the Friday and only on Verstappen's car.

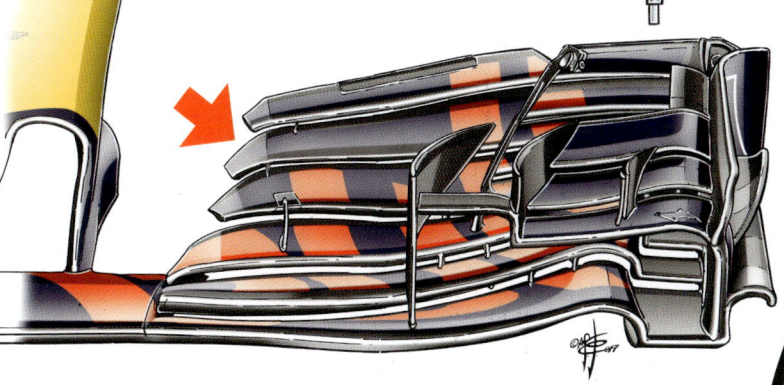

INTERLAGOS

A new front wing end plate with the leading edge of the outside tunnel with a distinct rise so as to be aligned with the car's positive rake and avoid grounding.

ABU DHABI

The last race of the season saw the introduction of a narrow internal front section of the cascade (arrow): note in this drawing the large portion of footplate inside the end plate.

RENAULT ENGINE

Another season for Red Bull penalised by the Renault engine's lack of power. In this drawing it is interesting to note the three large and sharply tapering ducts cooling the French Power Unit.

2018

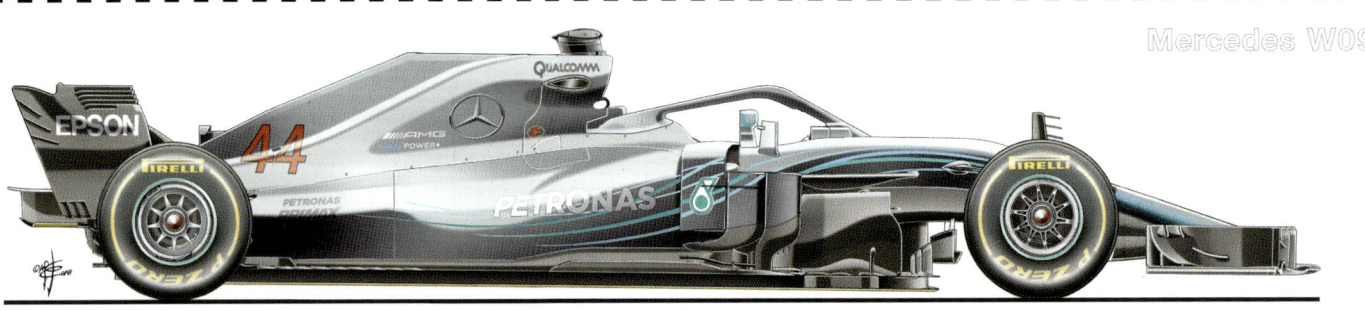

Mercedes W09

Ferrari SF71H

Red Bull RB14

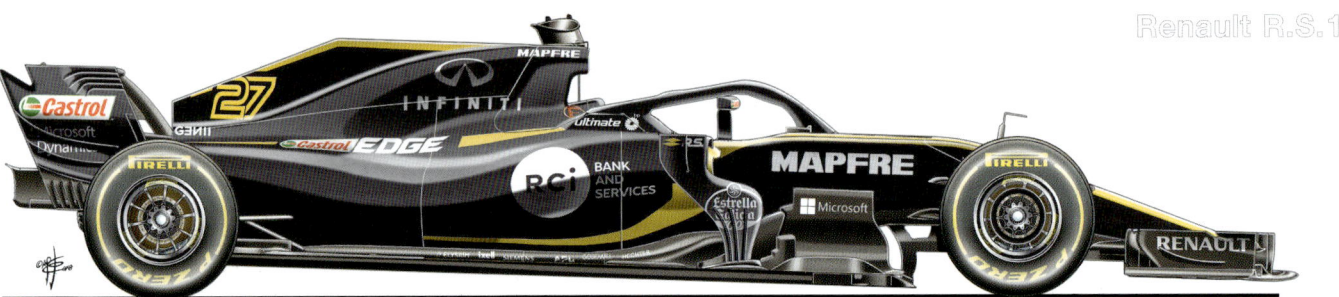

Renault R.S.18

Haas VF-18

McLaren MCL33

The 2018 season, which opened in an atmosphere of relative stability (apart from the reduction of the engine cover fin and the high T-wing and the abolition of the monkey seat), saw the official introduction to Formula 1 of the Halo and the consecration of the configuration conceived by Ferrari with the safety side structures located outside the sidepods, a feature that had created a sensation in the 2017 season and had now become a generalised trend. The configuration allowed the sidepod inlet to be set further back, away from the turbulence generated by the front wheels. Only four teams decided against adopting this stratagem (Mercedes, Force India, Renault and McLaren), which then became three as from the Austrian GP the W09 also adopted the trend imposed by the SF70H in 2017. A move that was officially defined as the most important development of the season on the W09.

Force India / Racing Point VJM11

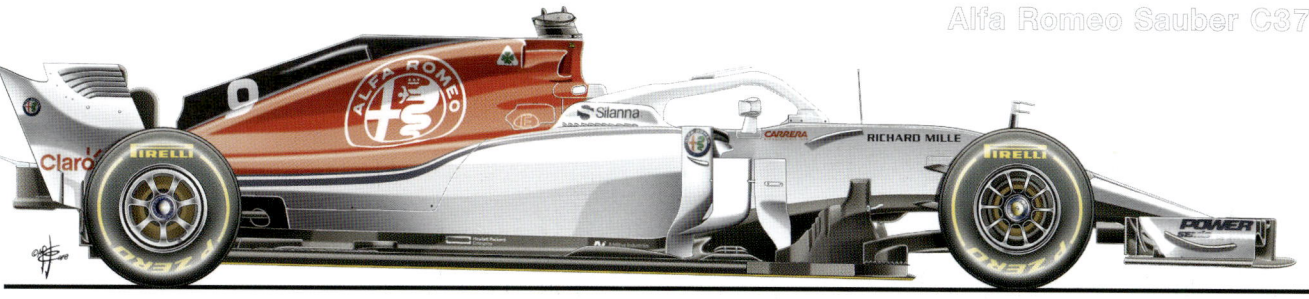

Alfa Romeo Sauber C37

Toro Rosso STR13

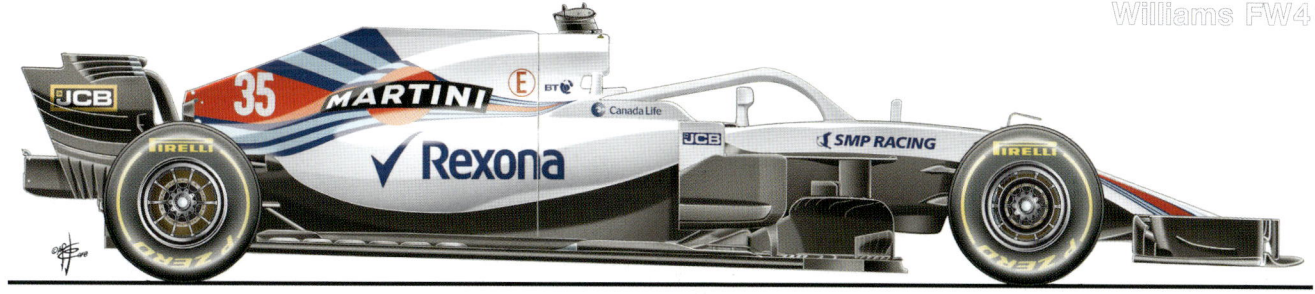

Williams FW41

Car TABLE 2018

	44-77 MERCEDES	5-7 FERRARI	3-33 RED BULL	27-55 RENAULT	
	W08	**SF70H**	**RB13**	**R.S.18**	
CAR					
Designers	James Allison Aldo Costa Andy Cowell	Mattia Binotto Enrico Cardile	Adrian Newey Rob Marshall Dan Fallows	Nick Chester Martin Tolliday	
Race engineers	Andrew Showling Peter Bonington (44) Tony Ross (77)	Matteo Togninalli Riccardo Adami (5) Dave Greenwood (7)	Paul Monagham Simon Rennie (3) Giampiero Lambiase (33)	Mark Slade (27) Karel Loos (55)	
Chief mechanic	Mattew Deane	-	Chris Gent Lee Stevenson	Robert Cherry	
CHASSIS					
Passo	3726 mm	3621 mm	3550 mm	3644 mm	
Front suspension	Push-rod 2+1 dampers and torsion bars	Push-rod 2+1 dampers and torsion bars	Push-rod 2+1 dampers and torsion bars	Push-rod 2+1 dampers and torsion bars	
Rear suspension	Pull-rod 2+1 dampers and torsion bars	Pull-rod 2+1 dampers and torsion bars	Pull-rod 2+1 dampers and torsion bars	Pull-rod 2+1 dampers and torsion bars	
Dampers	Sachs	Sachs	Multimatic	Penske	
Brakes calipers	Brembo	Brembo	Brembo	A+P	
Brakes discs	Brembo Carbon Industrie	Brembo CCR Carbon Industrie	Brembo	Hitco	
Wheels	BBS	BBS	O.Z.	AVUS	
Radiators	Secan	Secan	Marston	Marston	
Oil tank	middle position inside fuel tank	middle position inside fuel tank	middle position inside fuel tank	middle position inside fuel tank	
GEARBOX	Longitudinal carbon	Longitudinal carbon	Longitudinal carbon	Longitudinal titanium	
Gear selection	Semiautomatic 8 gears	Semiautomatic 8 gears	Semiautomatic 8 gears	Semiautomatic 8 gears	
Clutch	Sachs	Sachs	A+P	A+P	
Pedals	2	2	2	2	
ENGINE	Mercedes AMG F1 M09EQ	Ferrari 063	RBR - TAG Heuer RB14 2018	Renault RS18	
Total capacity	1600 cmc	1600 cmc	1600 cmc	1600 cmc	
N° cylinders and V	6 - V90°	6 - V90°	6 - V90°	6 - V90°	
Electronics	Mercedes	Magneti Marelli	Magneti Marelli	Magneti Marelli	
Fuel	Petronas	Shell	Total	Total	
Oil	Petronas	Shell	Total	Total	
Dashboard	Mercedes	Magneti Marelli	Red Bull	Renault F1	

GIORGIO PIOLA

8-20 HAAS	2-14 McLAREN	11-31 FORCE INDIA	9-16 ALFA ROMEO SAUBER	9-28 TORO ROSSO	18-35 WILLIAMS
VF-18	**MCL33**	**VJM11**	**C37**	**STR13**	**FW41**
Rod Taylor Ben Agathangelou	Peter Prodomou	Adrew Green Akio Haga	Simone Resta Ric Gandelin	James Key Jody Eggington	Paddy Lowe
Ajo Komatsu Gary Gannon (8) Giuliano Salvi (20)	Tom Stallard (2) Will Joseph (14)	Tim Wright (11) Bradley Joice (31)	Julien Simon Chautemps (9) Jorn Becker (16)	Mattia Spini (10) Pierre Hamelin (28)	James Urwin (18) Dave Robson (35)
Mattew Scott	Karl Lammenranta	Michael Brown Will Wikery	Reto Camenzind	Domiziano Facchinetti	Mark Pattinson
3621 mm	3568 mm	3640 mm	3551 mm	3650 mm	3600 mm
Push-rod 2+1 dampers and torsion bars	Push-rod 2+1 dampers and torsion bars	Push-rod 2+1 dampers and torsion bars	Push-rod 2+1 dampers and torsion bars	Push-rod 2+1 dampers and torsion bars	Push-rod 2+1 dampers and torsion bars
Pull-rod 2+1 dampers and torsion bars	Pull-rod 2+1 dampers and torsion bars	Pull-rod 2+1 dampers and torsion bars	Pull-rod 2+1 dampers and torsion bars	Pull-rod 2+1 dampers and torsion bars	Pull-rod 2+1 dampers and torsion bars
Sachs	McLaren	Sachs	Sachs	Koni	Williams
Brembo	Akebono	A+P	Brembo	Brembo	A+P
Brembo	Carbon Industrie Brembo	Hitco	Brembo	Brembo	Carbon Industrie
O.Z.	Enkey	BBS	O.Z.	O.Z.	O.Z.
Calsonic	Calsonic - IMI	Secan	Calsonic	Marston	IMI Marston
middle position inside fuel tank	middle position inside fuel tank	middle position inside fuel tank	middle position inside fuel tank	middle position inside fuel tank	middle position inside fuel tank
Longitudinal carbon	Longitudinal carbon	Longitudinal carbon	Longitudinal carbon	Longitudinal carbon	Longitudinal titanium
Semiautomatic 8 gears	Semiautomatic 8 gears	Semiautomatic 8 gears	Semiautomatic 8 gears	Semiautomatic 8 gears	Semiautomatic 8 gears
A+P	A+P	A+P	A+P	A+P	A+P
2	2	2	2	2	2
Ferrari 063	Renault RS18	Mercedes AMG F1 M09EQ	Ferrari 063	Honda RA6 18H	Mercedes AMG F1 M08EQ
1600 cmc	1600 cmc	1600 cmc	1600 cmc	1600 cmc	1600 cmc
6 - V90°	6 - V90°	6 - V90°	6 - V90°	6 - V90°	6 - V90°
Magneti Marelli	McLaren el.sys.	Mercedes	Magneti Marelli	Magneti Marelli	Mercedes
Shell	Mobil	Petronas	Shell	Total	Total
Shell	Mobil	Petronas	Shell	Total	Total
Ferrari	McLaren	P.I.	Magneti Marelli	Toro Rosso	Williams

2018 **REGULATIONS**

Following the epochal revolution introduced by the FIA for the 2017 season, the 2018 edition of the F1 World Championship did not entail technical modifications with the exception of the great visual impact of the definitive introduction of the Halo system after two years of debate and tests with various solutions, illustrated in the book dealing with the 2015/2016 season. This was a decision taken in the name of safety that has caused and continues to cause considerable controversy among F1 fans. As well as the aforementioned aesthetic impact, Halo also brings an increase in weight: with its supports, the system "costs" around an extra 15 kg; for this reason the minimum weight of the cars was increased by 5 kg, rising from 728 to 733 kg.

The fin on the engine cover and the upper T-wing that first appeared thanks to a loophole in the aerodynamic restrictions introduced with the 2017 regulations, along with the monkey seat, also disappeared. The engine exhaust was longer and lastly there was an improvement in safety with the addition of a third wheel retention cable (through to 2017 there were just two).

There were further very strict limitations on the engine: just three rather than four V6 internal combustion units were allowed, with each therefore being required to last seven races. Just two batteries, electronic control units and MGU-H (the turbo connected to the exhausts) were available over the course of the season.

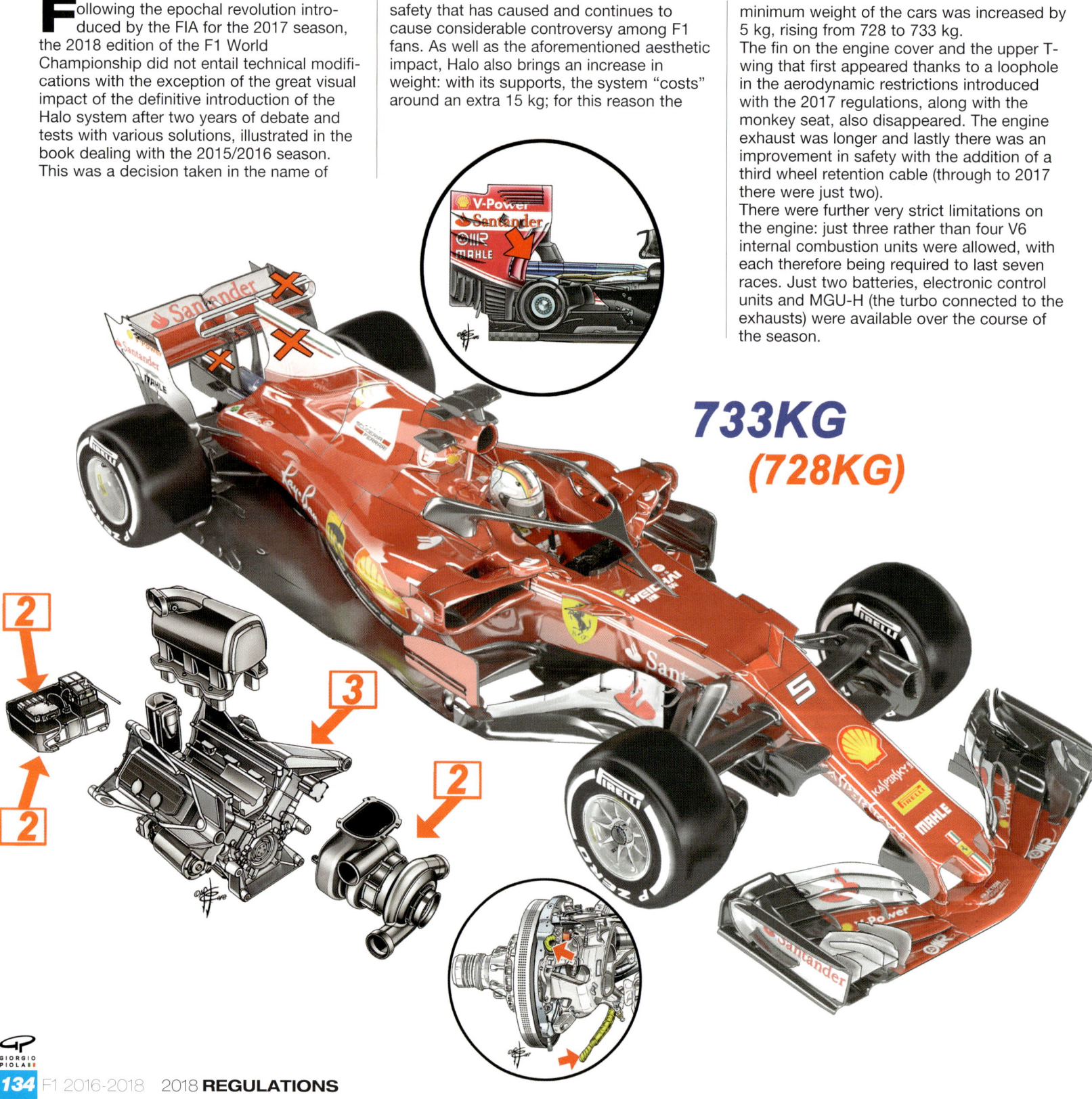

733KG
(728KG)

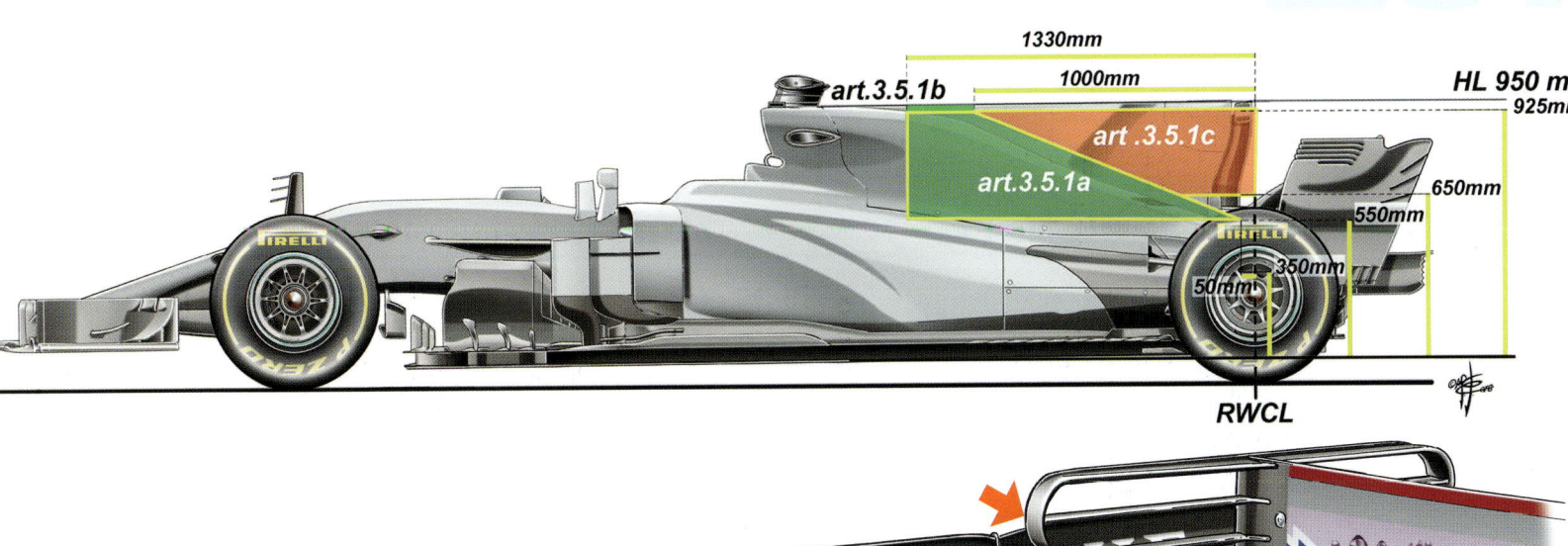

art.3.5.1b
1330mm
1000mm
art .3.5.1c
art.3.5.1a
HL 950 mm
925mm
650mm
550mm
350mm
50mm
RWCL

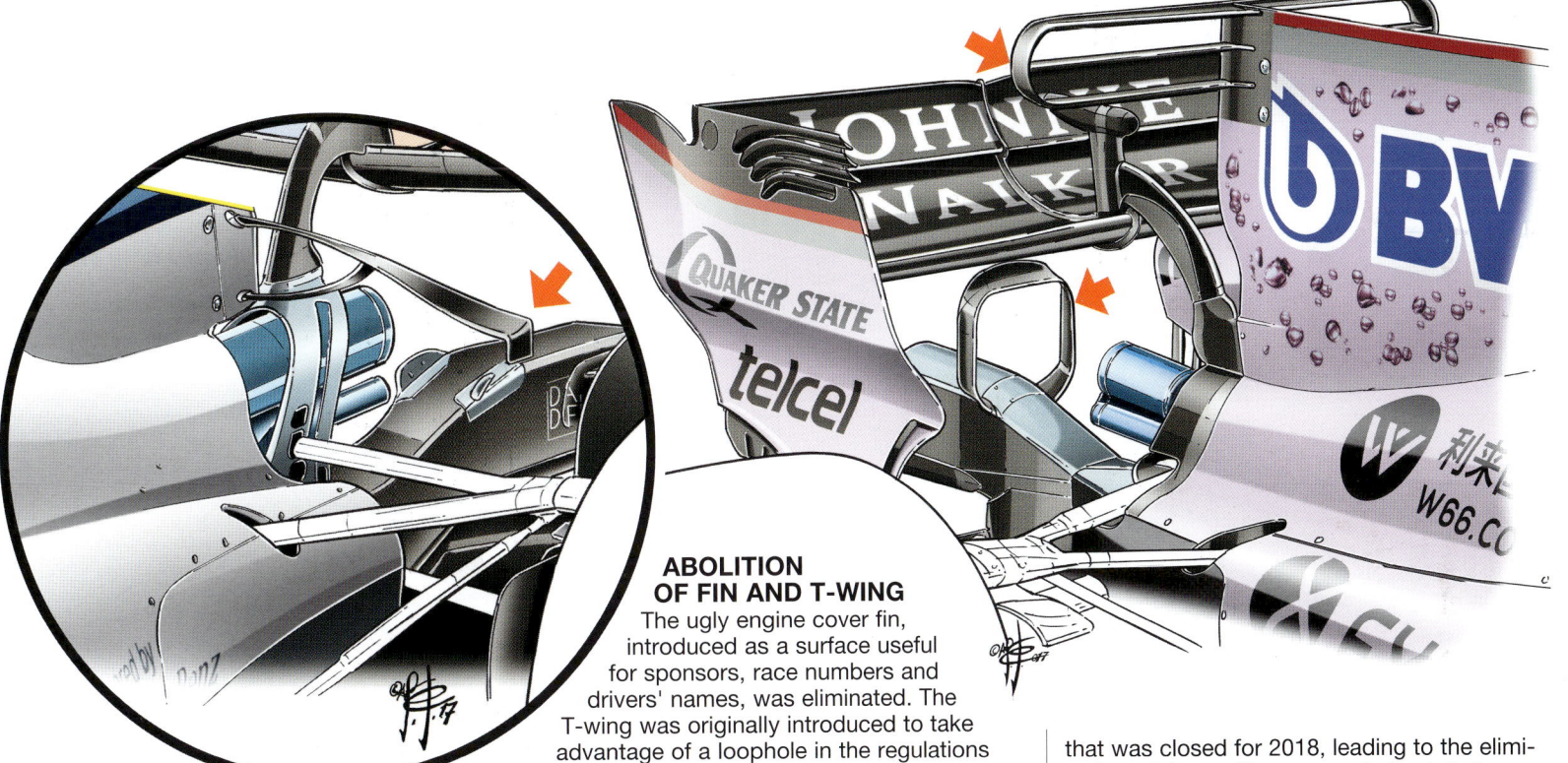

ABOLITION OF FIN AND T-WING

The ugly engine cover fin, introduced as a surface useful for sponsors, race numbers and drivers' names, was eliminated. The T-wing was originally introduced to take advantage of a loophole in the regulations that was closed for 2018, leading to the elimination of the profile, as seen in the left drawing. In green the surface permitted for the engine cover area and small fin, a configuration tested by Sauber in Austin, and in red the prohibited surface that includes the area previously left free. It should be remembered that the T-wing allowed downforce to be increased by better directing the air flow towards the rear wing. The Monkey Seat was also abolished. During the course of 2017, ever more complicated configurations had been seen for both areas. In the Force India drawing, a triplane T-wing and a "simple" Monkey Seat.

The 2018 regulations did however leave a very limited free zone lower down, above the 650 mm limit from the Reference Plane. In practice, the Williams FW40 configuration remained legal.

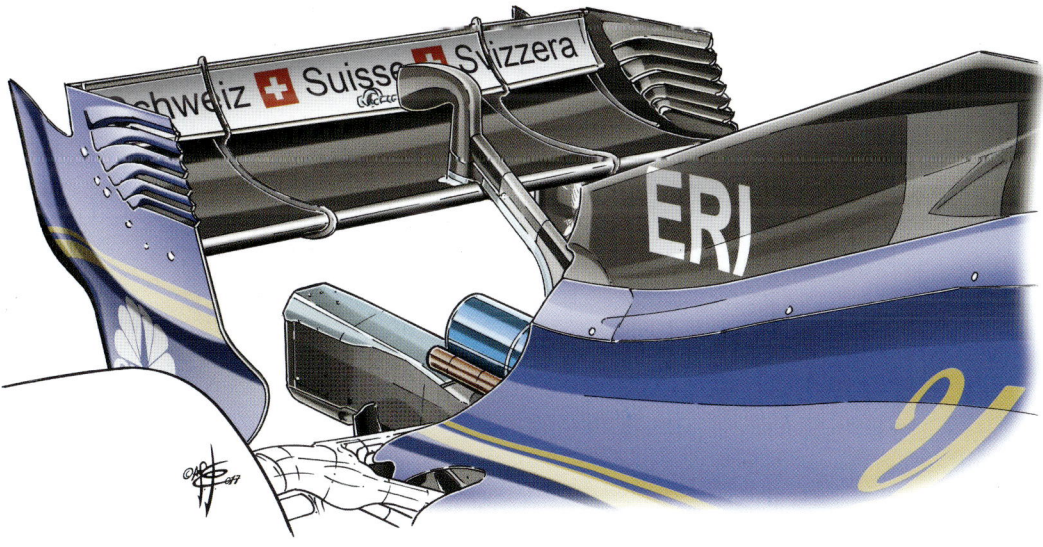

Bahrain 2015

HALO

The Halo system, first unveiled in Bahrain in 2015, put the F1 purists' noses out of joint; after two years of arguments and testing it was finally introduced with forms not very different to those conceived by Mercedes on behalf of the FISA (left). The new head protection structure brought costs in terms of weight, aerodynamics and fuel consumption. It had to resist severe crash tests, indicated in the drawing on the right, and had a notable impact on the chassis designs. The areas of the chassis around the Halo mounting points had to be reinforced as in the case of overturning the would be subjected to significant stresses. The increased weight was to have an effect: the system, complete with supports accounted for an extra 15 kg or so, while the minimum weight of the cars was increased by just 5 kg, from 728 to 733 kg, a limit that became particularly difficult to reach.

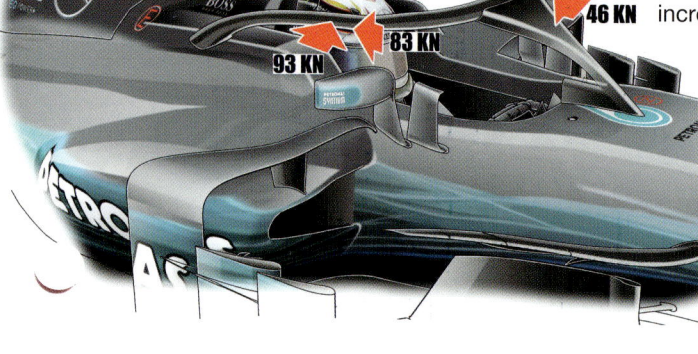

HALO REGULATIONS

In an attempt to reduce the negative effects on the air flows towards the engine air intake and the rear section of the car, the Federation left 200 mm of "free" space, immediately seized on by the teams from the test sessions following the Abu Dhabi GP, as iun the case of these three small profiles seen on the McLarens.

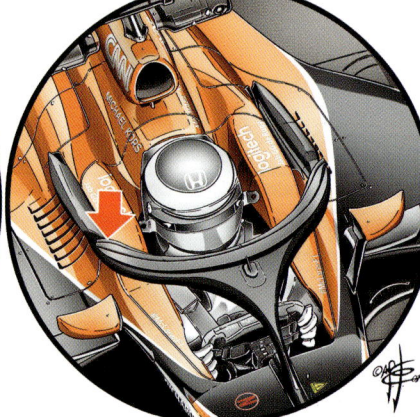

IMPACT BETWEEN WHEELS

There is no doubt that Halo is very effective in deviating large objects like wheels from the cockpit area. Lengthy tests were conducted with an impact speed of 220 kph, as can be seen in these two animation frames.

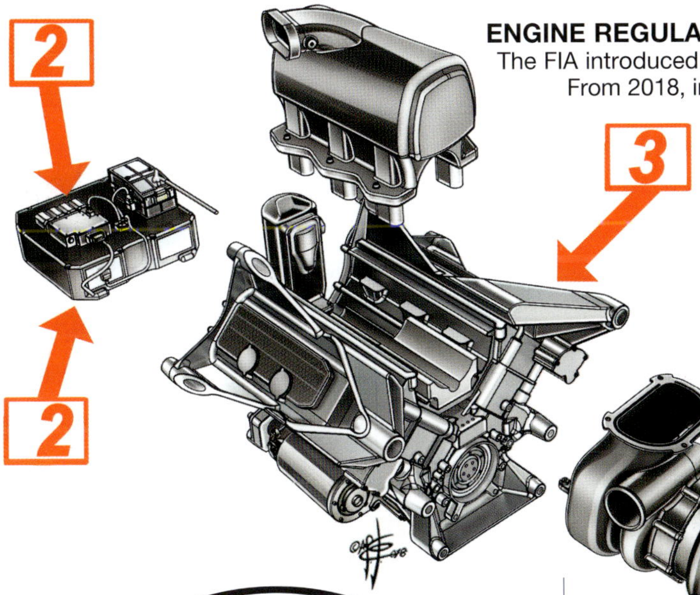

ENGINE REGULATIONS 2018

The FIA introduced a drastic reduction in the number of engine components that can be used. From 2018, in fact, teams were required to complete the full 21-race championship with 3 engines (ICE), 3 motor generator units-heat (MGU-H), 3 turbochargers (TC), 2 energy stores (ES), 2 control electronics (CE) and 2 motor generator units-kinetic (MGU-K). The exhaust pipes were lengthened by 10 cm to reduce the advantages deriving from any vents in the diffuser areas. The engines were also subject to other significant provisions: the FIA introduced strict limits on oil consumption. At the beginning of the 2017 season there were six-cylinder that "burned" 1.2 litres of lubricant per 100 km, while for 2018 this was halved to 0.6 litres per 100 km. Moreover, just one type of oil was to be available for the entire weekend and was to be declared before the start of the event.

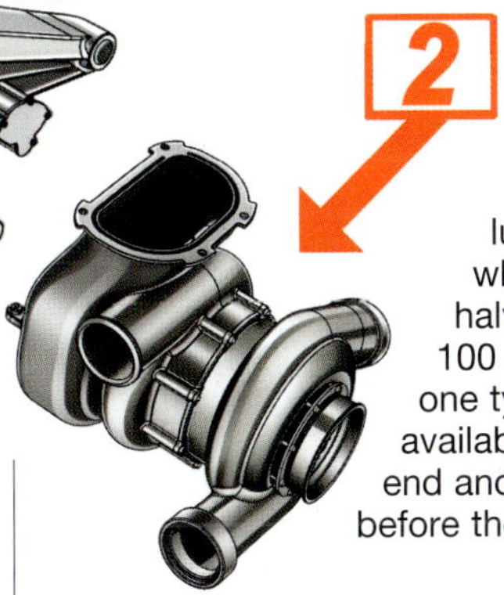

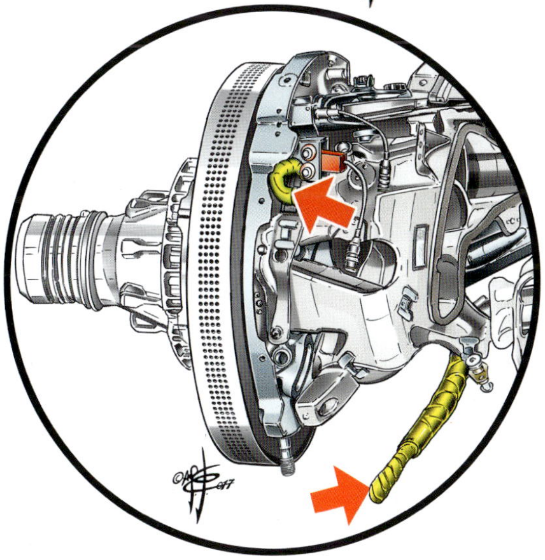

RETENTION CABLES

Safety concerns were also behind the introduction of the third wheel retention cable. No longer two cables as seen in the drawing but three. Moreover, the frontal crash test was to be more severe, requiring greater impact absorption by the nose and the chassis.

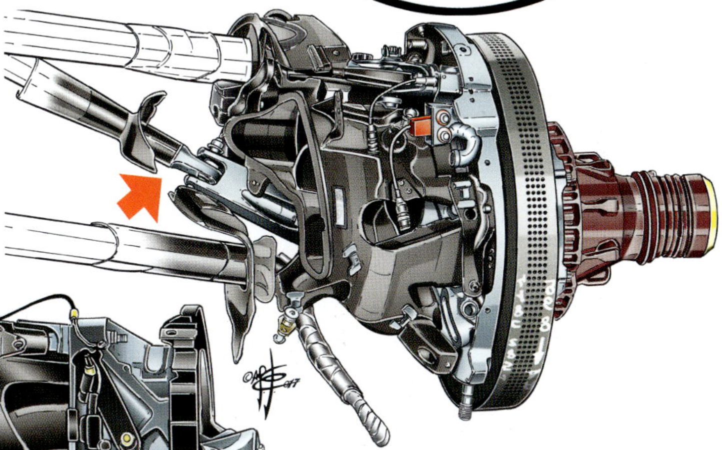

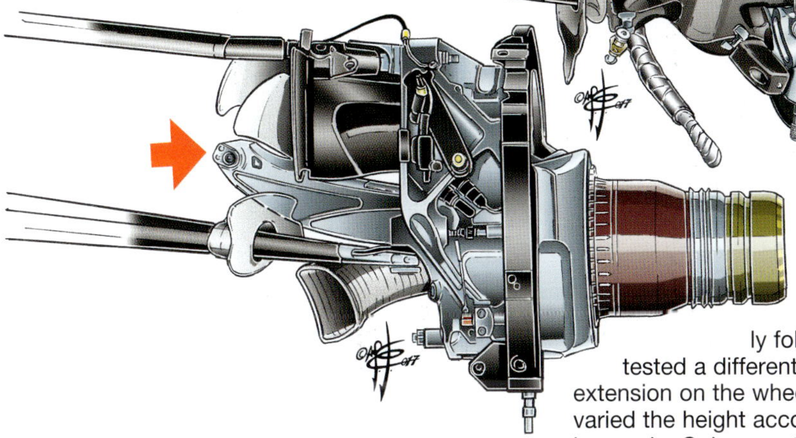

ABOLTION OF THE STRUT BRACKET

From the Belgian GP Ferrari, closely followed by Red Bull, tested a different strut mount extension on the wheel hub. This design varied the height according to the steering angle. Only once the season had finished, on the 12th of December, did the FIA ban this feature in response to a request for clarification by Ferrari.

PIRELLI TYRES

In order to guarantee greater flexibility in the choice of race sets and consequently to formulate more strategies with respect to the 2017 season, Pirelli expanded the range of dry tyres with compounds softer by a step with respect to those used in 2017.
The greatest novelty lay in the introduction of a new tyre that undercut the Ultrasoft and was named Hypersoft, distinguished by a pink band. This was the softest tyre ever produced by Pirelli for F1.
The Hard compound was distinguished by a blue band, while another even harder new compound was designated as the Superhard and given an orange band; this tyre took the full range to nine: seven slicks, the intermediates and the full wets.

orange=hard

pink=hypersoft

purple=extrasoft

red=supersoft

white=medium

yellow=soft

New **DEVELOPMENTS 2018**

The greatest revolution in recent Formula 1 history introduced in the 2017 season saw a great number of important new features and above all the pleasant surprise that, in starting with a blank sheet of paper, this time it was not Adrian Newey who was again stealing a march on his rivals, but rather Ferrari with the most innovative car in the field, with a wealth of extreme features. The 2018 season was characterised by absolute technical stability and saw the consecration of the new sidepod design devised by Maranello for the revolutionary SF70H and then adopted by almost all the teams, including Mercedes. This last introduced the design late, at the Austrian GP, and saw a notable increase in the W09's competitiveness. The New Solutions chapter presents similarities with the Controversies because certain novelties were the object of disputes that in order to avoid unnecessary repetition have been omitted here.

FERRARI MIRRORS

The extreme aerodynamic research in the area of the leading section of the sidepods led Ferrari to equip the SF71H with a new and sophisticated blown rear-view mirror housing with the objective of improving the flow towards a second flush cooling intake, which was also new (highlighted in the side view). The mirror design was also copied by Red Bull and Toro Rosso.

SAUBER SIDEPODS

Sauber also adopted the duel intakes for cooling the radiators like Ferrari. It did so however not with a flush intake like on the SF71H, but rather with a ear-like intake hidden by the horizontal plane, immediately behind the mirrors; the intake is highlighted in the cutaway in the drawing on the right.

FERRARI SIDEPODS

The drawing shows the stratagem used by Maranello to exploit a loophole in the regulations regarding the disposition of the anti-intrusion protection structures. The cones were to have an identical form on all cars and a certain position with respect to the cockpit. On the SF70H Ferrari had placed them not at the edge of the sidepod intake mouth, but rather within an aerodynamic profile that effectively allowed the sidepods to be set considerably further back, distancing them from the toxic turbulence generated by the front wheels. The detail drawing shows the upper deformable structure incorporated in the horizontal turning vane, placed ahead of the sidepod mouths.

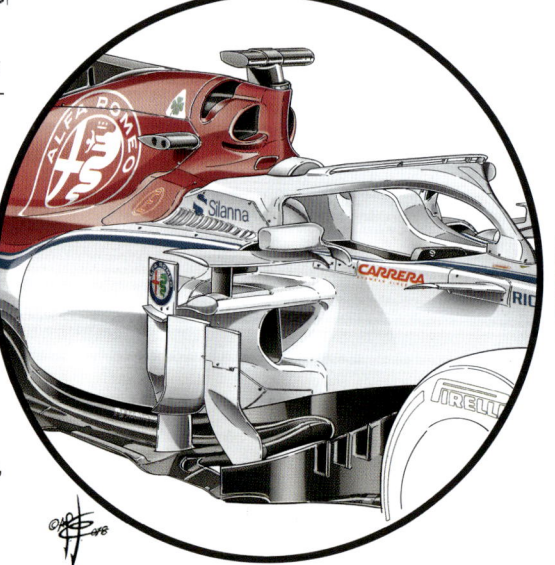

SAUBER AIR SCOOP

The Sauber air scoop was very sophisticated, a combination of the Ferrari 2010 design, with two rear-set ears, and that of the Mercedes, with a vertical roll bar and two lateral intake mouths. The design was reprised in 2011 by Lotus and Force India after the Federation had imposed dimensional limits on the single pylon roll bar.

Mercedes

Ferrari 2010

Force India 2011

Lotus 2011

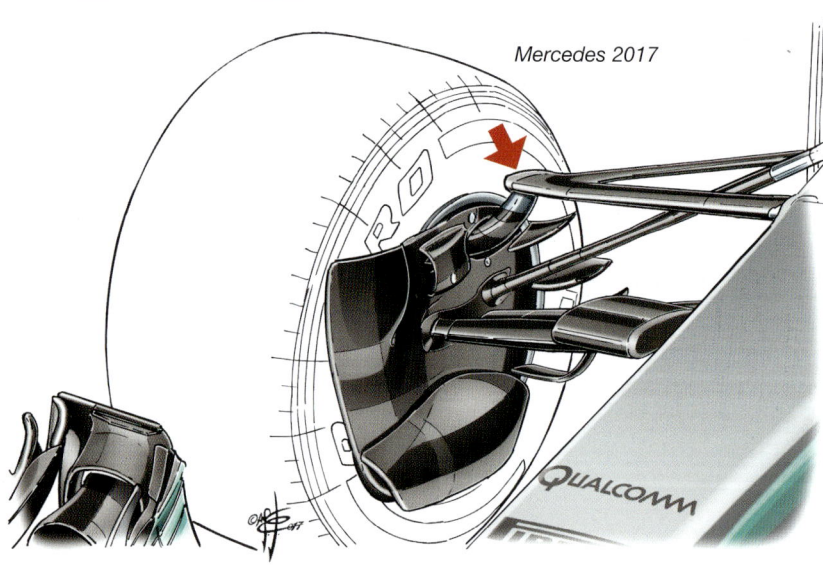

Mercedes 2017

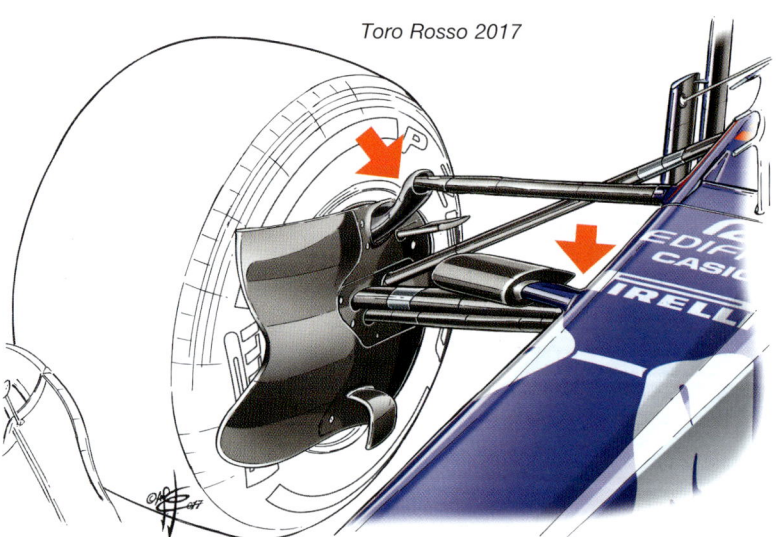

Toro Rosso 2017

MERCEDES REAR SUSPENSION

In the 2017 season the new front suspension configuration introduced simultaneously by Mercedes and Toro Rosso had caused a sensation with the new cars featuring this new and significantly higher upper wishbone mount that protruded above the wheel rim.

In 2018, Mercedes transferred the same concept to the rear, taking to the extreme to obtain great aerodynamic efficiency. The rear suspension upper wishbone mount was raised with a bracket that raised it much more than seen on the front end of the W08 in 2017.

The detail drawing shows a very stiff box-section structure that acted as an extension of the hub carrier. The extended part was designed to permit a wishbone positioned almost horizontally so that it could also act as a turning vane. This in part simulated the bean wing, the lower profile banned in 2014 but useful in that it works in synergy with both the rear wing and the diffuser.

The brake intake was raised so as to ensure that the lower wishbone, which was specially shaped in proximity to the brake duct, could interact with the faired arm in wing profile containing the drive shaft, creating a blow.

With the brake duct completely closed, the hot air from the brakes is vent not outside the wheel but from an aperture opened at the tail of the fairing that forms a carbonfibre hood for the bracket (drawing of the rear section).

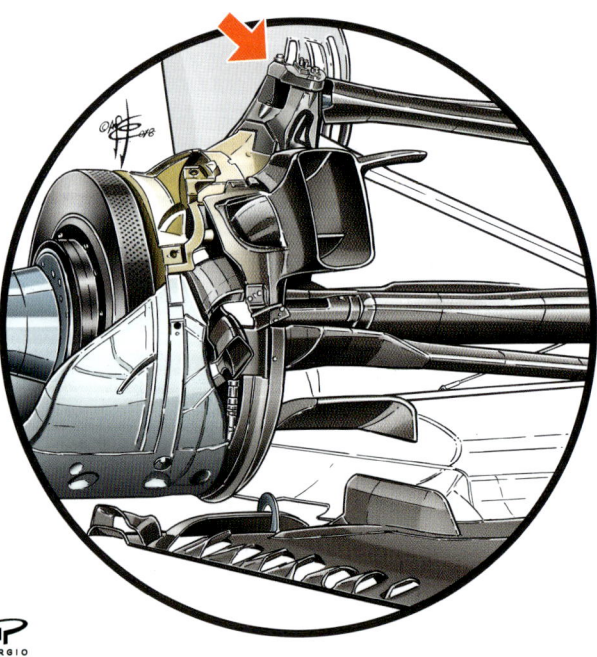

FERRARI BARCELONA

In the first free practice session for the Spanish GP, Ferrari fitted Sebastian Vettel's SF71H with modified rear suspension inspired by the Mercedes configuration.
In the comparison with the previous layout (insert) note how the upper wishbone mounting bracket on the hub carrier was raised. The new rear suspension gave the results hoped for and was also fitted to Kimi Räikkönen's SF71H from the Saturday: the two drivers used the configuration in qualifying, the race and then for the rest of the season.

McLAREN BARCELONA

McLaren introduced a new aero package at the Spanish GP with the first true development of the MCL33. There was an all-new lower nose (with very short front wing support pylons) that was very narrow with respect to the section of the MCL33 chassis, but had two very long vents at the sides (4).
The nose (1) was very small and characterised by a hole similar to the one seen on the Red Bulls to reduce the aerodynamic obstruction generated at the tip.
At the sides, two additional intakes reprised the concept already seen on the Force India in 2017 and, in particular on the Sauber C37, with a low vent (1) divided into three sections (2), visible in the view from below. The structure of the nose narrowed to a keel (3) in the lower part where there was a large "manta-fin" profile (5) that reprise the front extractor concept introduced by Mercedes in Spain in 2017 and copied by Williams in 2018.

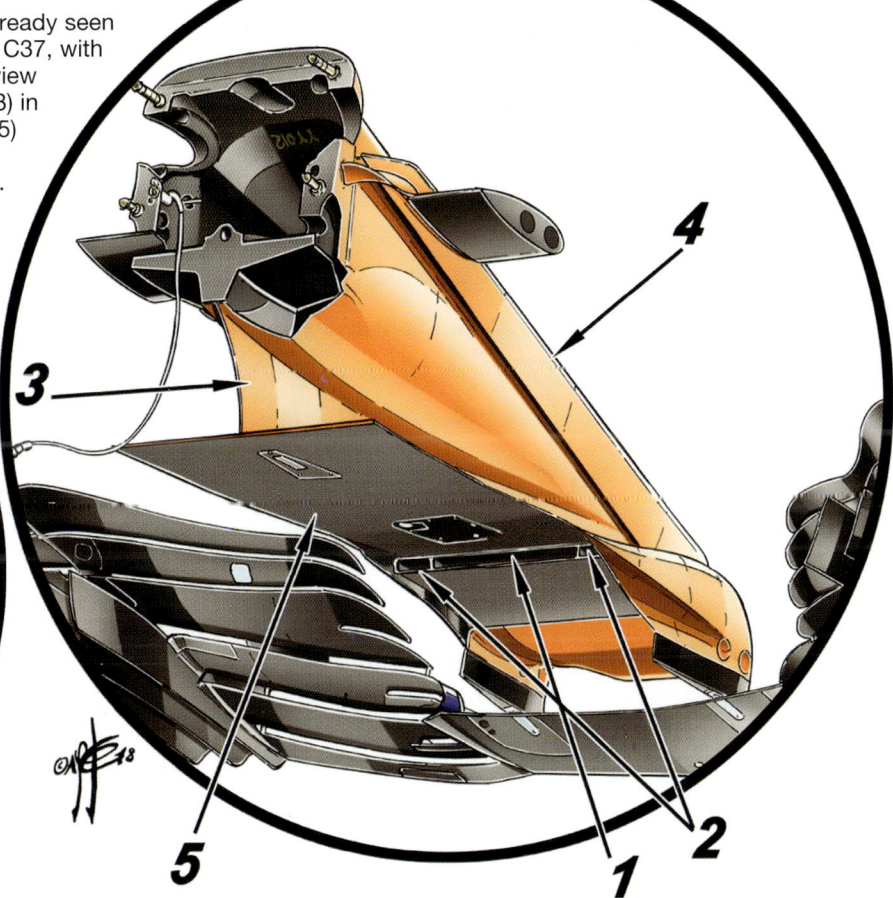

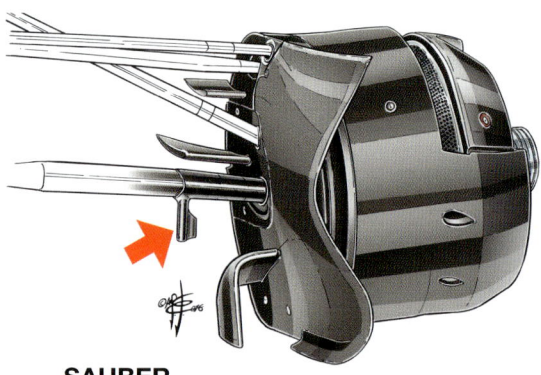

SAUBER

Sauber exploits the gaps in the regulations to introduce, from the start of the season, two turning vanes in the upper wishbone within the area 125 mm from the hub carrier, which was to be considered as part of the brake duct, even though it was an aerodynamic device mounted on levers.

In the drawing, indicated by the red arrows, you can see the two winglets that were designed to groom the flow so as to reduce the negative effects of the vortices generated by the front wheels.

In 2016, Toro Rosso (insert) had introduced a profile at Monaco in proximity to the strut, but within the lower wishbone, that was not as sophisticated as that on the Sauber. From the Spanish GP, Mercedes too used these small turning vanes.

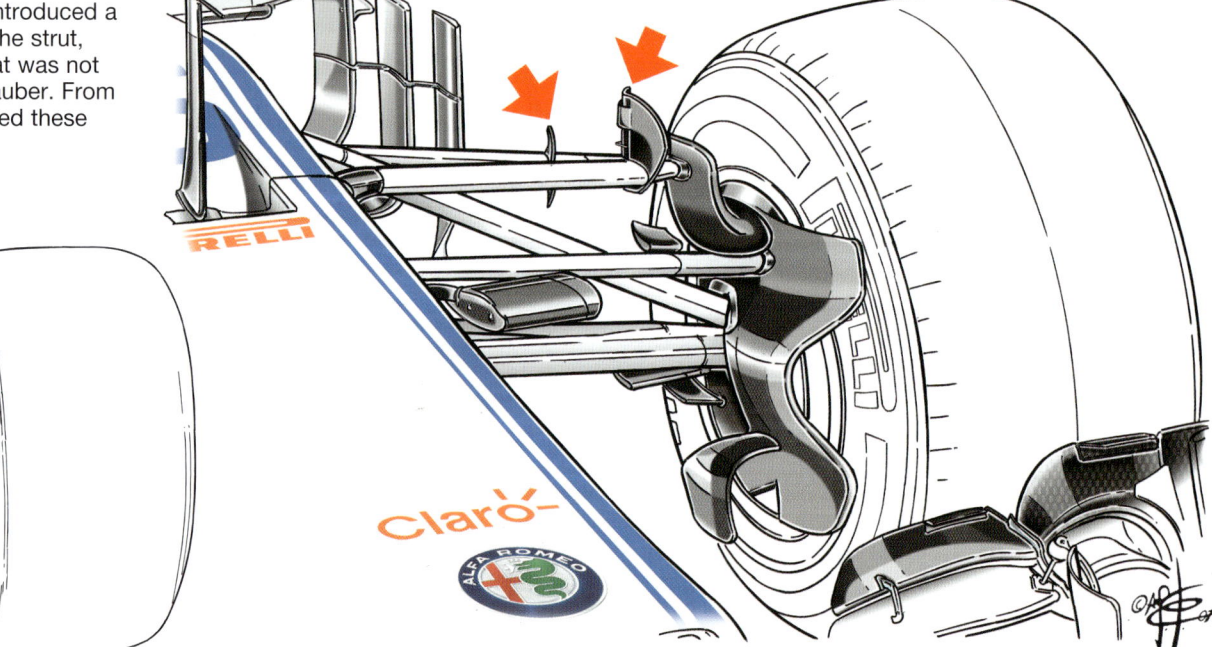

MERCEDES FRONT SUSPENSION

The W09 had a new and very sophisticated front suspension configuration with a layout that was asymmetrical in many points. The strut in the push-rod system was pivoted fairly high on the rocker (1) that itself acted on the third hydraulic element (2) that was much more voluminous than the one fitted to the W08.

At the front of the third damper there was a variable lung (2) controlling ride height without the need for the outlawed electronic system.

Another interesting feature was the system linking the two asymmetric torsion bars (3) with the right-hand one (visible on the left) protruding further than the left. This was a fabricated structure (4) that was infinitely stiffer than the 2017 one, but also light, thus achieving two objectives in a single feature.

There were those who claimed that the two levers united by a short connecting rod at the centre acted as an anti-roll bar: this was not the case. This important element remained concealed within the chassis and was not visible.

The front view clearly shows the steering box (5) set lower to allow the steering arm to be slightly higher than the lower wishbone, thus guaranteeing a blow useful in the definition of the flows towards the body of the car. In (6), the no less than three retaining cables per wheel.

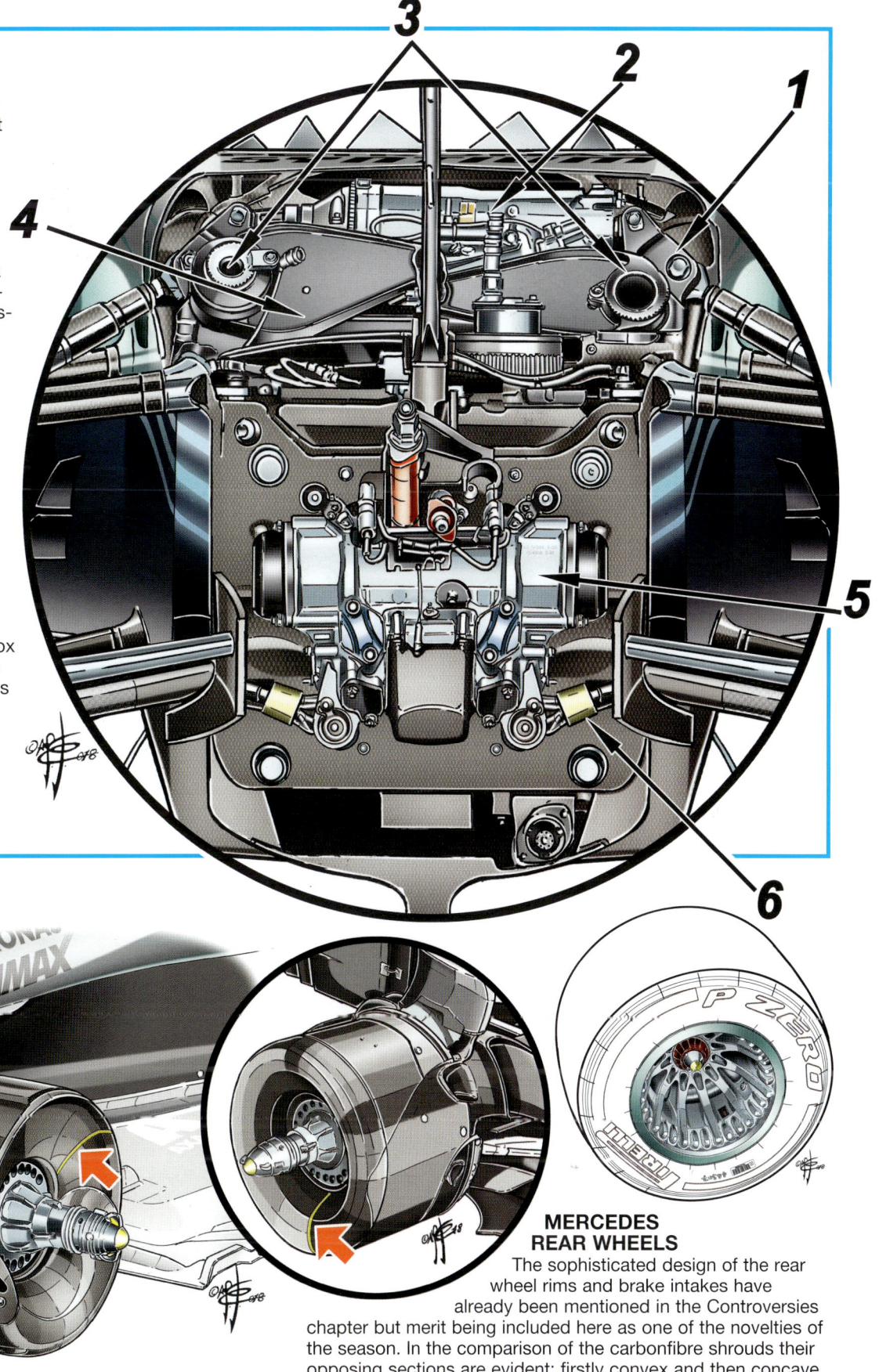

MERCEDES REAR WHEELS

The sophisticated design of the rear wheel rims and brake intakes have already been mentioned in the Controversies chapter but merit being included here as one of the novelties of the season. In the comparison of the carbonfibre shrouds their opposing sections are evident: firstly convex and then concave from the Belgian GP.

STEERING WHEELS 2018

In the 2018 season, the range of steering wheels used by the various teams was virtually unchanged, with all of them utilising the twin clutch paddles actuated with the fingers in moulded niches, a feature that had been adopted across the board after bing introduced by Mercedes early in 2016. After experimenting with the feature on Vettel's car alone from the 2017 Spanish GP (see the Cockpits 2017 chapter), Ferrari returned to the single offset paddle in Japan that had proved to be more effective with Räikkönen who got away to excellent starts. In the 2018 season there was a further evolution of the system, again reserved for the German driver: a small supplementary lever at the rear, jealously concealed until the moment in which Vettel, euphoric for the pole position conquered in Bahrain revealed it to the photographers' lenses when he propped the steering wheel against the Halo device. This momentary lapse provoked some disputes and a degree of suspicion regarding its true function and has been analysed in the "Controversies" section.

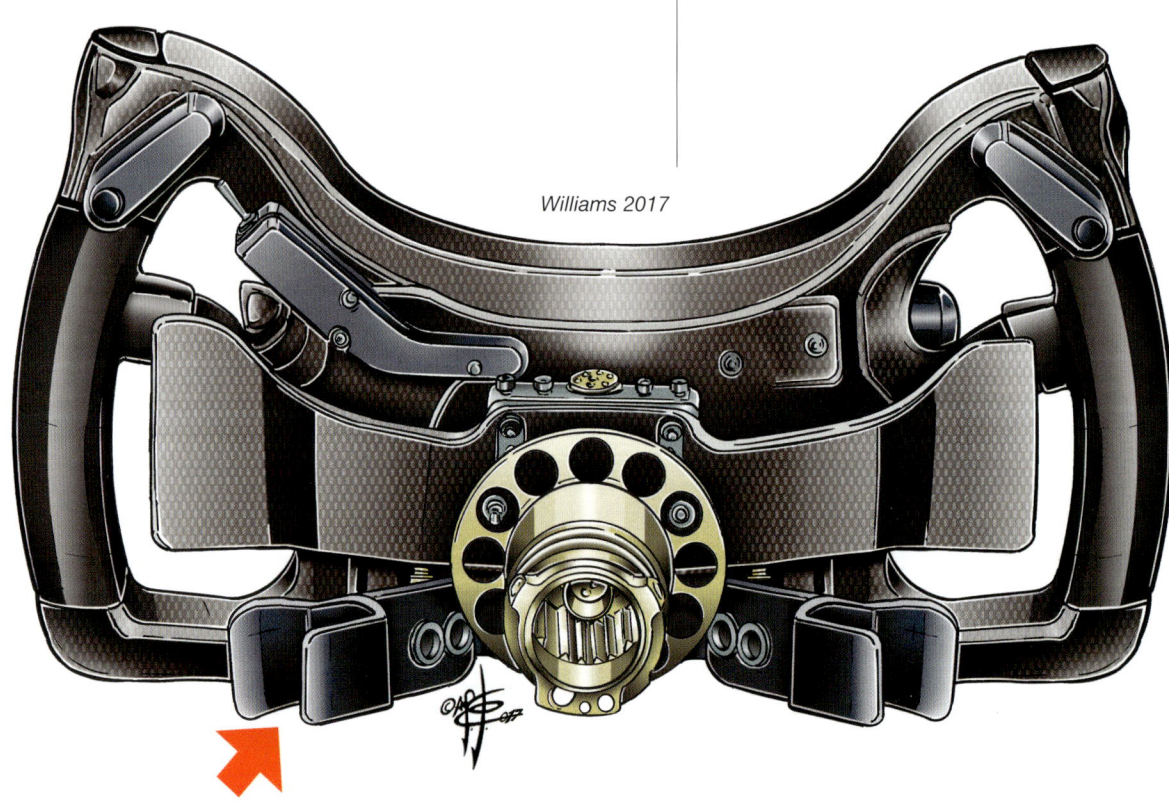

Williams 2017

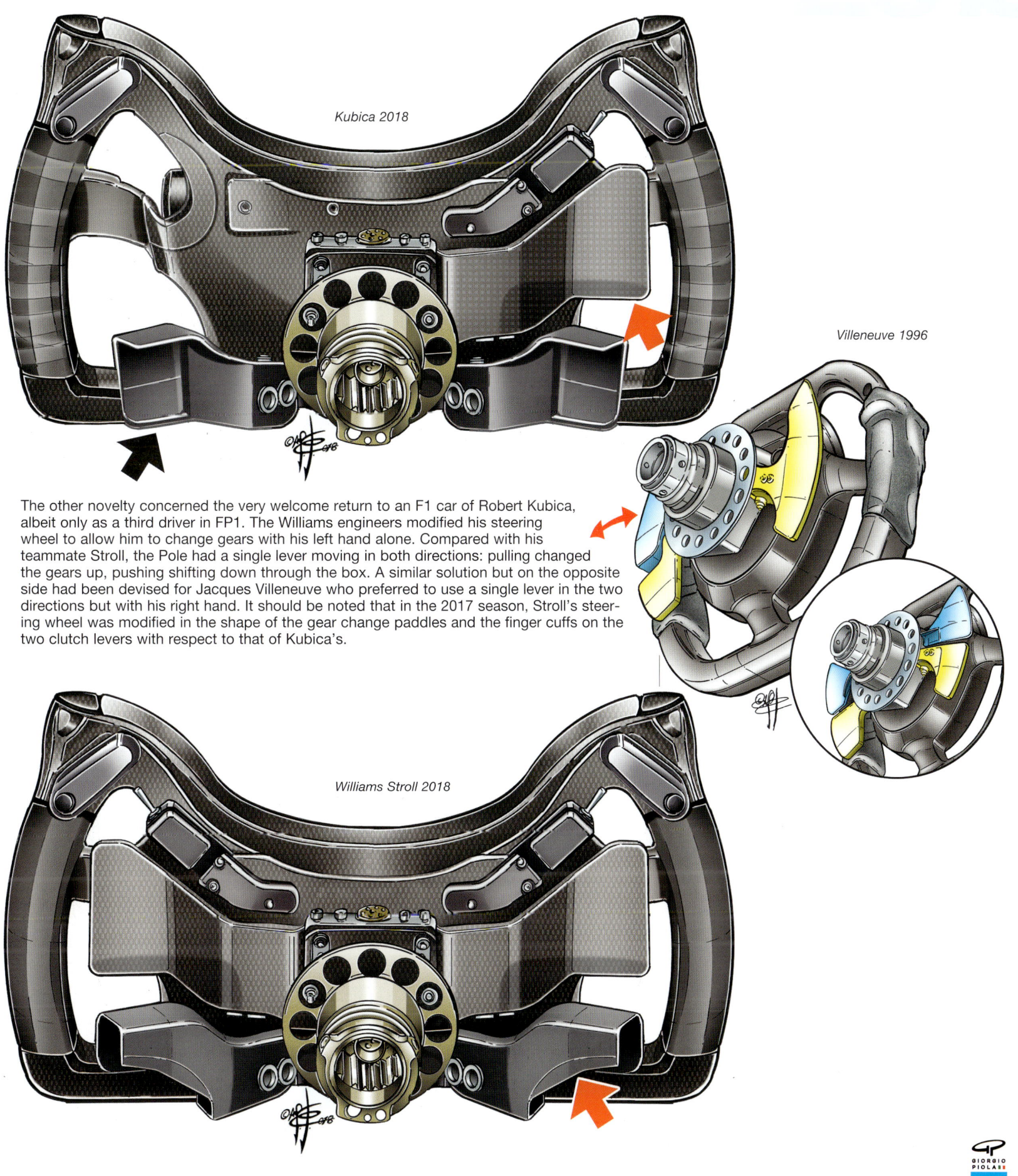

Kubica 2018

Villeneuve 1996

Williams Stroll 2018

The other novelty concerned the very welcome return to an F1 car of Robert Kubica, albeit only as a third driver in FP1. The Williams engineers modified his steering wheel to allow him to change gears with his left hand alone. Compared with his teammate Stroll, the Pole had a single lever moving in both directions: pulling changed the gears up, pushing shifting down through the box. A similar solution but on the opposite side had been devised for Jacques Villeneuve who preferred to use a single lever in the two directions but with his right hand. It should be noted that in the 2017 season, Stroll's steering wheel was modified in the shape of the gear change paddles and the finger cuffs on the two clutch levers with respect to that of Kubica's.

Top VIEWS 2018

In this last edition of the book there is a new feature with top view comparison reveals all the cars. This has entailed painstaking picture research given the difficulty of accessing suitable viewpoints compared to the side views. This has provided us with a further means of evaluating the differences between the various cars, not only because it gives a greater perception of the

different wheelbases, with Mercedes at the top and Red Bull bringing up the rear, but above all it gives a better idea of the shape of the bodywork.

The comparison was made with all the cars in early season form, hence the Mercedes W09 is depicted with the traditional sidepods and not the version introduced in Austria. The difference between the cars that did not follow

the Ferrari path is immediately evident and in this review there are four: Mercedes, Force India, Renault and McLaren. It was extremely interesting to note the narrowing of the so-called Coke bottle area, with very different configurations and almost no cars that could be said to resemble one another. Adrian Newey went completely against the trend with sidepods that were not only very narrow

but above all with a section that was almost the opposite of the general shape of the other cars and no undercut in the bottom part. This difference has been emphasised in a comparison between the Red Bull RB14 on the left and the Mercedes W09 on the right, on the Red Bull the flow of air to the rear end is not fed in the undercut of the lower part of the sidepod but rather laps around the

descending almost trapezoidal form of the sidepods. The Coke bottle area is particularly extensive. Seen from above, we can see all the various and ever-more numerous slots in the floor of the car, not only in the front section and ahead of the rear wheels, but also as a kind of longitudinal track along the sides of the sidepods. Slots that were constantly developed over the season.

In the comparison between the various cars we can also see the more or less extreme megaphone vent in the end part of the engine cover, with the Williams characterised by an exaggerated feature that betrays one of the weakness of this car, that is its inadequate cooling.

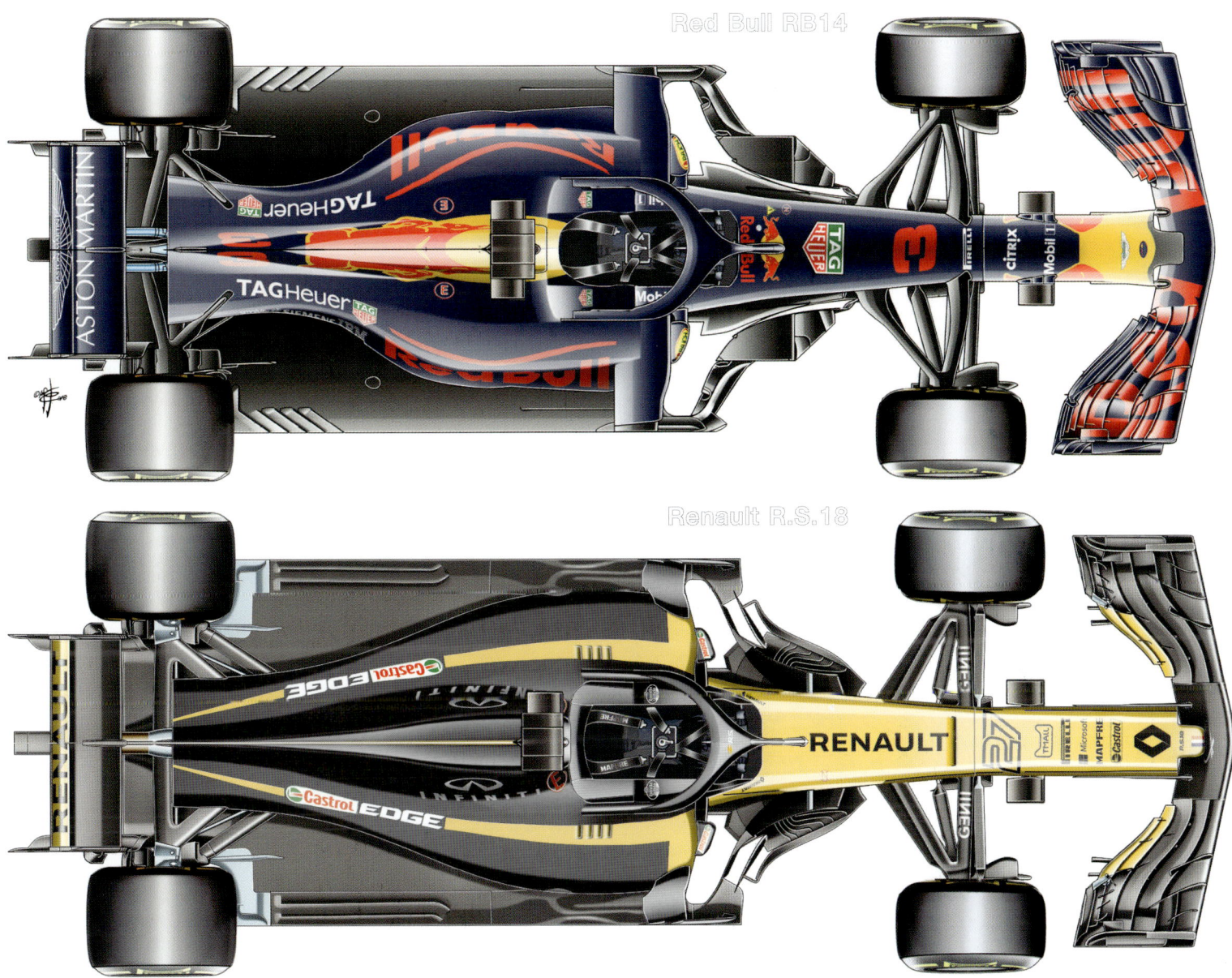

Red Bull RB14

Renault R.S.18

Haas VF-18

McLaren MCL33

Force India / Racing Point VJM11

Alfa Romeo Sauber C37

Toro Rosso STR13

Williams FW41

Talking about **WHEELBASES**

The tables that follow highlight the wheelbase and rake values of the cars from 2015, the last year examined in the previous editions of the Technical Analysis. It is immediately clear that the car that has dominated the Power Unit era, the Mercedes, has also been the one subjected to the greatest variations in terms of the wheelbase dimension. In 2015, in fact, it was the shortest car in the field only to become the longest (+300 mm) in the 2017/2018 seasons, after a slight increase of 15 mm that had taken it to the penultimate position in this list in 2016. At the top of the table in 2015/2016 were Renault and Toro Rosso, overtaken by the Silver Arrows in the next two-year period. This switch also involved Force India, subjected to an even greater variation (+279 mm) given that in 2016 the English car was shorter than its German "cousin". The team based at Silverstone adopted both the Mercedes engine and the gearbox casing with a spacer that was enlarged following the revolution in the regulations introduced by the FIA in 2017. After slipping back slightly in 2016, Ferrari saw a constant increase in wheelbase length over the last two years with +57 mm for the SF70H and a further 70 mm for the SF71H. Both Mercedes and Ferrari then increased their wheelbase by modifying the distance from the body of the car of both the front and rear axles; on the car from Maranello to a lesser extent with the front axle in part because the clever feature of inserting the deformable structures outside the entrance to the sidepods allowed these last to be distanced from the turbulence of the front wheels. Williams and Red Bull instead went against the tide and in 2017 actually reduced the wheelbase length of their cars. The first because, in contrast with Force India, they adopted only the Mercedes engine and pro-

2015

	Wheelbase		Rake
Renault	3.602 mm		1,0° (4)
Toro Rosso	3.533 mm		1,5° (2)
Ferrari	3.508 mm		1,4° (3)
McLaren	3.491 mm		1,9° (1)
Williams	3.483 mm		1,4° (3)
Red Bull	3.459 mm		1,9° (1)
Force India	3.411 mm		1,9° (1)
Mercedes	3.411 mm		1,0° (4)

2016

	Wheelbase (diff. 2015)		Rake
Renault	3.638 mm (+36 mm)		1,0° (4)
Toro Rosso	3.616 mm (+83 mm)		1,5° (2)
Williams	3.589 mm (+106 mm)		1,4° (3)
McLaren	3.534 mm (+43 mm)		1,9° (1)
Ferrari	3.494 mm (+50 mm)		1,4° (3)
Red Bull	3.432 mm (-27 mm)		1,9° (1)
Mercedes	3.426 mm (+15 mm)		1,0° (4)
Force India	3.412 mm (+1 mm)		1,9° (1)

MERCEDES WHEELBASES

The Mercedes was the car that was subjected to the greatest variations in wheelbase over the past four seasons, from one extreme to the other: the shortest csar in the field in 2015 and the longest in 2017-18. Moreover, it is the only one that did not exploit the high rake configuration introduced by Red Bull in 2011 and gradually adopted by all the other teams. It was no coincidence that the notable increase in wheelbase (a full 226 mm) was associated with the new regulations introduced in the 2017 season and the decision not to exploit the aerodynamic advantages brought by positive rake. In order to highlight the variation, the rear axle has been taken as a reference, while in fact both axles were distanced from the centre of the car.

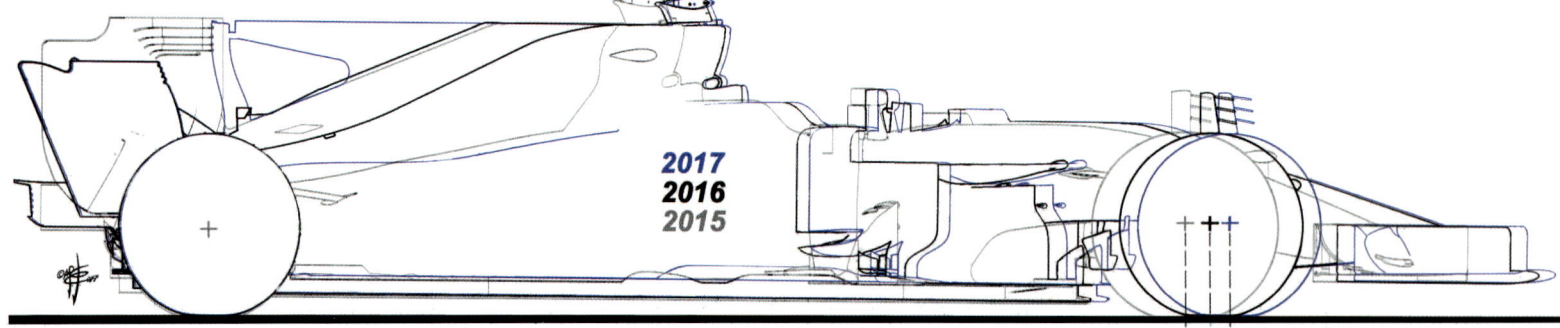

2015=3411mm   2017=3726mm
2016=3500mm

2017

	Wheelbase (diff. 2016)		Rake	
Mercedes	3.726 mm (+226 mm)		1,2°	(4)
Force India	3.691 mm (+279 mm)		1,9°	(1)
Renault	3.630 mm (-8 mm)		1,7°	(2)
McLaren	3.584 mm (+50 mm)		1,7°	(2)
Ferrari	3.551 mm (+57 mm)		1,5°	(3)
Toro Rosso	3.550 mm (+56 mm)		1,1°	(5)
Williams	3.545 mm (-44 mm)		1,7°	(2)
Red Bull	3.410 mm (-22 mm)		1,9°	(1)

2018

	Wheelbase (diff. 2017)		Rake	
Mercedes	3.726 mm (+0 mm)		1,25°	(6)
Toro Rosso	3.650 mm (+100 mm)		1,1°	(7)
Renault	3.644 mm (+14 mm)		1,7°	(3)
Force India	3.640 mm (-51 mm)		1,9°	(1)
Ferrari	3.621 mm (+70 mm)		1,53°	(4)
Williams	3.600 mm (+55 mm)		1,5°	(5)
McLaren	3.568 mm (-20 mm)		1,75°	(2)
Red Bull	3.550 mm (+140 mm)		1,9°	(1)

duced their own gearbox that was much shorter; the second because, the rake apart, they used an aerodynamic configuration that was very different with respect to all the other cars.

It is interesting to note the evolution of the positive rake configuration introduced by Adrian Newey in 2011 with the RB6. A configuration that instead of having a car virtually parallel to the ground entailed a very low front end and a higher rear, thus obtaining a set-up with a decisive "dive". A configuration that allowed the angle of incidence between the extractor profile and the ground to be increased. In practice it was means of effectively increasing the extractor dimensions that were otherwise restricted by the technical regulations. In practice, a loophole had been found by exploiting an extractor profile that used a much greater flow of air. The consequence was the generation of greater downforce that in 2011 had largely been determined by the blown exhaust which were instead abolished in 2012.

However, the tendency to exploit this positive rake was the objective of many times with the

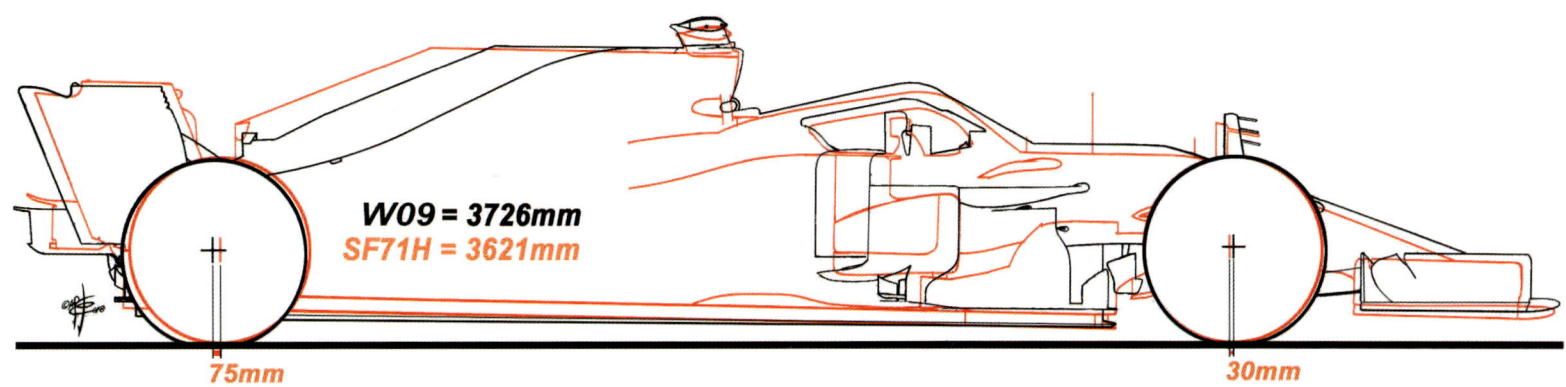

W09 = 3726mm
SF71H = 3621mm

75mm   30mm

MERCEDES/FERRARI

In the 2018 season Mercedes retained the same wheelbase dimension for the first time, with the car once again being the longest in the field. Ferrari instead increased its wheelbase by 70 mm with respect to the 2017 car, coming to within 105 mm of the Mercedes. In this case the comparison focuses on the centre of the cockpit, highlighting how both axles of the W09 were distanced with respect to the SF71H.

MERCEDES/RED BULL

The comparison between the W09 and the RB14 reveals a macroscopic distancing of the rear axle from the cockpit, with no less than 176 mm of difference between the two cars. It was no coincidence that the W09 was the longest in the field and the RB145 the shortest.

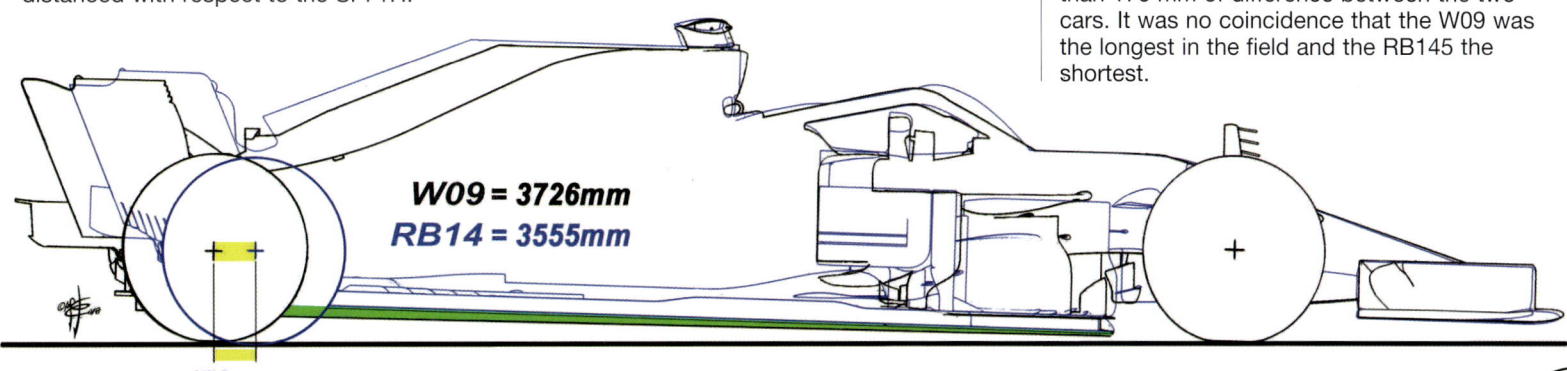

W09 = 3726mm
RB14 = 3555mm

176mm

exception of, among the majors, Mercedes, which continued to employ a less accentuated rake on its cars, even in the 2018 season; Force India instead exploityed an extreme rake configuration with a wheelbase that was almost identical. In effect, Andry Green, the very talented Force India designers, claimed that his car could have employed an even more extreme rake angle had it not been for the limits dictated by the Mercedes transmission. The fact that Force India, with its very tight budget managed in 2017 to be the fourth force in the World Championship ahead of wealthier teams suggested to the Brackley engineers that they should analysed the advantages of a set-up with greater positive rake. Instead in 2018 continued to follow the path of a more "neutral" configuration with the W09.

A low rake set-up tends to produce less maximum downforce but greater aero efficiency than high-rake cars and makes it easier to keep the airflow attached at the rear at low speeds when the rear ride height is at its maximum. However, the increased width of the floors stipulated in the 2017 regulations changed the balancing point between low and high rake concepts. The downforce produced by the underbody is a multiple of the underfloor area and the speed of flow. High rake produces a greater speed of flow because the greater expansion area behind the leading edge of of the floor' (T Tray) creates a low-pressure area that the oncoming air rushes to fill. The faster the flow, the greater the downforce produced. With the wider floor allowed in 2017, that multiple naturally increased more for the high-rake than

low-rake configuration, so increasing the advantage of a high rake. To compensate for this, Mercedes maximised the area of its floor by way of a very long gearbox casing that sited the rear wheels well back. It also sought to maximise the gap between the front axle and the leading edge of the floor in order to maximise the power of its airflow-accelerating vanes. This was to compensate for the smaller expansion ratio of the floor. Having set the rear wheels further back and the front axle further forwards created a very long wheelbase car which in turn meant it was always slightly heavier than ideal and could not use enough ballast to take full advantage of the fore-aft weight distribution tolerance allowed by the regulations. This narrowed its set up window, contributing to a car that was trickier than the rival Ferrari to get working over the full range of steering angles, tyre compounds and track temperatures.

RAKE COMPARISON

In the period in question the Red Bull always had the most accentuated rake, 1.9°, unexpectedly matched by Force India which, despite having a wheelbase very similar to the Mercedes, took full advantage of the configuration. The German company's car was instead the one with the most traditional set-up, reaching a maximujm rake angle of 1.25° in 2018. Ferrari also gradually increased its rake angle up to the 1.53° of the SF71H, behind Red Bull, Force India, McLaren and Renault.

Mercedes W09
Rake = 1.2°

Ferrari SF71H
Rake = 1.53°

Red Bull RB14
Rake = 1.9°

Controversies **2018**

With the technical revolution introduced in the 2017 season, the FIA adopted a system of dealing with borderline interpretations of the regulations without applying disqualifications but by banning excesses (on occasion with immediate effect), with clarifications being sent to all the teams.

For their part, the teams committed to not making official protests but rather asking for clarification from the FIA regarding any "borderline" features introduced by their rivals. This agreement was soon broken by Renault, which was fighting Haas for 4th place in the Constructors' Championship, and at the end of the Italian GP presented an official protest against Grosjean's car, which had finished 6th, regarding the irregularity of its underbody (article 3.7.1, concerning failure to respect the radius of 50 mm for the front part of the T Tray). This was moreover an irregularity ratified by a precise technical directive of the 25th of July, inserted after a request for clarification made by Renault. Note that the underbody in question had been introduced in Canada, but the FIA's ban only came into force on the eve of the Hungarian GP.

The FIA had given the American team time to get their cars in order ahead of Monza, but the modification could not be made for the Italian GP given that the team had to respect the summer break imposed by the regulations. For this reason, the technical commissioners allowed the Haas cars to enter the Italian GP, while they should have contested the irregularity at scrutineering prior to the practice sessions. The inevitable appeal from the American team was then definitively rejected by the FIA tribunal that met early in November and confirmed the exclusion of Grosjean's Haas from the standings.

It was actually Renault that during the 2018 winter tests introduced a feature that contradicted the spirit of the regulations, given that the exhaust pipe had been lengthened by 10 cm and the monkey seat had been eliminated to reduce the aerodynamic advantages of their blowing. On the R.S.18, from the first pre-season test session, the wastegate exhausts were located below the central exhaust, with the maximum inclination permitted by the regulations (5°) and therefore notably higher with respect to the competition. The configuration

exploited the blowing of the gases, despite their low energy, to improve the efficiency of the lower face of the rear wing main plane (protected by a heat-proof material) while helping to speed up the extraction of the air from the central area of the diffuser.

Surprisingly, the design was considered to be legal according to the letter of the regulations, was destined to be banned by the 2019 draft.

Again looking at exhausts, at Shanghai attention focused on the Ferraris that had a different engine sound than normal when the drivers lifted off, suggesting a mapping of the Power Units that permitted the wastegate exhausts to blow. Ahead of Baku, a communication was sent to the teams pointing out

RENAULT BLOWING

For the 2018 season the exhaust pipes were lengthened by 10 cm to reduce the advantages deriving from any vents in the diffuser area, as shown in the side view. The legalization of the blown exhausts introduced by Renault from the first test sessions caused a sensation. The venting was obtained by raising and inclining the exhausts as much as possible, clearly shown by the covering in ceramic material applied to the main plane. Even more absurd was the news that from the 2019 season this feature was to be banned. Would not have been more consistent to do so immediately? A comparison of the standard exhausts on the Red Bull which also used Renault power.

Red Bull

that engine mappings specifically conceived to increase the flow of exhaust gases in corners were prohibited. The suspicions were fed with the introduction of an additional lever on Vettel's steering wheel, discovered to the disappointment of the Ferrari engineers, following the German's pole position run in Bahrain when he left the wheel in clear view on the Halo support.

Ferrari again, in free practice for the German GP, had brought to the track a further evolution of the exhausts with the two wastegate pipes mounted vertically (one above the other) on the principal exhaust. A feature that not only served to narrow the engine cover but also influenced the aerodynamic effect of the exhaust gases under the rear wing, which was no longer seen but was the object of clarifications by the FIA on behalf of Mercedes. At the end of the championship all these features were definitively banned for the 2019 season.

The sophisticated rear-view mirror mount directly on the Halo introduced by Ferrari was rejected even before the start of practice for the Spanish GP. Accusations were levelled in particular at an additional fin that was eliminated from the following Monaco GP.

A further two features became regular objects of discussion in the second half of the season: Ferrari's dual battery and the holes in the rear wheels of the Mercedes, introduced in Belgium to better manage tyre temperatures, a weakness of the Silver Arrows during the races.

The dual battery controversy only broke out in 2018, but the feature devised by Ferrari had been already mounted in the cars in the previous years and had actually been patented. In effect, on the cars from Maranello the battery pack was divided into two distinct elements rather than being a single unit as on all the other cars. This configuration evidently brought advantages that justified its greater complexity; it was declared to be legal by the FIA but following the Italian GP it demanded the fitting of a second sensor. A legitimate request given that there were two separate cells, while for all the other cars with a monolithic battery clearly just one was needed.

The Ferrari engineers therefore cooperated with the FIA, fitting an additional sensor to verify that that there was no second energy output not monitored and unverifiable by the FIA.

The final controversy concerned the holes in the flanges of the sophisticated new rear wheel rims introduced by Mercedes at Spa. In Belgium, attention focused on the new shape of the O.Z. wheels, equipped between one spoke and the next with conspicuous "teeth" that had the same shape as the final part of the spoke. This feature had been designed to increase extraction of hot air from the wheel during rotation, permitting a real reduction in the temperature of the tyres and functioning as small additional "radiators". In Singapore, a further synergic novelty was discovered: the new concave shape of the carbonfibre shrouds, but the final piece in the sophisticated design to reduce the temperatures of the rear tyres was only revealed at Suzuka, on the start line for the Japanese GP: small blown holes inside the mating flange. A feature prohibited from 2012 on the Red Bull and the McLaren but held to be legal due to the microscopic dimensions of the holes compared to the earlier cases. However, in order to avoid sterile polemics and above all a possible protest, from the USA GP in Austin, Mercedes decided to close the holes with the exception of the last race at Abu Dhabi where Hamilton's car raced in what was almost a "rolling laboratory" form, crammed with sensors gathering data useful for the 2019 car.

FERRARI HOCKENHEIM

Ferrari tested this new wastegate exhaust configuration in FP1 at the German GP both to narrow the engine cover and to obtain a greater venting effect. The front view also shows a new main plane on the rear wing acting in synergy with the exhaust positions. An experiment that was taken no further.

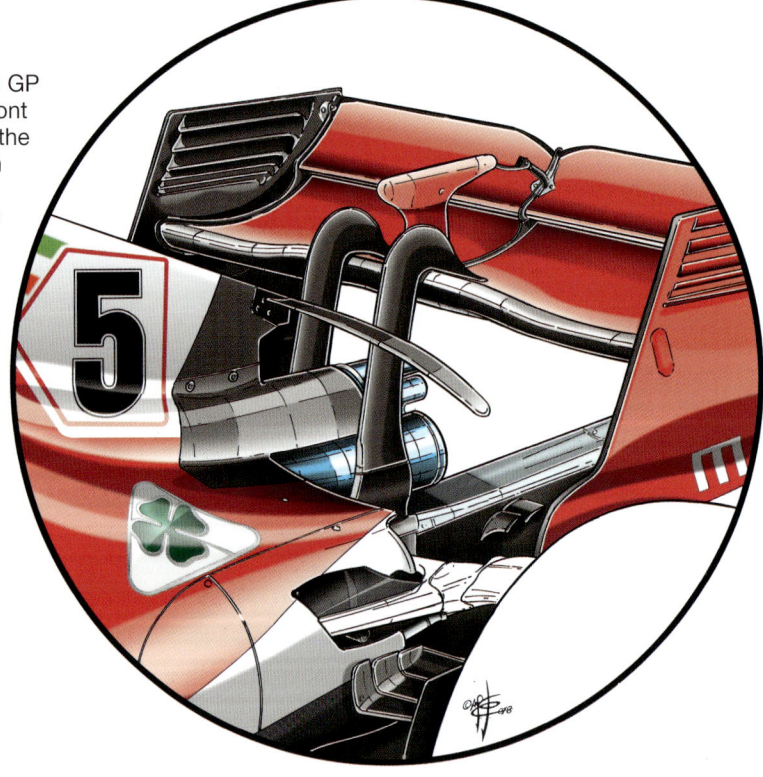

MIRRORS FERRARI

The SF 71H introduced another sophisticated aerodynamic feature with the rear-view mirrors equipped with a new vent. At the Spanish GP a notable evolution had already been introduced in this area with the mirrors mounted not on the sidepods but directly on the Halo, a feature which the FIA considered to be to close to the limit, although it should be noted that on almost all the cars the mirror mountings had become true aerodynamic appendices. What was under accusation was the conspicuous arched fin indicated by the arrow, considered to be a pure aerodynamic device and which Ferrari was obliged to remove before the successive Monaco GP. The prancing horse's aerodynamicists had in fact added a profile with a bracket that officially was there to prevent the mirror vibrating but which in reality helped reattach the flow of air towards the rear wing.

Barcelona

Monaco

FERRARI TWIN BATTERY

Ferrari's twin battery feature came to the fore only in 2018, but had been introduced with the arrival of the Power Units in 2014 and had been patented. A different configuration to all the others, in practice twin batteries in place of a single one, although both were placed in a single enclosure which is perhaps why it did not emerge for three years. This became one of the leit motifs of the season, above all following the requests for clarification by Mercedes. After having recognised the legality of the configuration, the FIA asked Ferrari to fit a second sensor (in addition to the one present on all the other cars) to monitor the output of the two separate cells. This was the root of certain rumours regarding a correlation between the Ferrari's fall-off in performance and the introduction of this more scrupulous control. Charlie Whiting himself explained at Suzuka that the drop in performance of the SF 71H did not depend on the second sensor, emphasising the complete legality of the feature adopted by Ferrari. In the drawing of the general layout of the Mercedes Power Unit you can see the single battery that in the case of the Ferrari featured two separate elements in a single container ostensibly in parallel longitudinally even though in the profile the division was made differently.

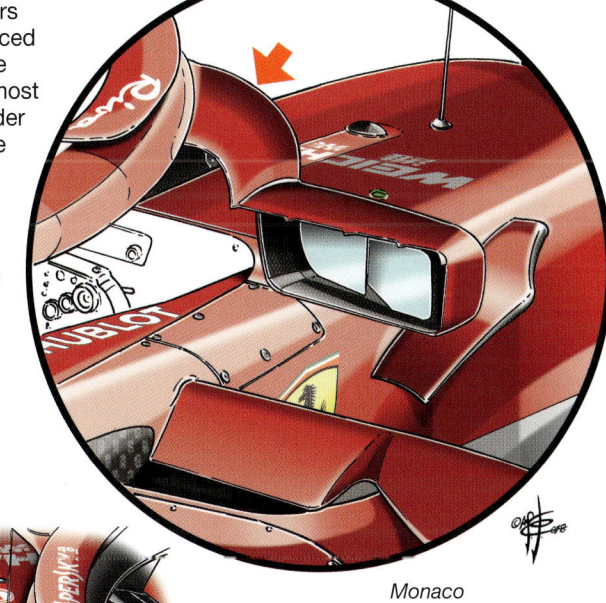

HAAS MONTRÉAL-MONZA

At Montr?al, Haas introduced a major evolution of the area of bargeboards either side of the monocoque, which were completely revised in all its various parts. At the time, attention had been focused on the new feature of the teeth on two levels in the lower part of the front of the sidepods. Renault had instead concentrate on another novelty: the conjunction of the vertical part of the new bargeboard with the central section of the T Tray that became the object of a letter of clarification due to its failure to respect the

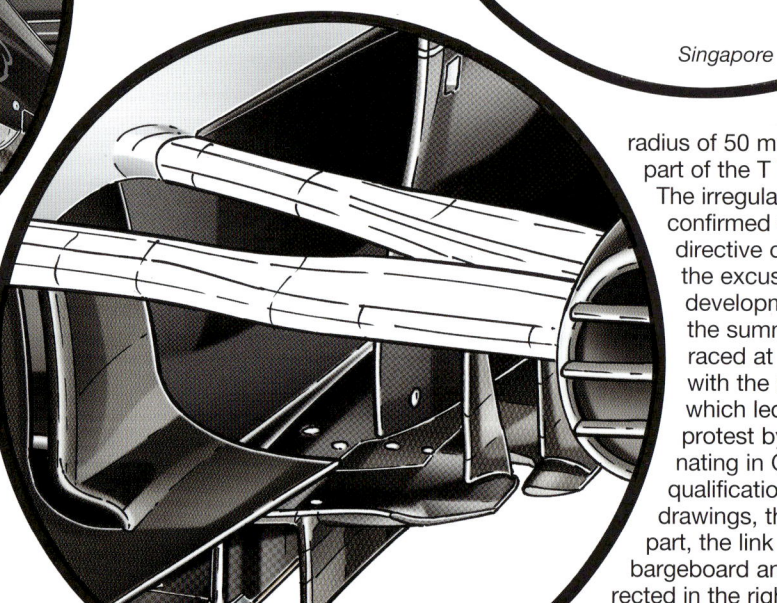

Singapore

Monza

Monza

radius of 50 mm for the front part of the T Tray (para. 3.7.1). The irregularity was officially confirmed with a technical directive dated 25 July. With the excuse of the ban on development work during the summer break, Haas raced at Spa and Monza with the irregular version which led to an official protest by Renault culminating in Grosjean's disqualification. In the two detail drawings, the incriminated part, the link between the bargeboard and the T Tray, corrected in the right-hand drawing.

MERCEDES REAR WHEELS SPA

At Spa, Mercedes introduced a new rear wheel developed in collaboration with O.Z. The unusual feature derived from the fact that between one spoke and the next there were two new and conspicuous teeth that were identical in shape but had two different functions; improving the extraction of hot air from the wheel and acting as small additional "radiators" helping to lower the temperature of the tyre carcass, thus better managing the duration of the rear covers, a weakness of the W09 in the races compared to the Ferrari.

MERCEDES HOLES

Only on the start line at Suzuka did the controversy over the holes in the mating flange of the Mercedes wheels come to the fore; they became the object of a request for clarification from Ferrari. They were considered to be mobile aerodynamic devices as their venting to the outside obviously occurred in rotation and not through the central part of the hub (non-rotating) as was the case at the front. In the 2012 season, conspicuous hole introduced by Red Bull and McLaren had been considered illegal. In the Mercedes case, the FIA considered the notably different size of the holes to be non-influential with regard to aerodynamics, accepting their heat dispersal function in synergy with the other more evident novelties introduced to the rear wheels from Spa. This episode did not lead to the issuing of a precise directive as had been the case with Haas. This gave rise to the polemics and doubts over the W09's lack of competitiveness in the US Grand Prix dominated by Räikkönen's Ferrari. In fact, Mercedes then preferred to close the holes in the mating flange, using them only in the final race in Abu Dhabi.

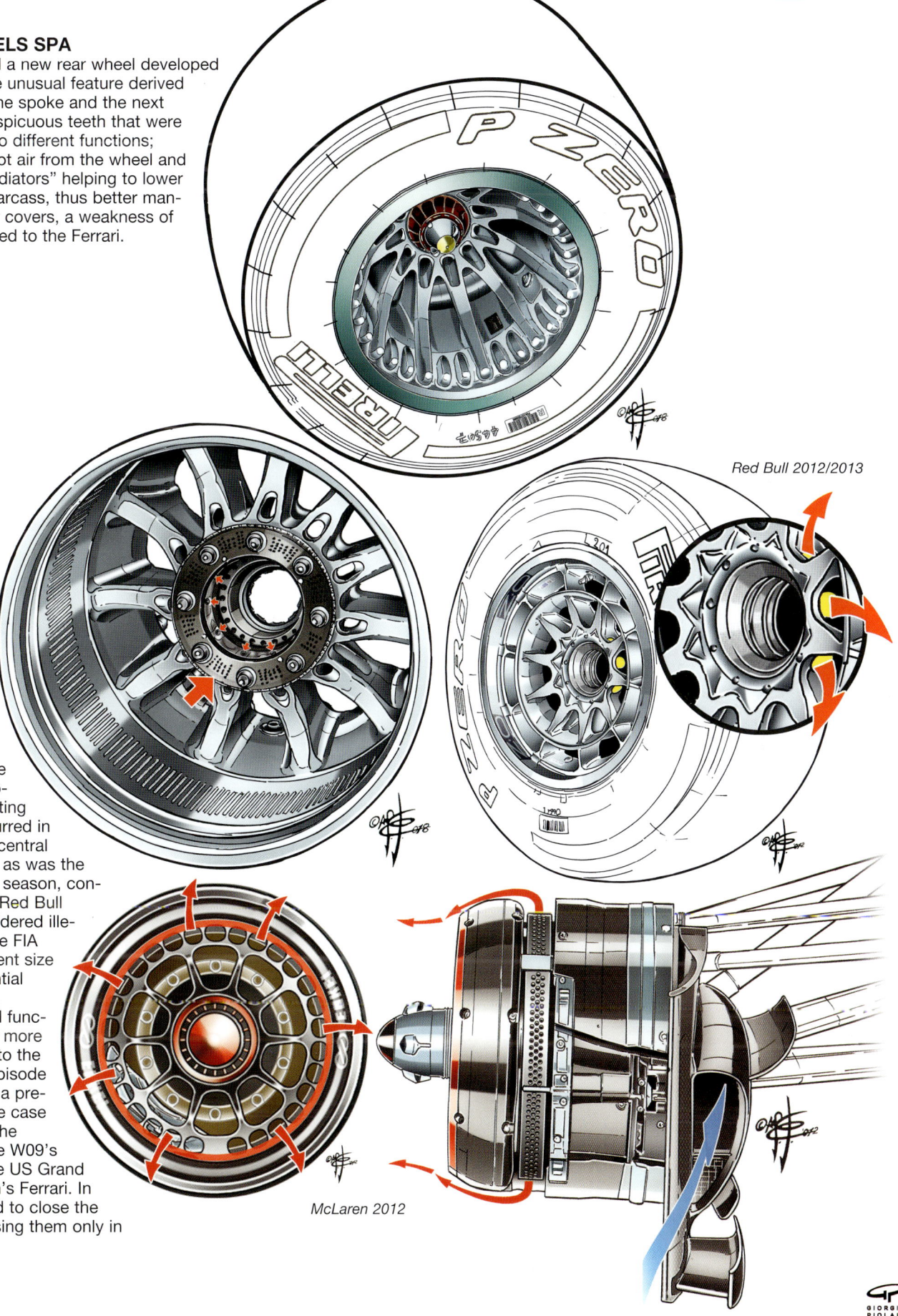

Red Bull 2012/2013

McLaren 2012

HALO

The 2018 season saw the obligatory introduction of Halo to all the cars after the various teams had tested the feature in the Friday morning practice sessions during the previous season. Its impact on the aerodynamics and weight of the cars (+15 kg) was significant and required careful analysis in the wind tunnel. Halo also led to a slight increase in fuel consumption. As highlighted in the Regulations 2018 chapter, the experiments had begun in the 2017 season and the first drawing shows the experiment made by Ferrari in FP1 at Singapore with the engine air intake equipped with Pitot tubes to verify the influence of the Halo on the flow to the engine.

At the end of the season during free practice following the Abu Dhabi GP, McLaren had fitted the first example of an aerodynamic device designed to groom the air flow and vortices around the Halo, introducing three small spoilers that then appeared on a number of 2018 cars, as in the drawing of the Toro Rosso. The regulations in fact allowed a degree of freedom within the 2 cm around the Halo, the structure of which had to be identical for all the teams.

Ferrari 2017

McLaren 2017

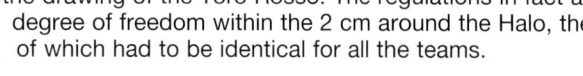

Toro Rosso 2018

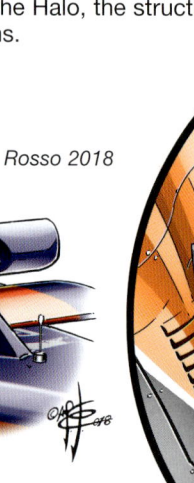

WILLIAMS

The safety structures outside the start of the sidepods became a trend, as we can see in the extreme example of the Williams FW41, characterised by very short sidepods and conspicuous boomerang-shaped turning vanes in the upper part of the chassis Mercedes 2017-style.

FERRARI REAR-VIEW MIRRORS

There was much interest from the outset in the Ferrari rear-view mirror configuration for the SF71H. After having caused a surprise in the 2017 season with the protection structures outside the sidepods, in order to improve the efficiency of the high intake cooling the radiators, the mirrors enclosed a sophisticated vent that helped make the cooling more efficient, in the comparison in the ovals the shape of the sidepod mouth air intake is emphasised.

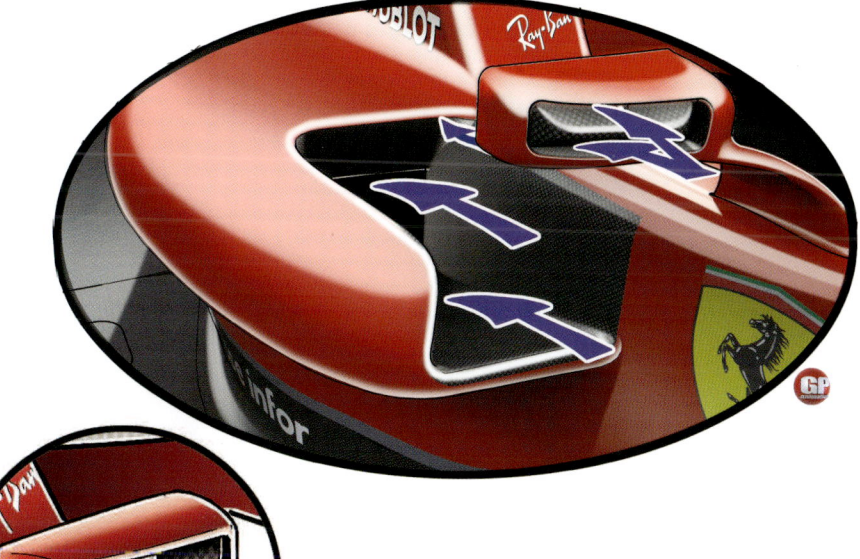

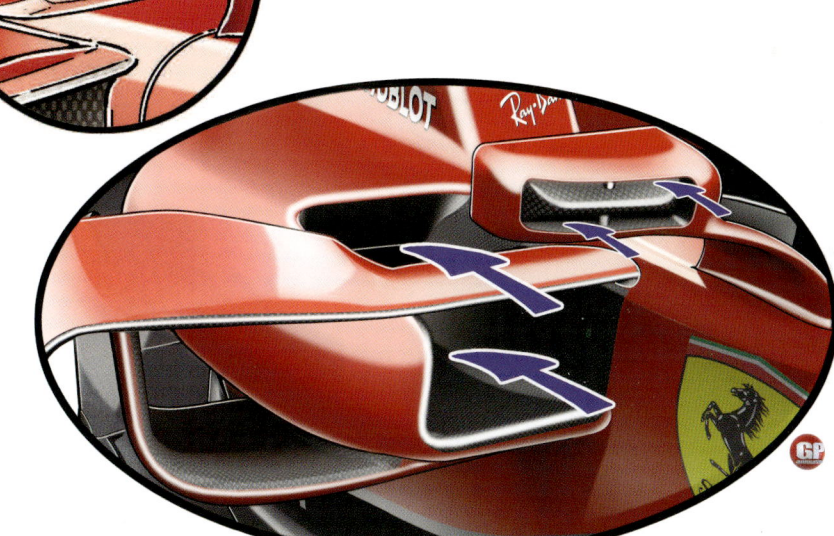

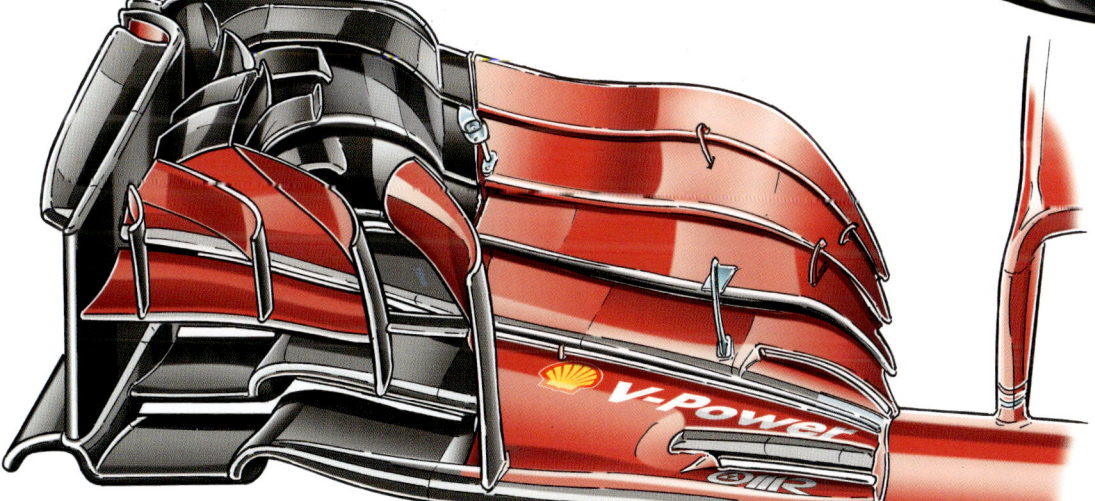

FERRARI FRONT WING

The area close to the sidewall was subjected to constant development throughout the 2018 season. The drawing highlights the increase of the flat area with the "kink" in the main plane moved inside. Note also the sequence of vents in all the different profiles and the three vertical turning vanes in the upper flaps designed to expel the air outwash the wheels.

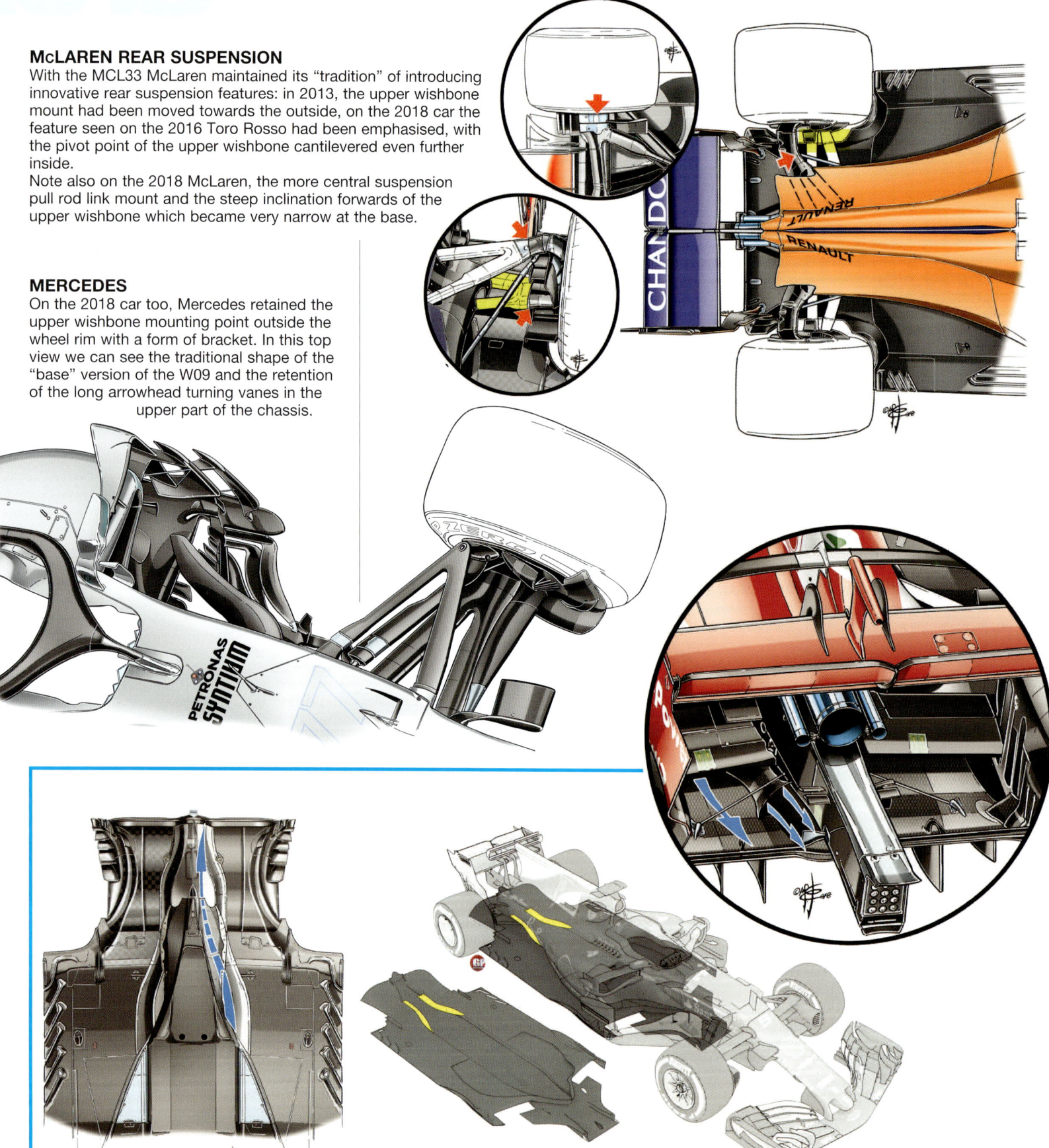

McLAREN REAR SUSPENSION

With the MCL33 McLaren maintained its "tradition" of introducing innovative rear suspension features: in 2013, the upper wishbone mount had been moved towards the outside, on the 2018 car the feature seen on the 2016 Toro Rosso had been emphasised, with the pivot point of the upper wishbone cantilevered even further inside.

Note also on the 2018 McLaren, the more central suspension pull rod link mount and the steep inclination forwards of the upper wishbone which became very narrow at the base.

MERCEDES

On the 2018 car too, Mercedes retained the upper wishbone mounting point outside the wheel rim with a form of bracket. In this top view we can see the traditional shape of the "base" version of the W09 and the retention of the long arrowhead turning vanes in the upper part of the chassis.

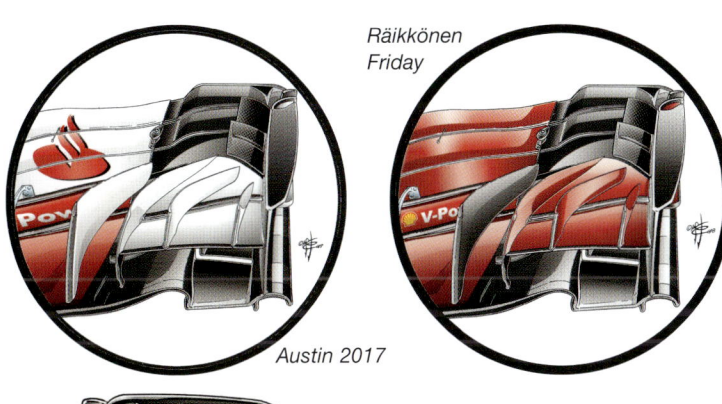

Räikkönen Friday

Austin 2017

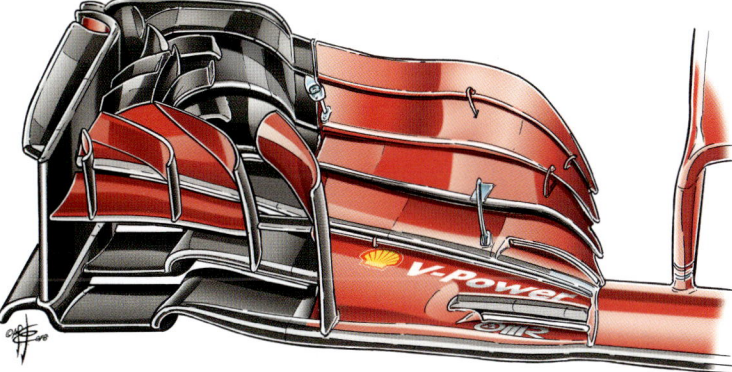

FERRARI MELBOURNE

The new Ferrari front wing at the first race of the season compared to the old one inspired by the design used in Austin in 2017.
The main plane has an almost straight configuration with discreet leading edge and a minimal "kink" in proximity to the footplate (narrower) linking it to the endplate.
The main plane also featured a vent, like the first flap which doubled up in proximity to the metal bracket supporting the three additional flaps which had a pointed shape facing down and inwards to favour the passage of the Y250 vortex that feed the flow destined towards the rear of the car. The small tooth sealing the external inverted U channel (arrow) was eliminated.

FERRARI "DUAL" DIFFUSER

The feature adopted by Ferrari in the central area of the diffuser from the first test sessions and improperly known as the "double diffuser" aroused considerable interest. It was in effect a megaphone vent from the lung that formed in the lower bodywork of the SF71H between the diffuser keel and the gearbox-safety rear structure assembly as highlighted in the diagram of the 2017 car. The feature had also been present on the 2016 car.

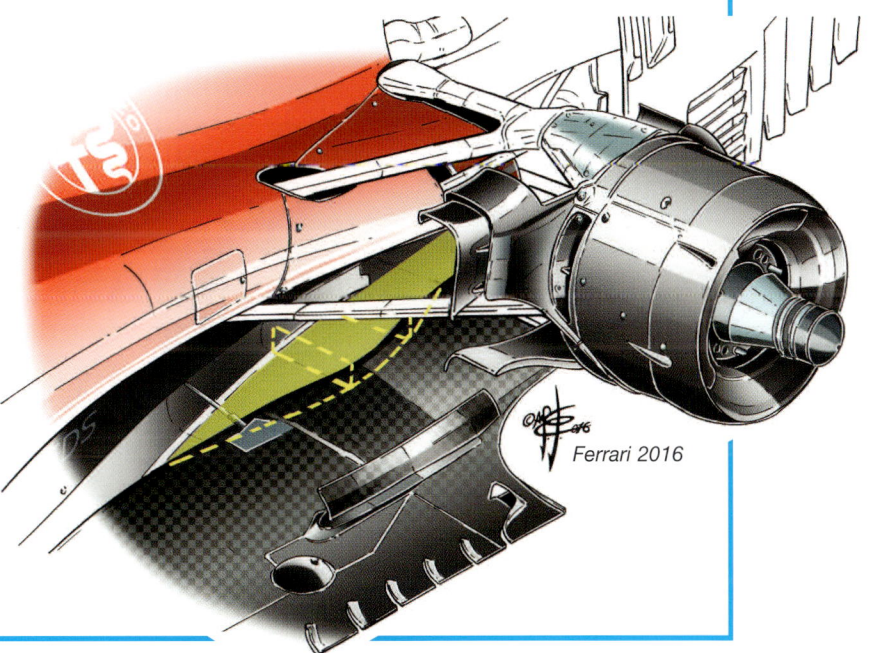

Ferrari 2016

RED BULL

Evolution of the RB14 bargeboard. In the insert with the base version, the placing of the safety structure inside the lowest profile, ahead of the sidepods, covered by the series of Ferrari-style lateral vanes. The version introduced at the last Barcelona test session and which debuted at Melbourne also had the long, arrow-shaped turning vane like the Mercedes.

MERCEDES MELBOURNE LATERAL VENT

At Melbourne, Mercedes added a long, vertical S vent to the horizontal slots either side of the cockpit, a feature seen on the 2017 Williams.

WILLIAMS

The inefficiency of the Williams FW41 cooling system was clearly revealed by this rear view with the enormous megaphone opening that obviously penalised the car's aerodynamic efficiency.

MERCEDES REAR SUSPENSION

Mercedes introduced the same important technical novelty it had adopted on the front end (New Features 2018 Chapter) to the rear suspension too.

The rear suspension upper wishbone mount was raised with a bracket that protruded much more than seen in 2017 on the front end of the W08.

The extended part was designed to permit a wishbone positioned almost horizontally so that it could also act as a turning vane.

The brake intake was raised so as to ensure that the lower wishbone, which was specially shaped in proximity to the brake duct, could interact with the faired arm in wing profile containing the drive shaft, creating a blow.

RED BULL

Adrian Newey applied to his RB14 a feature already seen on its Toro Rosso cousin in the 2017 season, this slight curve in the footplate area ahead of the rear wheels.

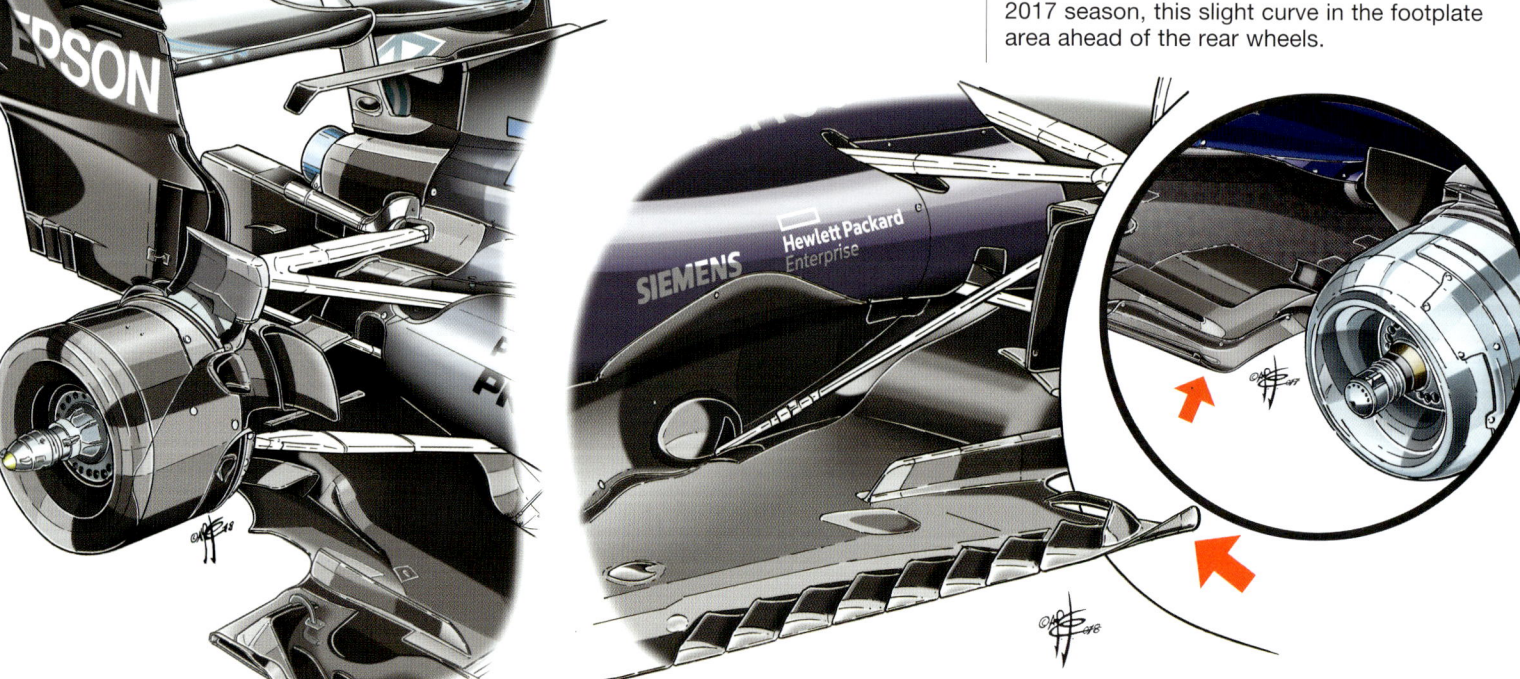

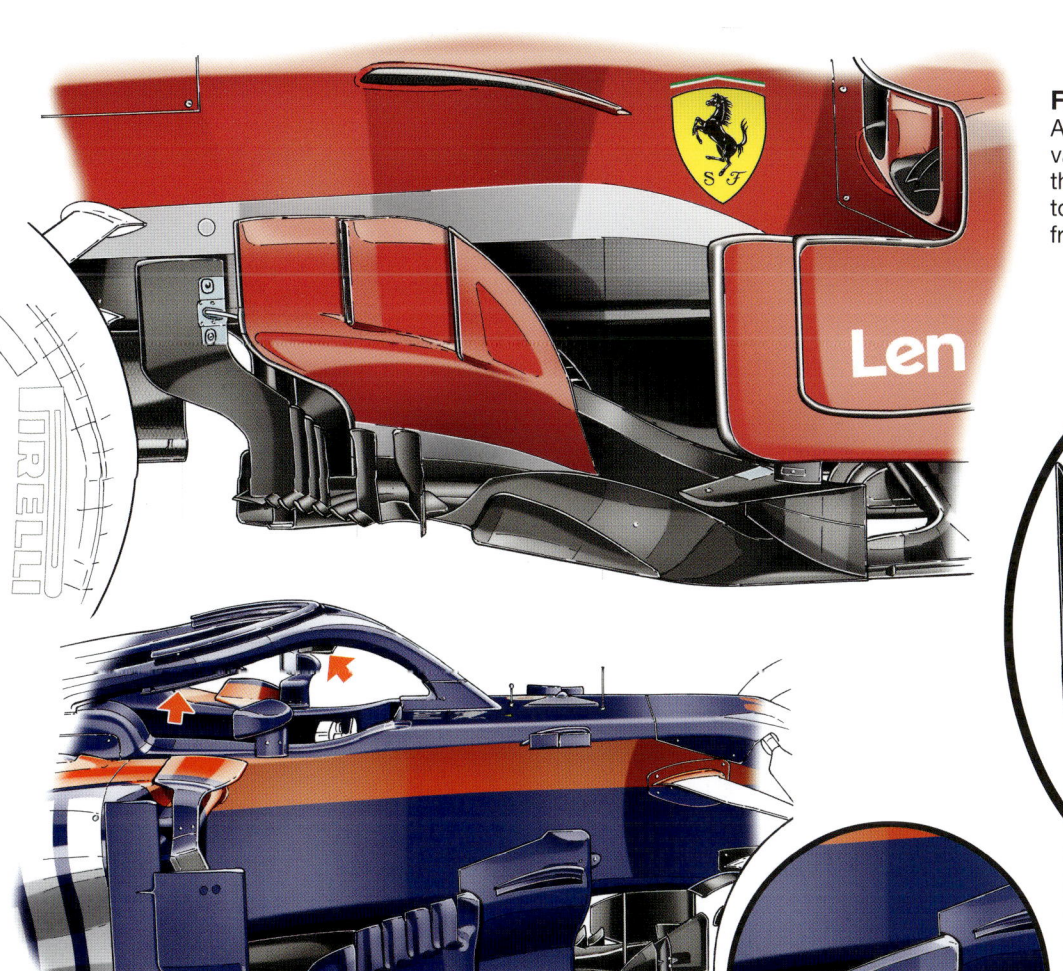

FERRARI BAHRAIN
A new bargeboard with an additional vertical vane in carbon fibre and a series of "fingers" in the lower area topped by vertical turning vanes to better groom the turbulence generated by the front wheels.

MERCEDES
In free practice on the Friday, Mercedes experimented again with the application of a serrated adhesive tape already seen in previous seasons, this time attached to the lower part of the main plane and not in the flap.

TORO ROSSO BAHRAIN
A new bargeboard for Toro Rosso with in the insert the old design that highlights the multiplication of the vertical slots in the second large vertical vane. A small fin was added to the lower part of the Halo, along with slim profiles in the upper part already seen in pre-season testing.

FERRARI SHANGHAI
The area of the floor ahead of the rear wheels was modified at almost every race. On the SF70H, this Mercedes-style horizontal fin appeared in addition to the four longitudinal slots.

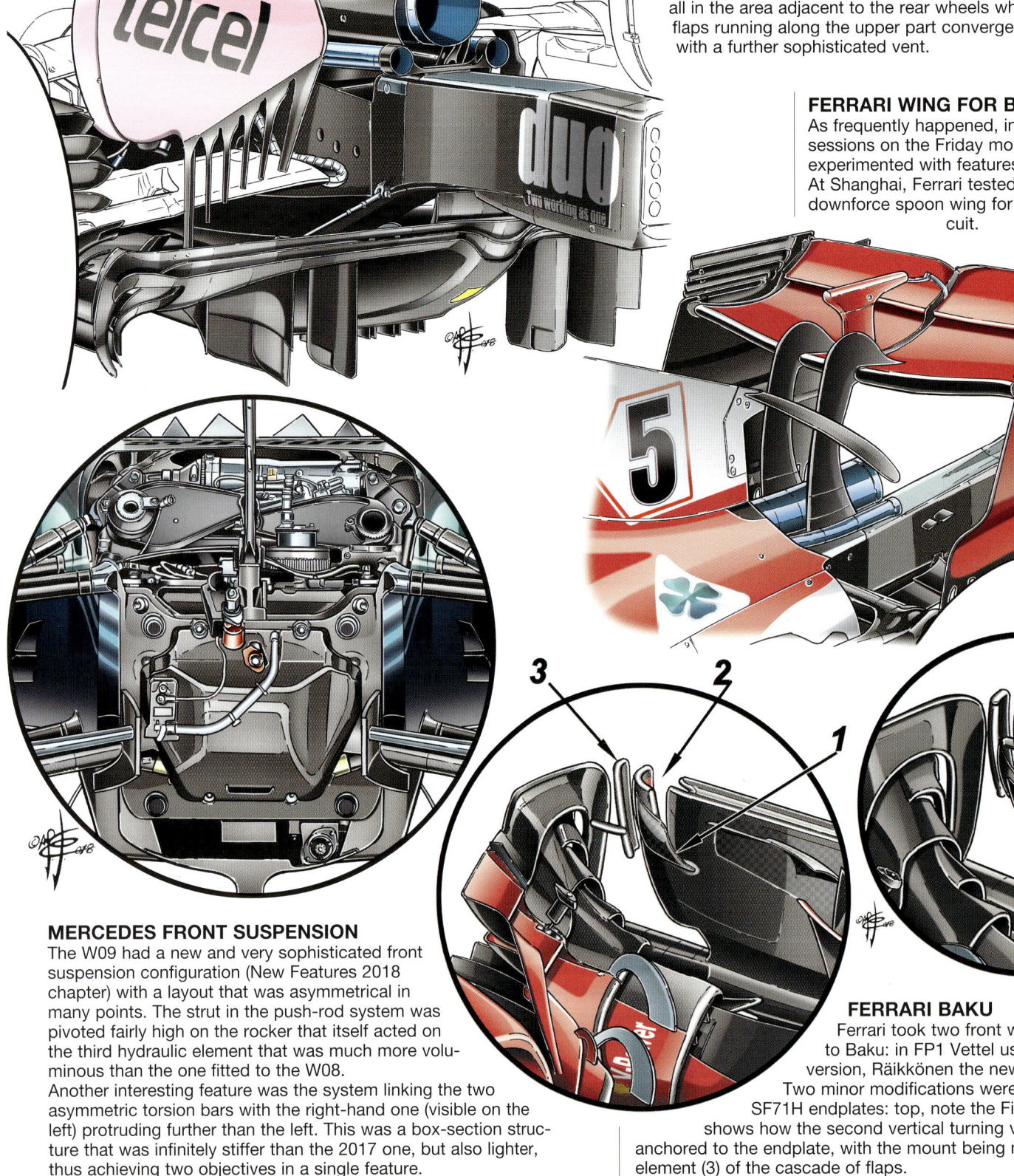

FORCE INDIA SHANGHAI

A new diffuser for the Force Indias at Shanghai, modified above all in the area adjacent to the rear wheels where the two small flaps running along the upper part converge and link vertically with a further sophisticated vent.

FERRARI WING FOR BAKU

As frequently happened, in the free practice sessions on the Friday morning the teams experimented with features for the next race. At Shanghai, Ferrari tested the medium downforce spoon wing for the fast Baku circuit.

MERCEDES FRONT SUSPENSION

The W09 had a new and very sophisticated front suspension configuration (New Features 2018 chapter) with a layout that was asymmetrical in many points. The strut in the push-rod system was pivoted fairly high on the rocker that itself acted on the third hydraulic element that was much more voluminous than the one fitted to the W08.

Another interesting feature was the system linking the two asymmetric torsion bars with the right-hand one (visible on the left) protruding further than the left. This was a box-section structure that was infinitely stiffer than the 2017 one, but also lighter, thus achieving two objectives in a single feature.

Bottom, the fairing that covers the steering box set lower but allowing the steering arm to be slightly higher than the lower wishbone and generating a blow contributing to the definition of the flows towards the body of the car.

FERRARI BAKU

Ferrari took two front wing configurations to Baku: in FP1 Vettel used the standard version, Räikkönen the new one.

Two minor modifications were made to the SF71H endplates: top, note the Finn's front wing that shows how the second vertical turning vane is no longer anchored to the endplate, with the mount being moved to the final element (3) of the cascade of flaps.

The endplate featured a vertical vent (2). The fins on the internal part were different, with Vettel's being longer (1). The cutaway in the lower part of the endplate was also different and larger.

RED BULL BAKU

Red Bull used the rear wing with the least downforce of the field: the main plane chord was very short and it was almost flat while the mobile flap was also significantly smaller given that the advantage offered by DRS was much less than at other circuits.

The endplates showed two horizontal vents and no vertical slots.

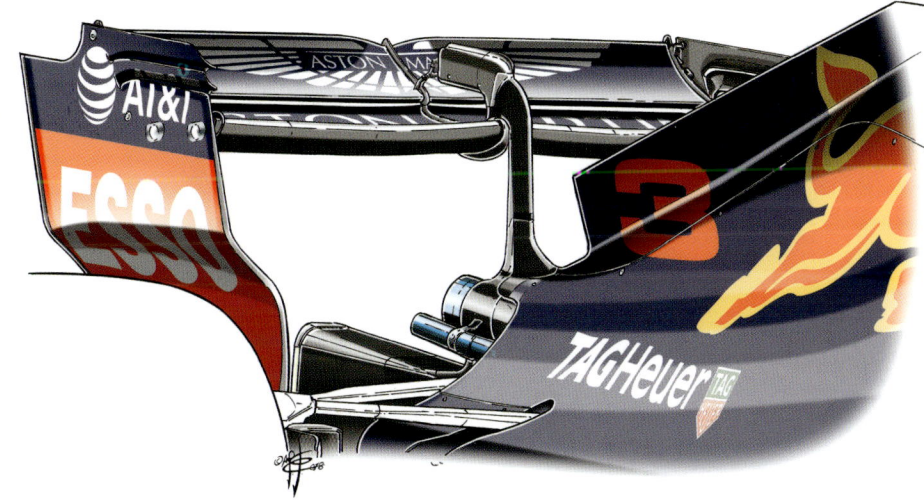

MERCEDES BAKU

Mercedes tested the low downforce spoon wing and the one with the straight main plane and more downforce before opting for the spoon version that was not as extreme as the Red Bull low downforce wing. The drawing clearly shows the greater angle of the profiles, also revealed by their position with respect to the endplate with four slots.

FERRARI MIRRORS

A micro-development by Ferrari: three small vortex generators were added to the upper part of the mirrors. They had a truly minimal impact, but everything contributed to the chase for performance.

There was also a curious doubling of the mirrors to provide the driver with the best possible rear visibility.

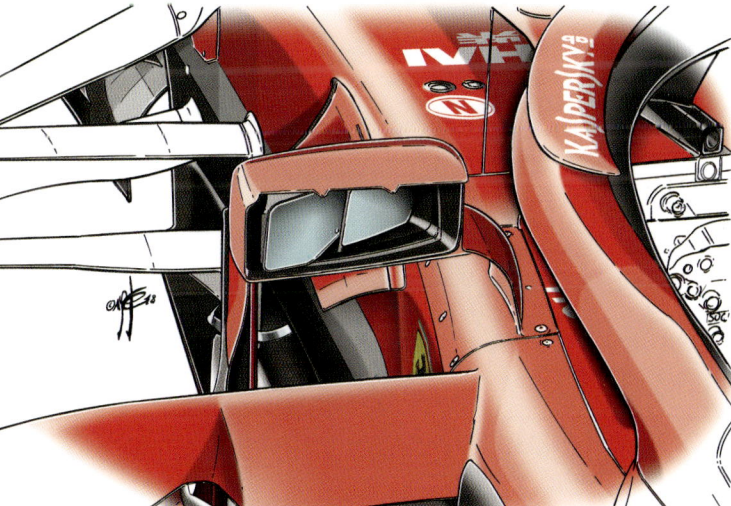

FORCE INDIA BARCELONA

A new front wing for the Force Indias with the main plane featuring a clear kink in conjunction with the Y250 zone. There were also new blown flaps in the part sloping towards the centre. However, this new wing was not used in either qualifying or the race.

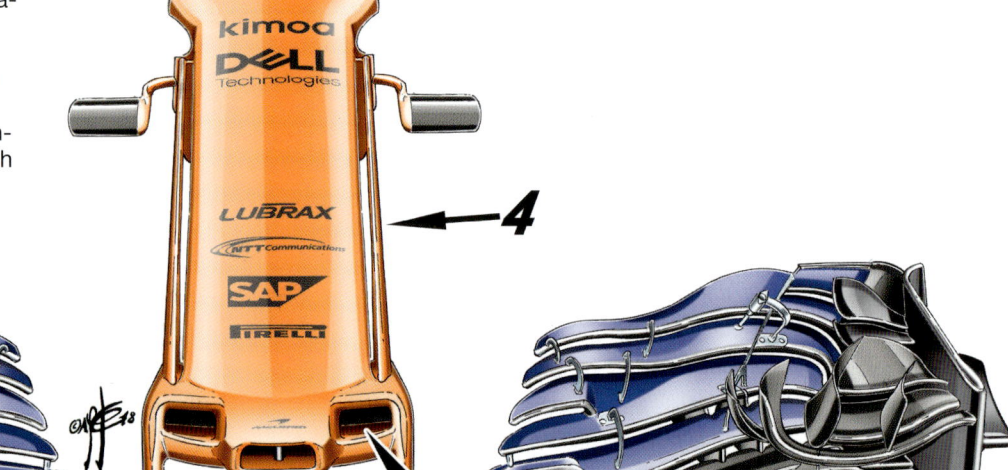

MCLAREN BARCELONA

A new nose debuted in Spain by McLaren with the triple intake (1-2) in Red Bull/F .India-style and above all a new vertical vent that ran along the sides of the nose cone.
The apertures at the tip of the nose avoided the aerodynamic stall created in this point and separate the flow of air towards the turning vanes that worked in perfect synergy with the large "manta ray" profile that reprised in part the idea of the front extractor seen on the Mercedes W09.

MERCEDES BARCELONA

At Barcelona, Mercedes adopted the vertical fin introduced by Sauber close to the upper wishbone mount close to the upright.
A feature designed to direct the wake from the front wheel to the outside and prevent it from negatively affecting the floor.

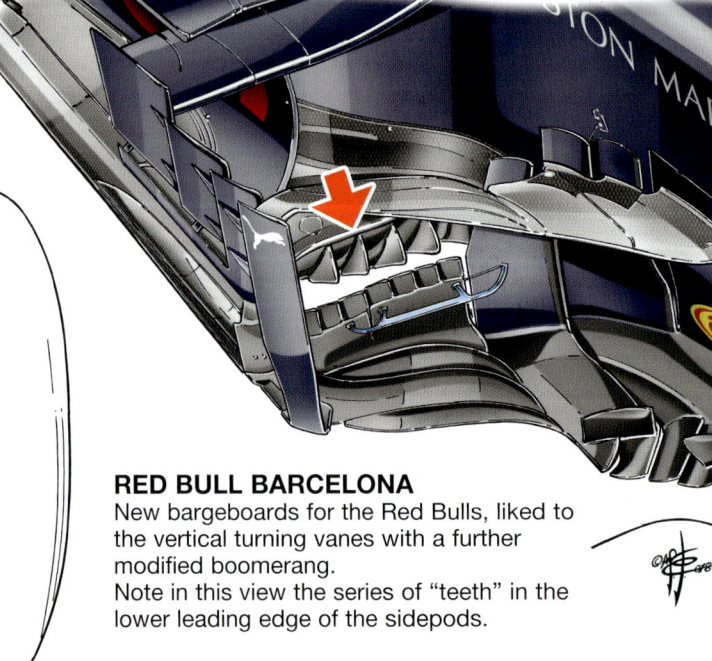

RED BULL BARCELONA

New bargeboards for the Red Bulls, liked to the vertical turning vanes with a further modified boomerang.
Note in this view the series of "teeth" in the lower leading edge of the sidepods.

FERRARI REAR SUSPENSION

In Spain Ferrari adopted the raised and faired bracket for the rear suspension upper wishbone mount, as seen earlier from the launch on the Mercedes. The comparison also reveals its aerodynamic function in forming what was almost a winglet in this raised area.

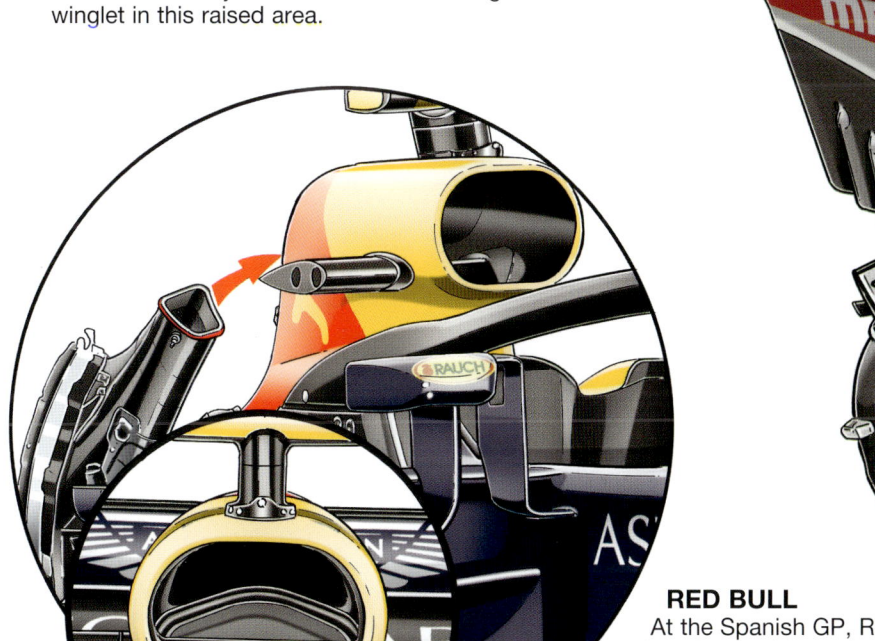

RED BULL

At the Spanish GP, Red Bull introduced a notable update package that comprised a new front wing, a better cooling system and a large engine air intake in order to try to extract as much power as possible from the disappointing Renault unit. This modification required a new crash test of the roll bar area. On the front wing the flap assembly was more detached from the endplate (2) equipped with a new small blown fin (1).

MERCEDES MONACO

Mercedes introduced two rows of three mini-flaps to the upper front part of the sidepods which served to orient the flow attached to the body towards the outside of the rear wheel. The solution was not new – the McLaren MCL33 for example reveals that the German aerodynamicists had drawn inspiration from the vortex generators seen on the car from Woking at the beginning of the season.

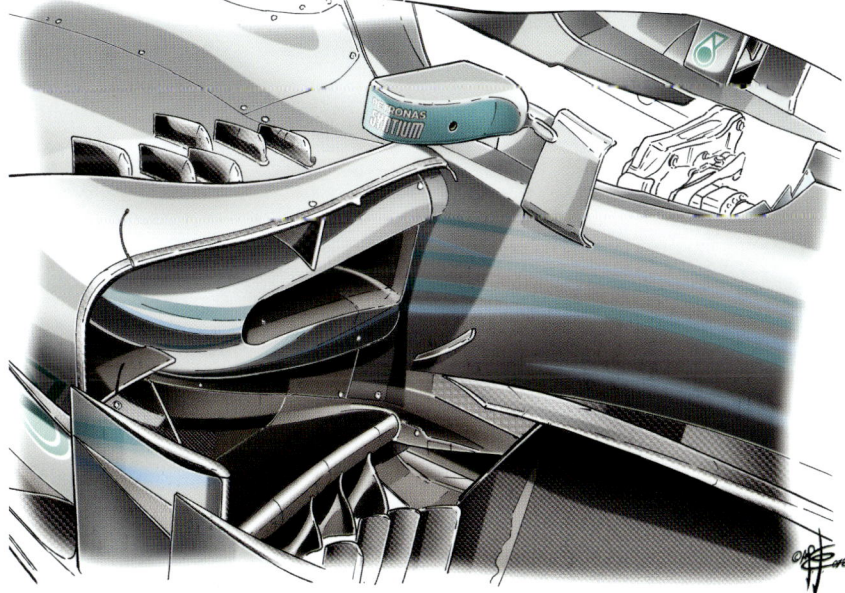

FORCE INDIA MONACO

As well as bring a high downforce rear wing, F.India also multiplied the flaps above the exhaust where the regulations permitted the fitting of wing profiles, while from the beginning of the season T-wings in the upper part had been banned. The Silverstone-based team attempted to increase the aerodynamic loading at the rear by proposing two dual profiles attached to the bottom of the vertical engine cover fin. Each pair of elements was linked by metal brackets to form a kind of "shutter".

RED BULL MONACO

At Monaco, Red Bull fitted a low T-wing for the first time, anchored to the support linking the vertical fin of the engine cover to the rear wing pylon.

This profile had a curved edge in the innermost part, vaguely following the line of the rear wing main plane. Also of note was the return of a kind of Monkey Seat: the regulations permitted the introduction of flaps as long as they were not cantilevered with respect to the main exhaust.

FERRARI MONACO BRAKES

There are three main braking points at Monaco (Sainte Devote, Massenet and the Chicane du Port), but the truly critical aspect of the circuit derives from the low average speed, which does not permit effective brake cooling. Ferrari brought new brake calipers and discs with more than 1400 holes in oblique rows and more friction material at the sides.

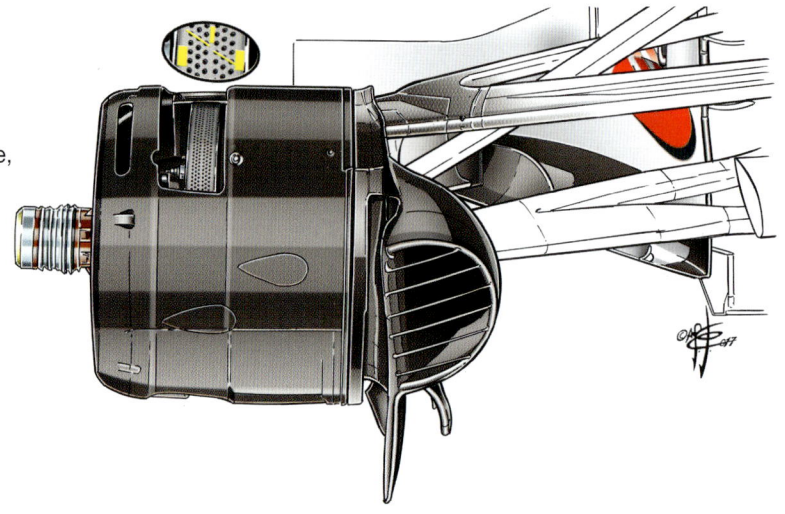

SAUBER SIDEPODS

Sauber revised the shape of the bridge-like turning vane mounted above the sidepod mouth to unite the element with the protection structure at the side of the cockpit. A channel was formed that directs the air at the lower part of the profile to ensure the necessary flow to the upper radiator intake above the side-impact protection structure.

McLAREN FRONT WING

At Monaco, McLaren introduced a new front wing for Alonso: the innermost upper flap was composed of two elements separated by a slot while previously it had been a single, shorter piece. The other modification concerned the final part of the endplate: between the last flap and the strap appeared a profile curving outwards the reprised concepts already seen on the Mercedes W09.

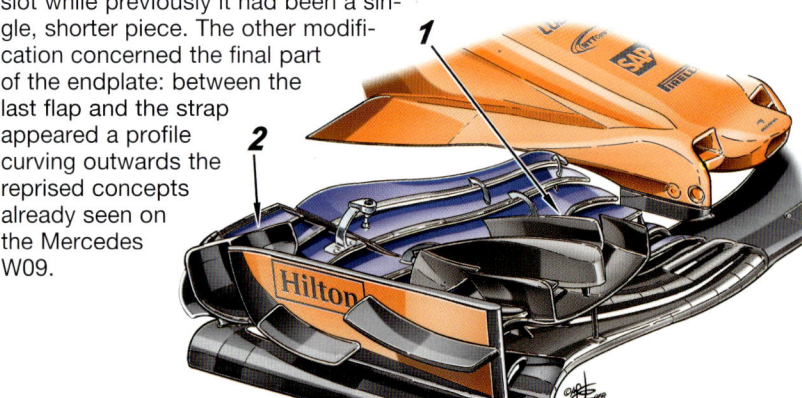

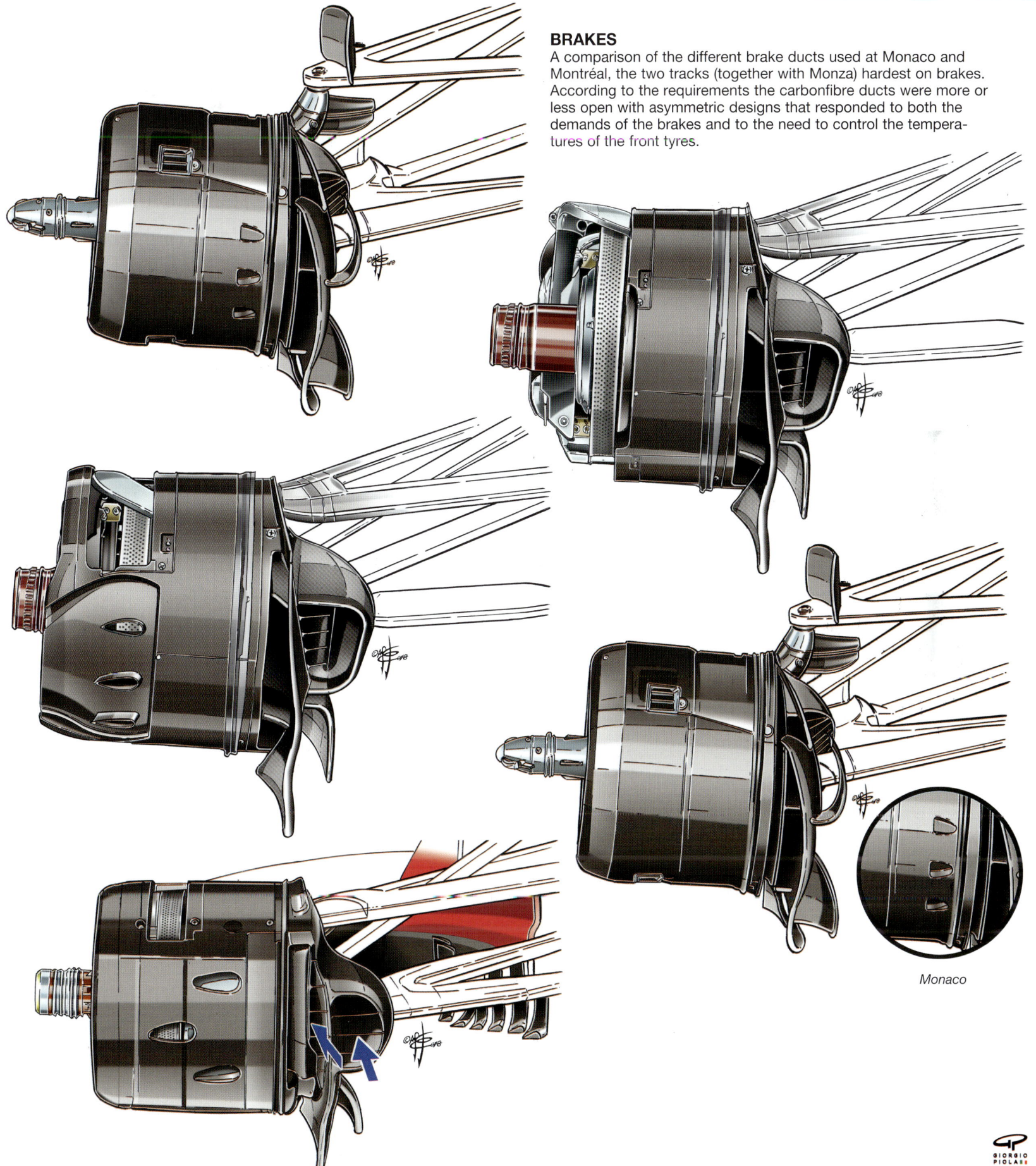

BRAKES

A comparison of the different brake ducts used at Monaco and Montréal, the two tracks (together with Monza) hardest on brakes. According to the requirements the carbonfibre ducts were more or less open with asymmetric designs that responded to both the demands of the brakes and to the need to control the temperatures of the front tyres.

Monaco

SAUBER DIFFUSER

In the 2018 season, the area of the diffuser inside the rear wheels became the object of significant and interesting developments; this drawing shows the version introduced by Sauber in Canada with two practically vertical turning vanes. There was also a blade extending forwards between the rear wheel and the rear extractor. This last presented a kind of narrow footplate.

WILLIAMS REAR WING MONTRÉAL

For the Canadian GP Williams developed a rear wing with a spoon main plane and a dual support similar to that of the Ferrari with a swan's neck shape. The FW41 was a car with a clear lack of downforce hence the fitting of a mini-flap even on a fast track like Montreal, mounted on the trailing edge of the rear protection structure.

RED BULL MONTRÉAL

As in Baku, at Montréal were the cars with the lowest downforce configuration in the field. The drawing shows how the chord of the supplementary flaps was reduced and the fourth element was removed in the search for maximum aerodynamic efficiency.

FERRARI

Modified bargeboards were fitted to the SF71H at Montreal: the three main elements were enlarged (now being taller) with more evident slots.

In the section that links the endplates there are three small flaps, vortex generators helping to feed the flow.

There were seven blades on the bargeboard floor grooming the flow disturbed by the front wheel turbulence.

Verstappen

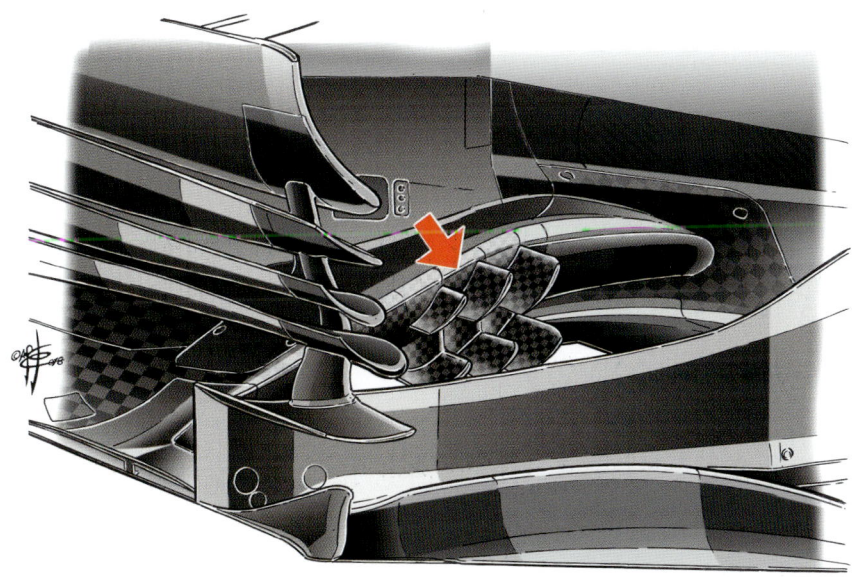

HAAS MONTRÉAL

In Canada, Haas introduced a major aerodynamic package. The drawing illustrates the bargeboards and the flow vanes mounted to the side of the radiator intakes.

The front part of the bargeboards featured horizontal blades on which were located a series of vertical winglets of various sizes and shapes (white arrow). The lower part featured double rows of serrations, not seen on any other car.

FERRARI PAUL RICHARD

A major aerodynamic package for Ferrari: this comparison shows the most concealed part, the footplate area to the sides of the T-Tray, revolutionised in its sophisticated vents.

The new solution was cleaner and more linear, with conspicuous longitudinal cuts that formed a kind of inverted L also known as the "golf clubs" and retained for the following races.

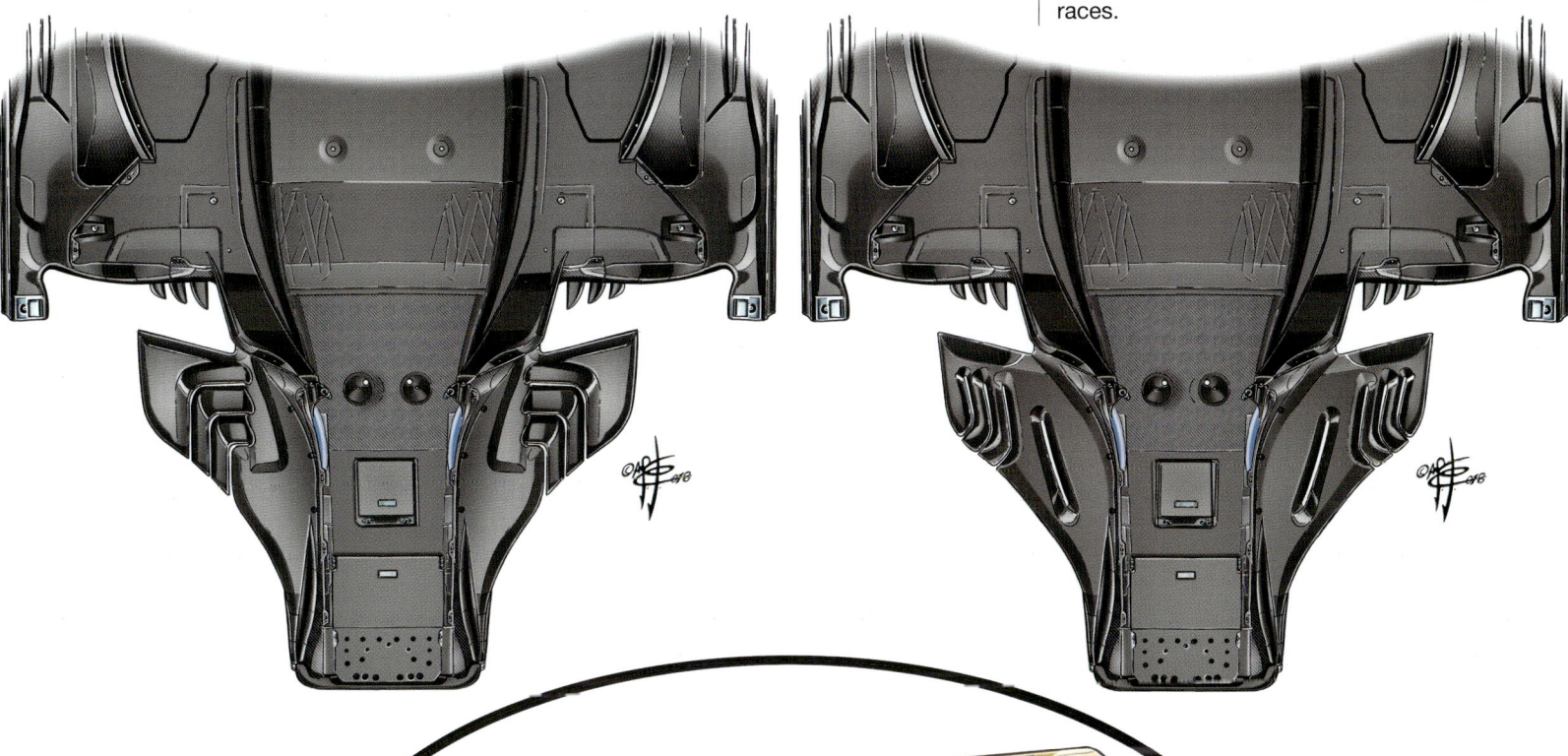

RENAULT S-DUCT

No less than three intakes channelled air into the Renault nose, but only two fed the S-Duct that opened in the upper part of the chassis.

MERCEDES BRAKE INTAKES

In France, the W09s had a double fin equipped with a vent anchored to the brake duct, behind the front suspension strut. These appendices served to groom the vortices with which it was possible to reduce the negative effects of the turbulent wake of the front wheel.

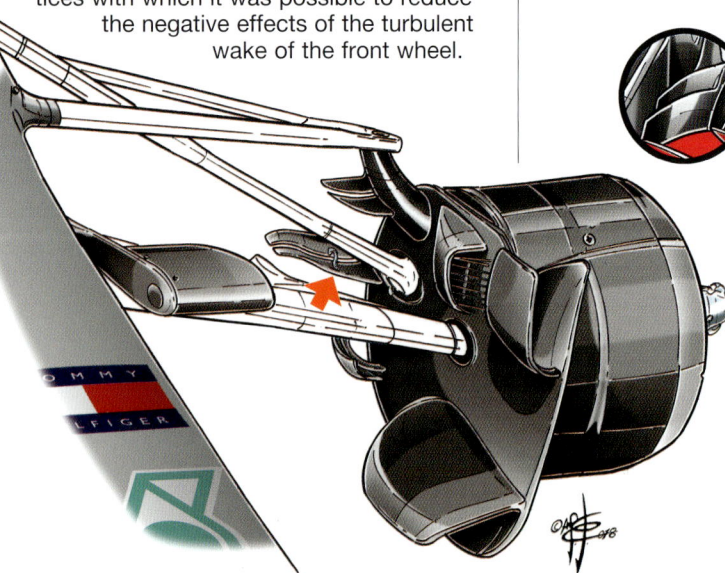

FERRARI FRONT WING

At the Paul Ricard circuit Ferrari presented a front wing that reproduced a number of old features in the red central area, combining them with new parts visible at the extremities: to begin with, the flap anchored to the inside of the end-plate is no longer rectangular with a vent, but triangular and left the supplementary flaps which were of a new design more visible.

In particular, the last profile that highlights the presence of a carbonfibre rib acting as a Gurney flap, shows how the part remaining in carbonfibre merges into the previous part. In the inserts, the earlier version.

SAUBER PAUL RICHARD

New bargeboards on the Saubers in France with the lanceolate portion larger, inside which the delta-shaped floor with no less than five oblique blades.

We are talking about, to take one example, the area where Ferrari located the unusual elements shaped like golf clubs of larger sizes moving towards the rear. This was an area of the car that had become strategic in the management of the flows under the car.

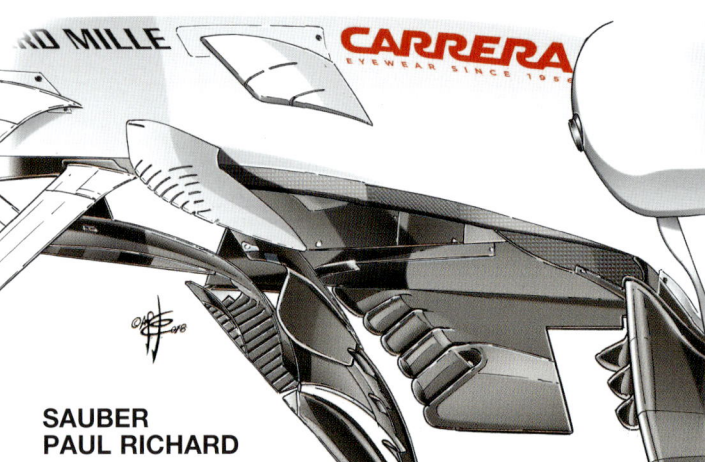

RED BULL PAUL RICHARD

The lower part of the large bargeboard were modified with a different arrangement and shape for the small turning vanes placed on the Mercedes-style longitudinal extensions (knives).

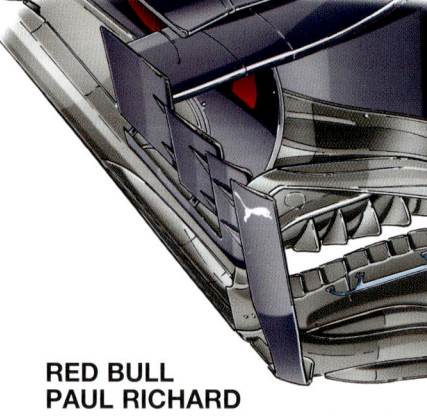

RED BULL DIFFUSER

Again in France, the latest modification the diffuser in the link between the two flaps that run the full width and the notably flared vertical peripheral area.

TORO ROSSO ZELTWEG ENDPLATES

The configuration introduced by Toro Rosso with the front wing endplate featuring a footplate area curiously divided into three sections with raised leading edges was new, a feature that however failed to convince and was not used in the race.

MERCEDES ZELTWEG

The other major novelty at Zeltweg was represented by the rear wing endplate inspired by the design introduced on the McLaren MCL33. The leading edge of the endplate retained the kink required by the regulations, while at the back the fringes, which descended mid-way, maintained a straight line, creating a vent that contributed to improving the efficiency of the rear wing which presented a main plane with a distinct raised section in the central part, while the mobile flap had a V-shape and presented a Gurney flap with a decreasing incidence from the inside out.

MERCEDES FERRARI STYLE SIDEPODS

At Zeltweg Mercedes introduced what was to all intents and purposes a kind of the "B" version of the W09. This was the most important evolution of the season and focused on the adoption of Ferrari-style sidepods in the W09s with the safety structure advanced and outside the sidepods (1) which were therefore set further back where they were less disturbed by the turbulence from the front wheels. The mirrors supports was also all-new.

McLAREN FRONT WING

McLaren experimented with an all-new front wing for Alonso in FP1 at the Austrian GP (bottom), but then in the race the old version was used. The new front wing lost the kink in the main plane that created the tunnel in the flaps directing the flow towards the outside of the front wheel.

It presented a main plane leading edge that was raised halfway along both sides, producing a conspicuous slot located between the support for the larger upper flaps and the endplate. The vent was retained in the delicate area that generates the Y250 vortex.

In contrast with the other designs that had supplementary flaps with the internal surfaces cut away, the McLaren wing presented a further three elements with almost the same length and only the final one having a very short chord and an undulating configuration.

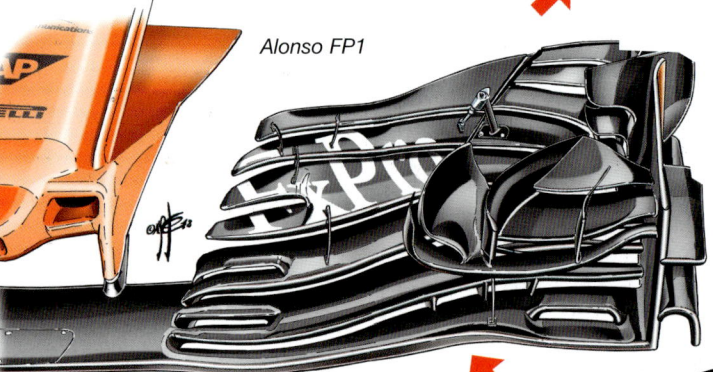

Q/R

Alonso FP1

Zeltweg

Toro Rosso 2017

Shanghai

FERRARI SILVERSTONE DIFFUSER

The main novelty on the SF71H that won at Silverstone was on the floor: a second rail was added to the one introduced in Spain running almost the full length between the two axles. In this way we can see that three vents were interrupted by two diagonal partitions. The tracks curved at the end towards the rear wheel, working in synergy with the more traditional diagonal blows.

The longitudinal arch introduced at Shanghai was replaced with a Toro Rosso-style shaped edge (insert).

In order to increase downforce, the profile extending from the rear brake duct was also enlarged: the small serrated endplate with two slots from Austria (in the top circle, blue arrow) was replaced for the British GP with two completely open vents (blue arrow). The end part of the engine cover was also different.

RED BULL SILVERSTONE

In FP1 Red Bull conducted a series of comparison tests: Verstappen lapped with the front wing from Austria (the same as in Canada), while Ricciardo used a version without the final flap.

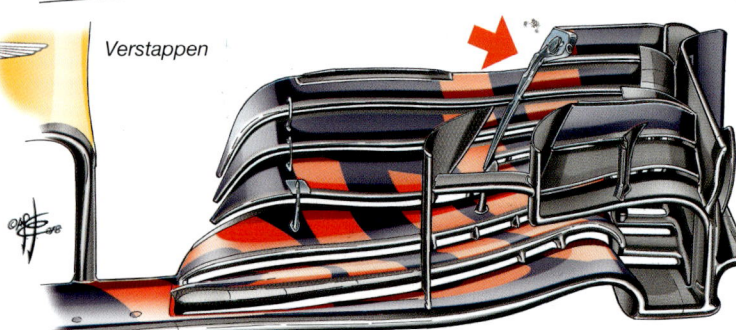

Ricciardo Q/R

Verstappen

GIORGIO PIOLA

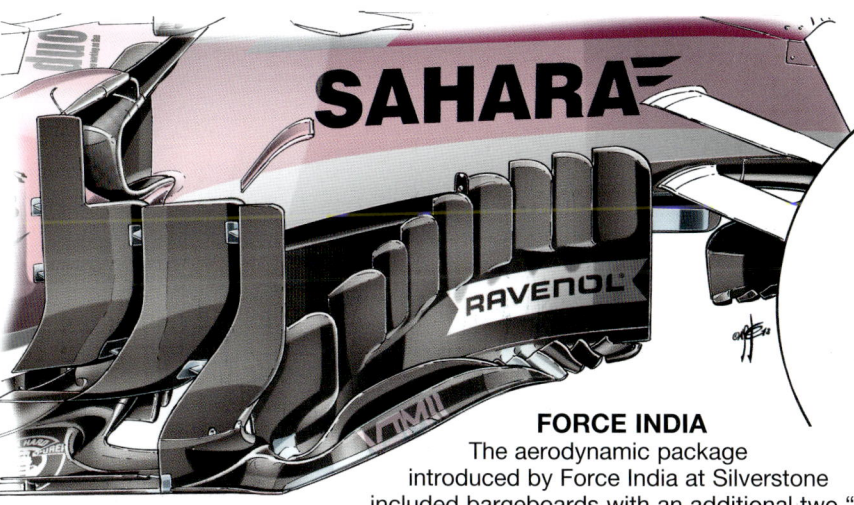

SAHARA

RAVENOL

FORCE INDIA

The aerodynamic package introduced by Force India at Silverstone included bargeboards with an additional two "fingers" in rapid prototyping that were larger than the previous ones and had the clear task of supporting two horizontal "sabers".

The complicated lateral turning vane divided into four parts (three in carbonfibre and one pink) was revised in the lower part with small horizontal deflectors that served to laminate the flow towards the rear diffuser.

MERCEDES DUAL MIRRORS

In an attempt to improve rear visibility for the drivers, Mercedes tested two stacked mirrors to identify the area of greatest efficacy.

TORO ROSSO SUSPENSION FAILURE

The front left suspension on Hartley's Toro Rosso failed spectacularly in FP3 with the push rod link mount on the upright. This was a serious failure that obliged the team from Faenza to investigate and then replace the suspension on Gasly's car too.

FERRARI SILVERSTONE REAR WING

Ferrari took a very low downforce to Silverstone that had been tested at Baku. The SF71H's rear wing was characterised by a spoon main plane with the leading edge curving upwards. The mobile flap presented a V at the centre, with its surface narrowing to the outside edges, in proximity to the endplates with a clear intent to reduce drag. The endplates instead had no less than five horizontal slots with small profiles that had a varying incidence and chords increasing as they rose. The end part of the engine cover was different, lower at the centre (insert).

V-Power

MAHLE

WILLIAMS SILVERSTONE

At Silverstone, Williams introduced a new front wing characterised by the absence of the kink at the ends of the main plane and with a smaller footplate area.

RENAULT HOCKENHEIM

The Renaults had a new front wing design in Germany. The four upper flaps in the internal part, painted yellow, were very short and, above all, where characterised by a straight edge, while there was also an arched version. There were instead no less than six "shutter" flaps in the external part, with the last three having a V-shaped king almost in the middle aligning with the four vertical elements (red arrow) that separated the ever-shorter endplate from the flaps.

There were then five vertical fins in the area of the upper flaps, the first of which creating a slot in the lateral strap assembly.

FERRARI HOCKENHEIM TWIN VERTICAL WASTEGATE EXHAUSTS

In Friday morning practice in Germany, Ferrari experimented with a different wastegate exhaust position without then using it in the race. The feature was the object of a dispute and we analysed it in the Controversies chapter.

RED BULL HOCKENHEIM DUAL RAILS

Red Bull introduced two long slots ahead of the rear wheels at Hockenheim, following a trend that had been launched by McLaren and then taken up by Renault and above all Ferrari.

The "rail" as the two slots had by now been nicknamed merges into the three diagonal slots in front of the rear wheels.

These features were designed to reduce the turbulence that affects the efficiency of the diffuser due to the deformation of the rear tyre shoulder.

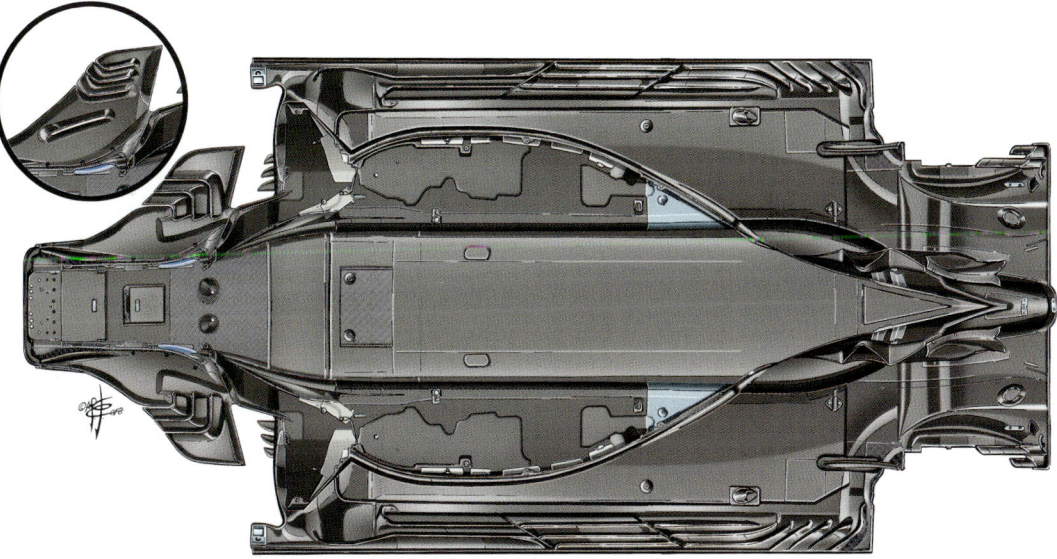

RENAULT BRAKE DISCS

The number of cooling holes in the carbonfibre brake discs increased greatly, in part because the discs themselves increased in thickness from 28 to 34 mm. The configuration of the C.I. brakes was curious, with the holes aligned neither horizontally nor obliquely but in a kind of combination of the two as can be seen through the open drum of the Renault.

FERRARI FLOOR EXPOSED

It required a constant stake out to manage to capture the lower bodywork completely exposed, highlighting the complexity of the vents in all the crucial areas to improve the efficiency of the lower part of the car. The "rails" along the sides were subjected to constant development, as were the diagonal vents ahead of the rear wheels. These very long cuts also replaced the "scimitar" from the previous year.

Hockenheim

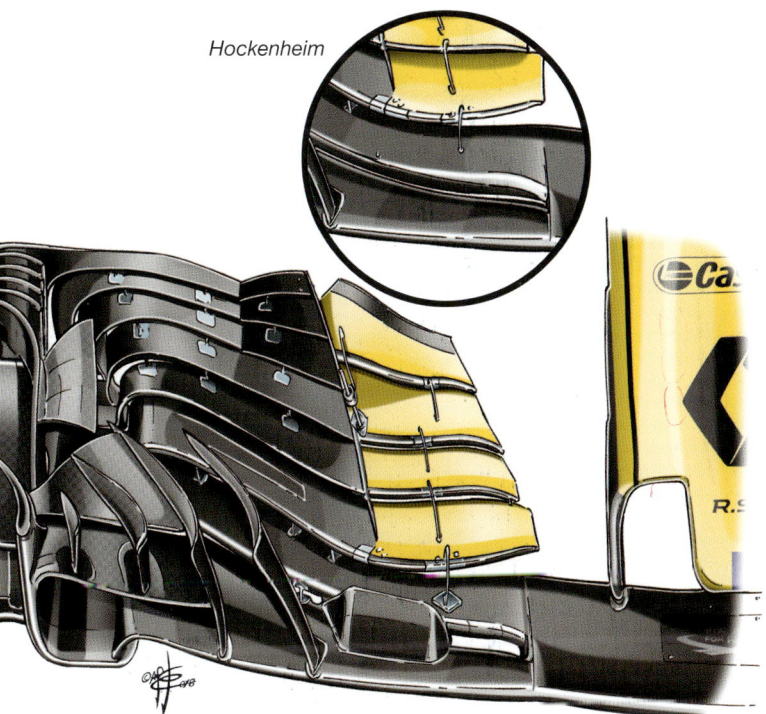

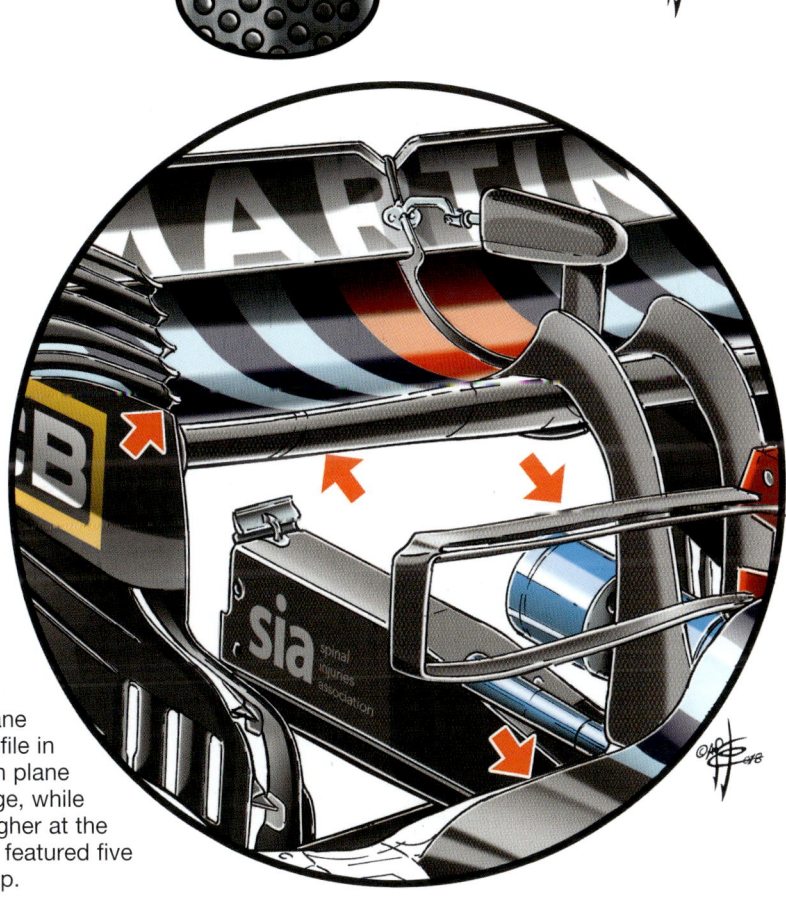

RENAULT BUDAPEST

In Hungary, Renault tested without any great success a modification to the front wing introduced in Germany, splitting the long slot in the main plane. In the race the version introduced at Hockenheim (insert) was chosen.

WILLIAMS BUDAPEST

At Budapest, Williams again fielded the high downforce wing already seen at Monaco. A biplane T-wing with a blown profile in the upper part. The main plane had a raised leading edge, while the engine cover was higher at the sides and the endplates featured five horizontal slots at the top.

HAAS SPA FLARED FLAP

At Spa, Haas fielded a very low downforce rear wing: the main plane was almost flat with a leading edge tipped upwards, especially at the centre between the two support pylons. The mobile flap configuration was curious with a very short chord and above all the trailing edge that was not straight but had two curvatures to reduce drag and improve the maximum speed.

FERRARI FRONT WING

Red Bull-style endplates on the SF70H for experimental purposes. The initial part of the external arched area was notably raised with respect to the horizontal.

MERCEDES FRONT WING

The novelty of the front wing fielded by Mercedes lay not so much in the reduced chord flaps as in the addition of this turning vane between the two already present on the W09 in an attempt to channel as much air as possible towards the outside of the front wheels.

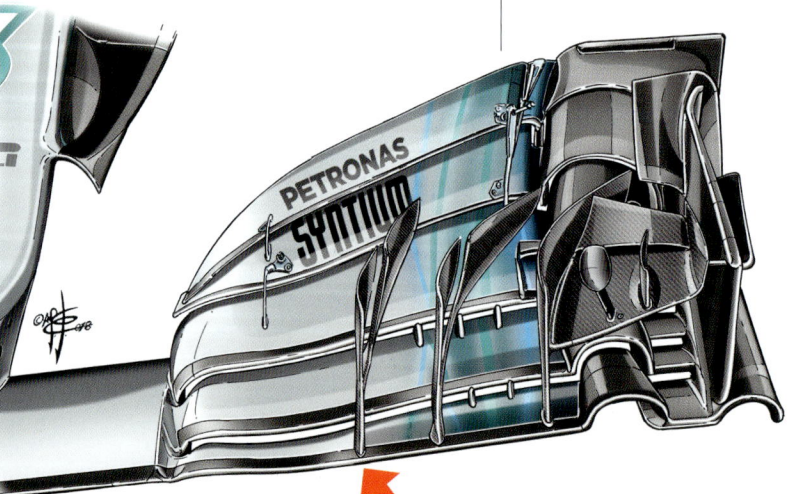

RED BULL LOW DOWNFORCE

At Spa too, Red Bull presented the lowest downforce rear wing in the field, with end-plates without horizontal slots in the upper part.

FERRARI SPA

Ferrari too presented a particularly low downforce rear wing at Spa, with a main plane that was almost flat, only slight flared upward in the central section of the leading edge.

The flap also had a fairly short chord because the efficacy of the DRS on a fast track like the one in Belgium is significantly reduce compared with other circuits. The SF71H nonetheless featured four horizontal vents in the upper part of the endplates, while they had all disappeared on the Red Bull.

To improve the aerodynamics and reduce drag, the end part of the sidepods was further cut away.

RED BULL MIRRORS

Red Bull also decided to create slots around the rear-view mirrors, adopting the concept launched on the Ferrari at the start of the season; a true turning vane was actually created around the mirror.

Red Bull adopted this complex feature after Toro Rosso had experimented with a blown mirror on the occasion of the British Grand Prix at Silverstone.

MERCEDES MONZA

At Monza, Mercedes made minor modifications to the lower part of the chassis behind the bargeboards, with four winglets of increasing size (red arrow) managing the flows under the car. The presence of the four flaps revealed by the corresponding cuts seen on the winglet (white arrow) over these small elements as every aperture on the floor had to be visible from above.

The shape of the board flanking the car was also different (blue arrow). There was also a modification to the delicate linking zone between the bargeboards and the turning vanes at the sides of the sidepods: a small flared profile appeared in this area.

TORO ROSSO

A new feature on the Toro Rosso in the area ahead of the rear wheels with a kind of small symmetrical twin spoiler designed to create vortexes in both the upper and lower parts of the stepped floor.

RED BULL MONZA

The RB14 was extreme at Monza too, with a rear wing that was almost flat as can bee seen in this side view. The endplates had no vents in a search for maximum penetration, while the front wing had been literally deconstructed.

MERCEDES MONZA

A low downforce front wing for Mercedes based on the version introduced in Belgium and characterised by the final flap with a further reduction in the chord, without however arriving at the extremes of the Red Bull.

TORO ROSSO DRS

Toro Rosso brought a very low downforce rear wing with the main plane practically neutral and a mobile flap with a minimal incidence, making the DRS much less efficacious than at other circuits where the difference in speed with the wing open or closed was much greater.

The fairing of the DRS was different to all the other and as well as grooming the flows at high speed, avoided the problems with the blocking of the flap that Ericsson had suffered in practice on the Saturday morning.

The fairing of the actuator prevented the mobile flap, subjected to loading at the very high speeds of Monza, from lifting more than it should, blocking the system's recall manoeuvre for the closing of the DRS, acting as a stop.

FERRARI SINGAPORE LIVERY

The new Ferrari livery debuted in Singapore, with the sponsor clearly visible on the engine cover and the bargeboards.

FORCE INDIA SINGAPORE

Force India presented what was almost a B version of the VJM11 at Singapore. The most evident changes were made to the barge-boards: the element was now attached to the chassis via a bridge-like profile as on the McLaren MCL33. The mirror no longer had three supports but just two elements that acted as turning vanes on the sidepods, while the more traditional one mounted on the chassis was eliminated. Lower down, below the raised leading edge of the floor can be seen six vertical elements (clearly visible once the lateral turning vanes are detached) in place of the usual three.

FERRARI REAR WING

The new rear wing shows the endplate design launched by McLaren the previous season and seen on the Mercedes and Renault in 2018 too.
The vertical fringes (arrow) did not follow the flared line imposed on the endplates by the regulations, instead creating vents that contributed to increasing downforce and grooming the wake.

MERCEDES SINGAPORE

Mercedes opted for a rear wing with a deep chord main plane with a an upwards kick in the central part and a high incidence mobile flap equipped with a Gurney flap.
As well as the five horizontal vents at the top, the endplate also featured an interesting novelty at the back: the McLaren-style fringing was exaggerated by a vertical slot that reached the upper section and required two metal brackets to ensued that the element affected by the flow was not torn off at maximum speed.

FERRARI SINGAPORE

Maximum brake cooling at Singapore, with the shrouds having new holes because with night race was held with high ambient temperatures on a street circuit with low average speeds.
It should come as no surprise that Ferrari fitted new carbonfibre Brembo discs with more than 1,400 cooling holes in inclined rows of six, visible through the aperture at the top.

SPA-SINGAPORE MERCEDES

We only noticed it at the Singapore GP, but along with the new rear wheels introduced in Belgium, Mercedes also introduced a new rear brake drum that unusually was concave and had a different extraction effect in this delicate area of the car, improving and rendering more constant the temperature of the rear tyres while also improving heat dispersal from the bearings.

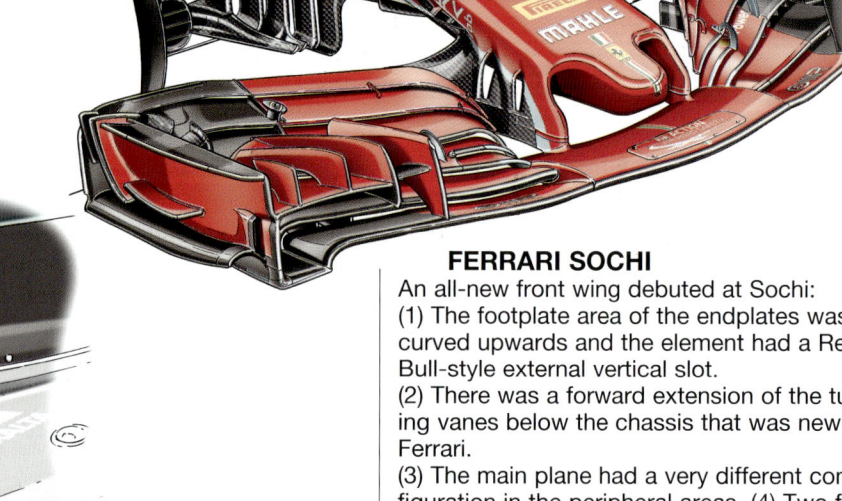

FERRARI SOCHI

An all-new front wing debuted at Sochi:
(1) The footplate area of the endplates was curved upwards and the element had a Red Bull-style external vertical slot.
(2) There was a forward extension of the turning vanes below the chassis that was new for Ferrari.
(3) The main plane had a very different configuration in the peripheral areas. (4) Two further fins were introduced to the outside section of the vertical vane, along with a vent (5) in the penultimate flap and a different shape (6) to the last flap.

RED BULL SOCHI

A new endplate for the Red Bulls, cut away at the rear to leave room for two vertical turning vanes designed to channel the air to the outside of the wheels. Note the small tunnel inclined upwards on the leading edge and cut obliquely to match the positive rake set-up.

SAUBER SOCHI

On the Saubers at Sochi, along with the modifications to the bargeboards and the floor, there were changes to the outermost area of the diffuser with the proliferation of a series of micro-flaps that were designed to groom the wake behind the rear wheels in order to improve the aerodynamic efficiency.

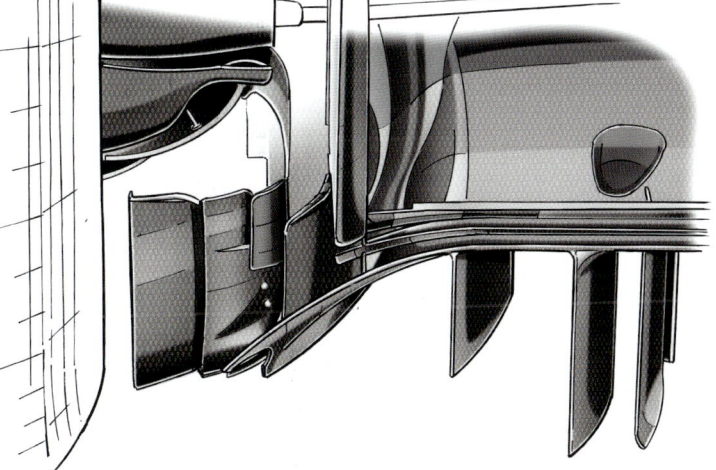

MERCEDES SOCHI

Mercedes introduced a feature already seen on the Ferrari. The Brackley-based team in fact abandoned the mono-pylon rear wing support, adopting the idea that characterised the SF70H, with the two swan's neck supports that were retained on the SF71H.

MERCEDES SUZUKA

A further modification to the front wing of the Mercedes with the addition on the inside of the endplates of a small flap in addition to the one fitted lower down and at the back. In the insert, the earlier version.

SAUBER BARGEBOARD

New bargeboards with a McLaren-type mount and a myriad mini vortex generators in the lower part of the Saubers under constant development. Note at the top, the three small fins grooming the flows towards the sidepod intake opening.

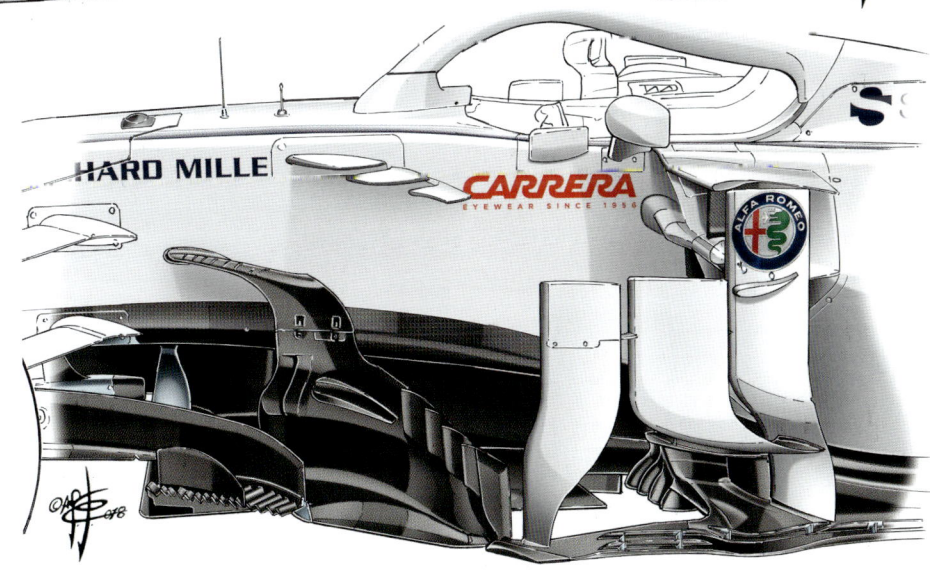

FERRARI FLOOR

At Suzuka a new floor was tested that shows the last slot, the one in front of the rear wheels, which was elongated through to the endplate. The aim was to channel the vortexes close to the vertical strap so as to limit the negative effect generated by the turbulence from the tyre rotating in contact with the asphalt.

However, the slot filled up with rubber debris during FP1, neutralising its efficacy and delaying its race debut.

FERRARI BRAKE DUCTS

Again at Suzuka, Ferrari presented a new brake duct feature with the addition of a further L-shaped fin (as on the McLaren), below, in the lower part of the brake duct lip to as to better direct the flow of air towards the bargeboards. In the insert, the earlier version.

FERRARI AUSTIN

At Austin, new bargeboards debuted on the SF71H with the floor characterised by no less than 11 vortex generators to improve the flow directed towards the rear diffuser: each element had a different chord and length. The first were wider and shorter, while they became increasingly narrower and undulating towards the rear. The vertical elements also increased in number, with a further two small turning vanes being added.

McLAREN SUZUKA

After having introduced in the 2017 season (insert) the straight fringes detached from the endplate that started a trend, at Suzuka McLaren presented a simplified version without the cantilevered elements.

FERRARI FLOOR

A new floor was fitted to Kimi Räikkönen's car for FP1 at Austin equipped along with the slot in front of the rear wheel three new rows of small turning vanes in the vicinity of the rails. This new solution was then discarded before being reprised, again unsuccessfully, in Mexico.

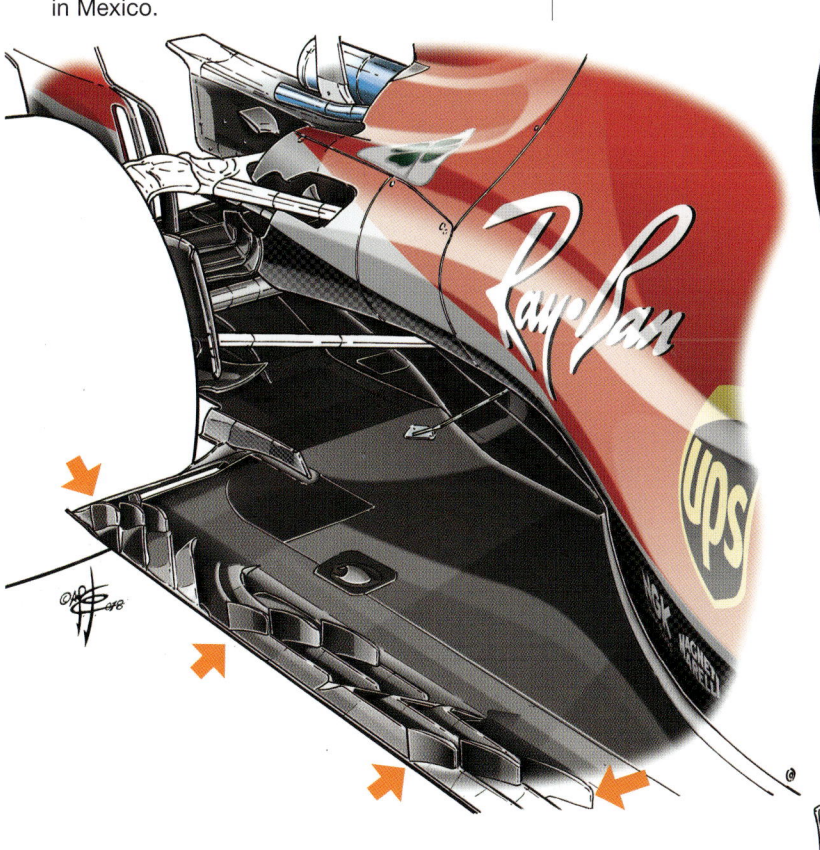

MCLAREN AUSTIN

At Austin, McLaren abandoned the longitudinal vent either side of the nose cone that had characterised the MCL33 through to that race.

TORO ROSSO

At Austin, Toro Rosso adopted a modification the front wing: the endplate was trimmed at the rear and a second turning vane was inserted, tasked with orientating the flows towards the outside of the front tyre in order to improve the aerodynamic efficiency of the car from Faenza.

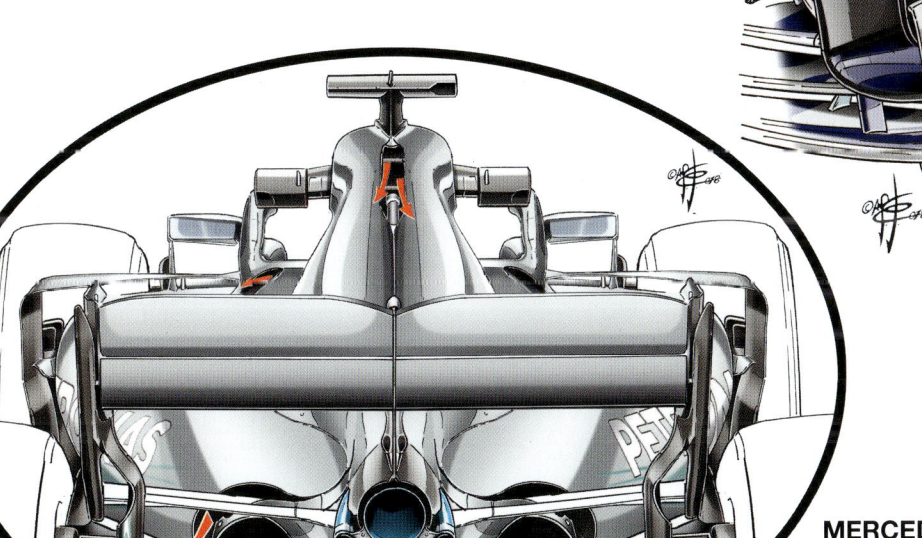

MERCEDES MEXICO CITY

In order to overcome the cooling loss associated with the altitude, the W09s saw a return of the "chimney" opened on the engine cover fin, while the lateral vents on the ends of the sidepods were subjected to a minor variation.

RED BULL MEXICO CITY

Just a few days later, Red Bull presented an idea almost identical to the mini turning vanes applied by Ferrari to the rails along the floor at Austin. Clearly they were produced using a rapid prototyping process and simply glued in place. They were not used in qualifying and the race.

WILLIAMS MEXICO CITY

The end section of the Williams in Mexico was a true chasm. An even more extreme version of the configuration seen at Budapest. Obviously it had a detrimental effect on the car's aerodynamics.

FERRARI BREMBO DISCS

In Mexico, Brembo previewed material for the 2019 season: Ferrari tested a new brake disc with seven holes on diagonal rows.
The Mexico City circuit is ideal for evaluating cooling given that it is located at an altitude of 2,240 metres where the air is less dense and a greater flow is required to guarantee an effective braking system. In the inserts, previous versions of the ventilation holes.

FORCE INDIA FUEL TANK/CHASSIS

In this drawing we have highlighted the widening of the chassis area to enable the fuel tank capacity to be increased without interfering with the length of the chassis itself.

TORO ROSSO INTERLAGOS

At Interlagos, Toro Rosso introduced new bargeboards with vertical element in two pieces at the front, linked by a flap mounted on the outside arching slightly downwards.

It also featured the three cuts corresponding to similar apertures seen on the floor, with inclined blades designed to direct the underbody flow. Above this structure there was also a flap arching towards the bottom. Either side of the cockpit could be seen a fin with five elements, the first larger and the other four square and of the same size.

The number of turning vanes was also increased from two to three elements, drawing inspiration from the latest configurations seen on the Red Bull and the Haas.

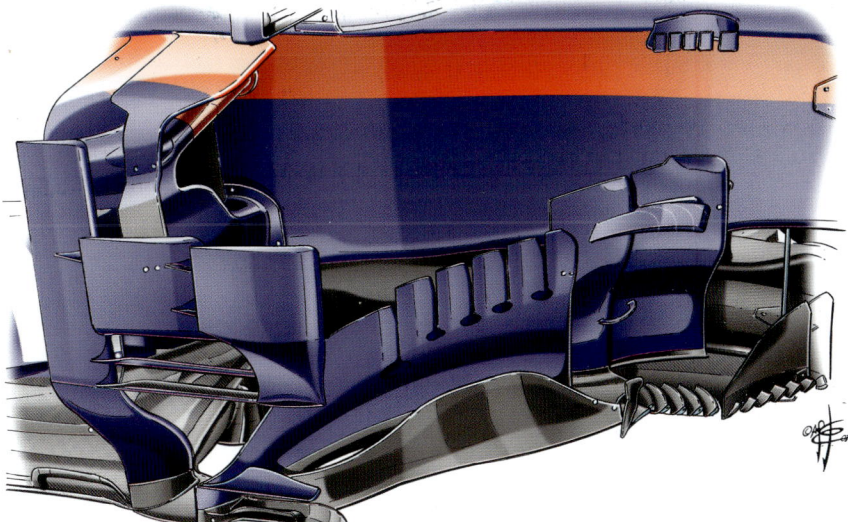

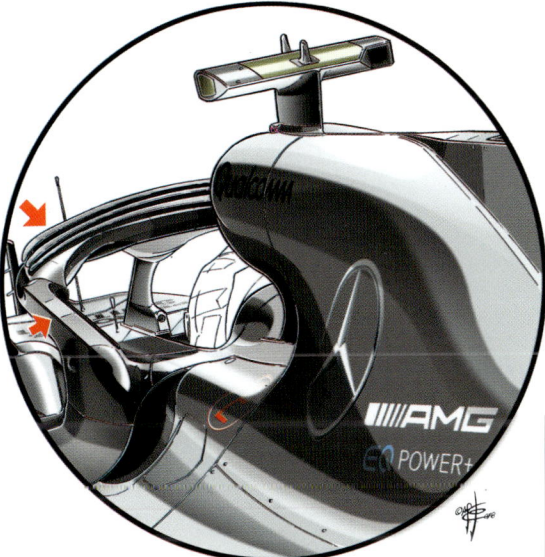

MERCEDES HALO

After having competed for the whole seen with the Halo in standard" form, Mercedes introduced three small aerodynamic profiles at Abu Dhabi as many other teams had already done in the front central area, while the lateral part had a more angular section at the top.

FERRARI ABU DHABI

In FP1 Räikkönen's Ferrari was fitted with a simplified front wing without all the upper flaps and turning vanes outside the endplate, in order to bring the design closer to the configuration to be used in the 2019 season. Ferrari did not produce a 2019 wing, but tried to adapt the 2018 version to the regulatory dimensions for the forthcoming season (the main plane could be 20 cm wider), in order to discover with the array of sensor located in front of the sidepods how the wake from the front wheel changed with the inevitable loss of downforce.

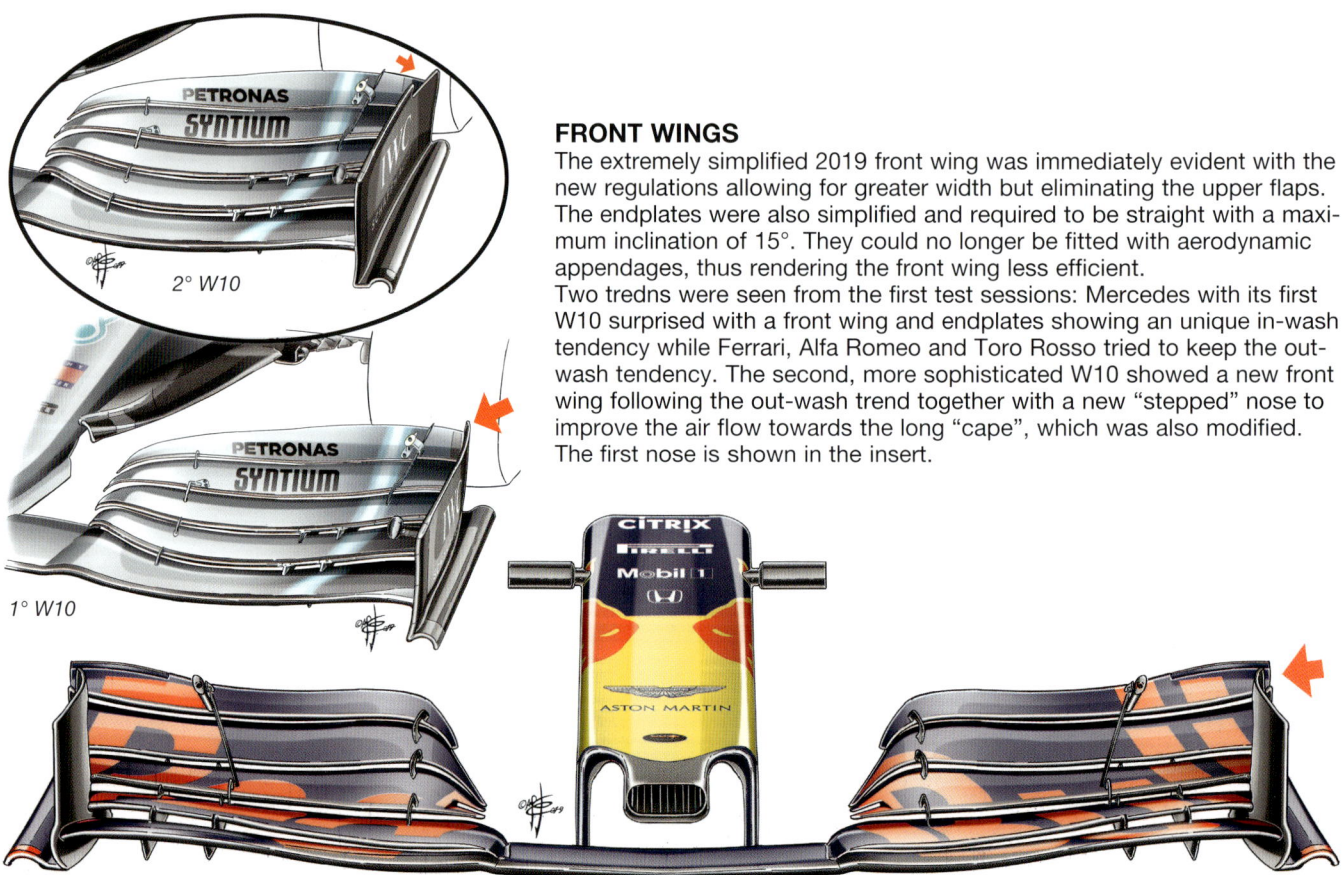

FRONT WINGS

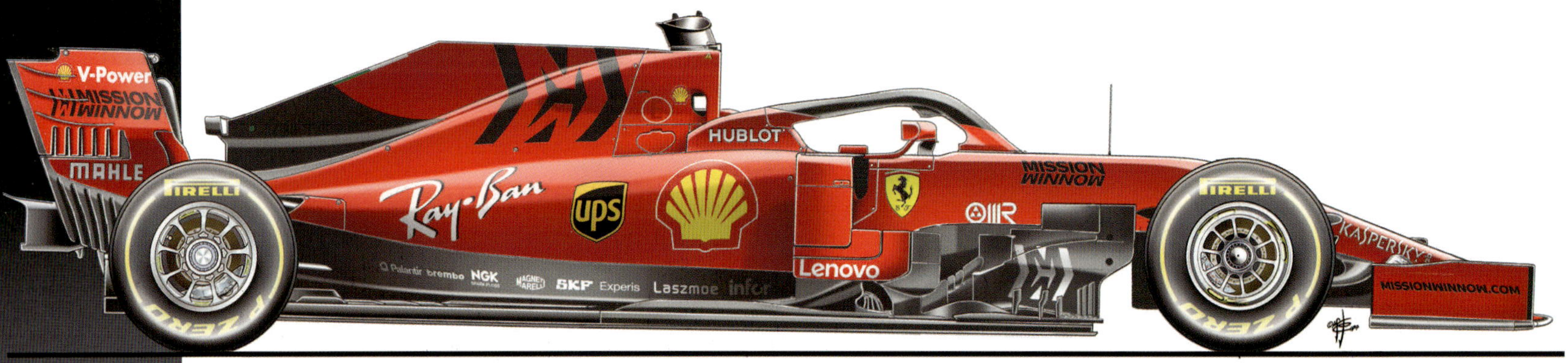

The extremely simplified 2019 front wing was immediately evident with the new regulations allowing for greater width but eliminating the upper flaps. The endplates were also simplified and required to be straight with a maximum inclination of 15°. They could no longer be fitted with aerodynamic appendages, thus rendering the front wing less efficient.

Two tredns were seen from the first test sessions: Mercedes with its first W10 surprised with a front wing and endplates showing an unique in-wash tendency while Ferrari, Alfa Romeo and Toro Rosso tried to keep the out-wash tendency. The second, more sophisticated W10 showed a new front wing following the out-wash trend together with a new "stepped" nose to improve the air flow towards the long "cape", which was also modified. The first nose is shown in the insert.

2° W10

1° W10

FERRARI

The 2019 regulations banned the blown axle (lower insert). Consequently, the position of the brake calipers on the on the Ferrari F90-H was modified and internal ducts were created to allow the air to be blown towards the outside of the front wheels. This feature was also seen on other cars, including the Mercedes.

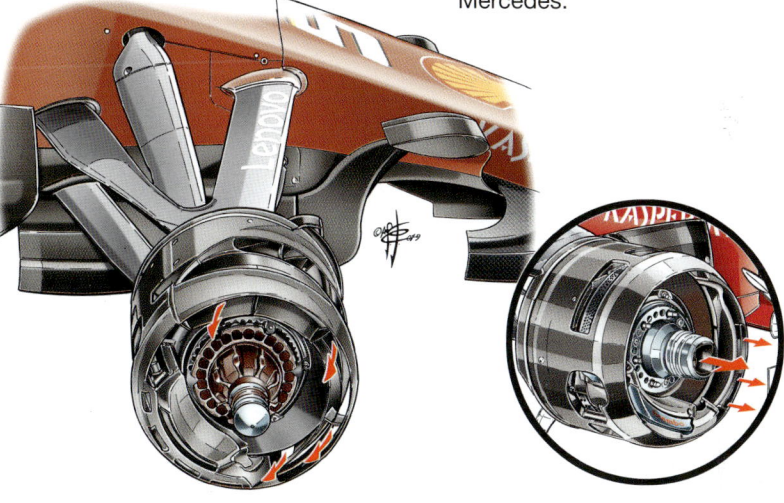

ALFA ROMEO

As happens with new regulations, from first test sessions we saw a number of new aero devices on the 2019 cars. Alfa Romeo introduced four additional fins on the upper part of the chassis, like Mercedes with its second W10, helping to extract air from the S -duct and direct the air flow outboard.

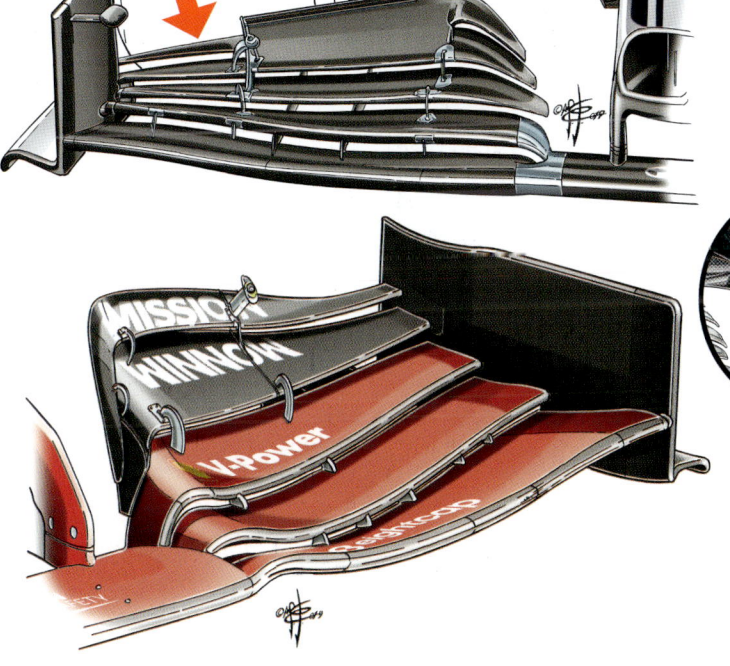

2019 **REGULATIONS**

For the 2019 season the FIA has again revolutionized the cars' aerodynamics so as to reduce the slipstreaming effect and improve overtaking; this wide-ranging intervention has affect almost all the components of the F1 cars' aerodynamics. The aim of the changes was to improving overtaking by acting on two fronts: a radical modification of the aerodynamics so as to make the cars less sensitive to the slipstreaming effect and the uprating of the DRS effect. All the main points of this revolution are shown in the front view.

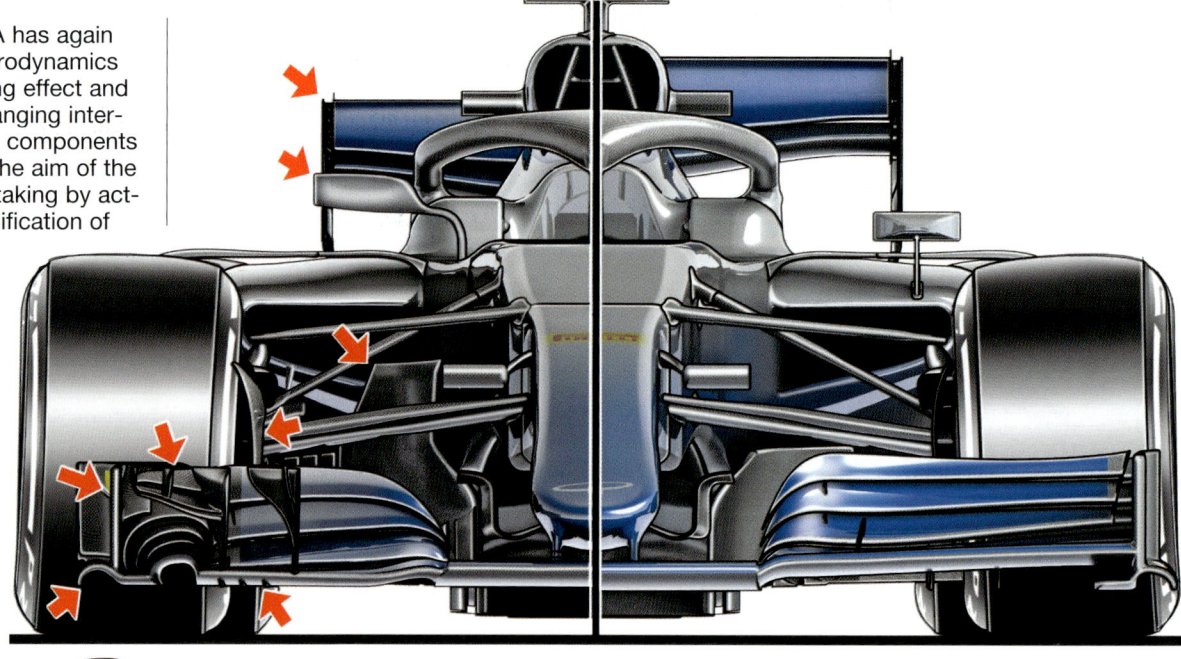

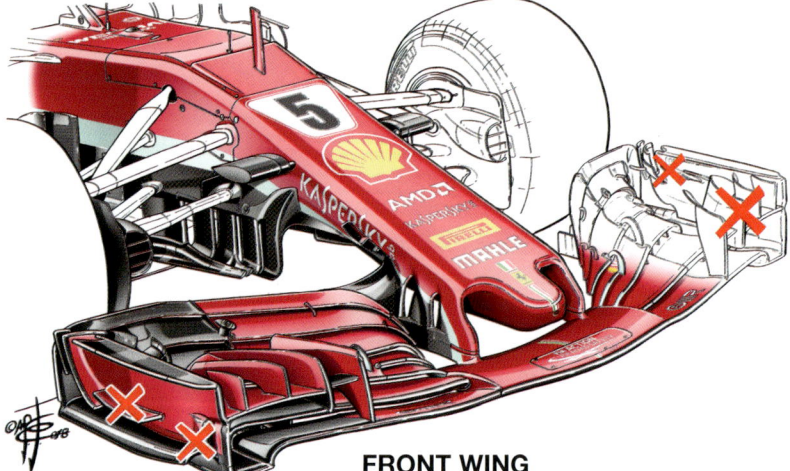

2018   2019

900mm   1000mm

+25mm

FRONT WING

The front wing has been widened from 1800 mm to 2000 mm, thus approaching the external dimensions of the tyres. The length of the profile and therefore the front overhang has been increased by 25 mm, as has the height, with the result that the front end plates are 25 mm longer and taller.

The upper flaps have been banned with the simplification of the front wing that can now have just five planes with the possibility of adjustable flaps on the last two.

The end plates have been simplified and must be straight with a maximum inclination of 15°. The end plates must also be contained within an area between 910 and 950 mm from the car's centre line.

This is in order to try to contain the deviation of the air flow towards the outside of the tyres, thus rendering the front aerodynamics less efficient.

Just two vertical elements of a maximum of 75 mm tall are tolerated in the lower part of the wing, set between 500 mm and 800 mm from the centre line.

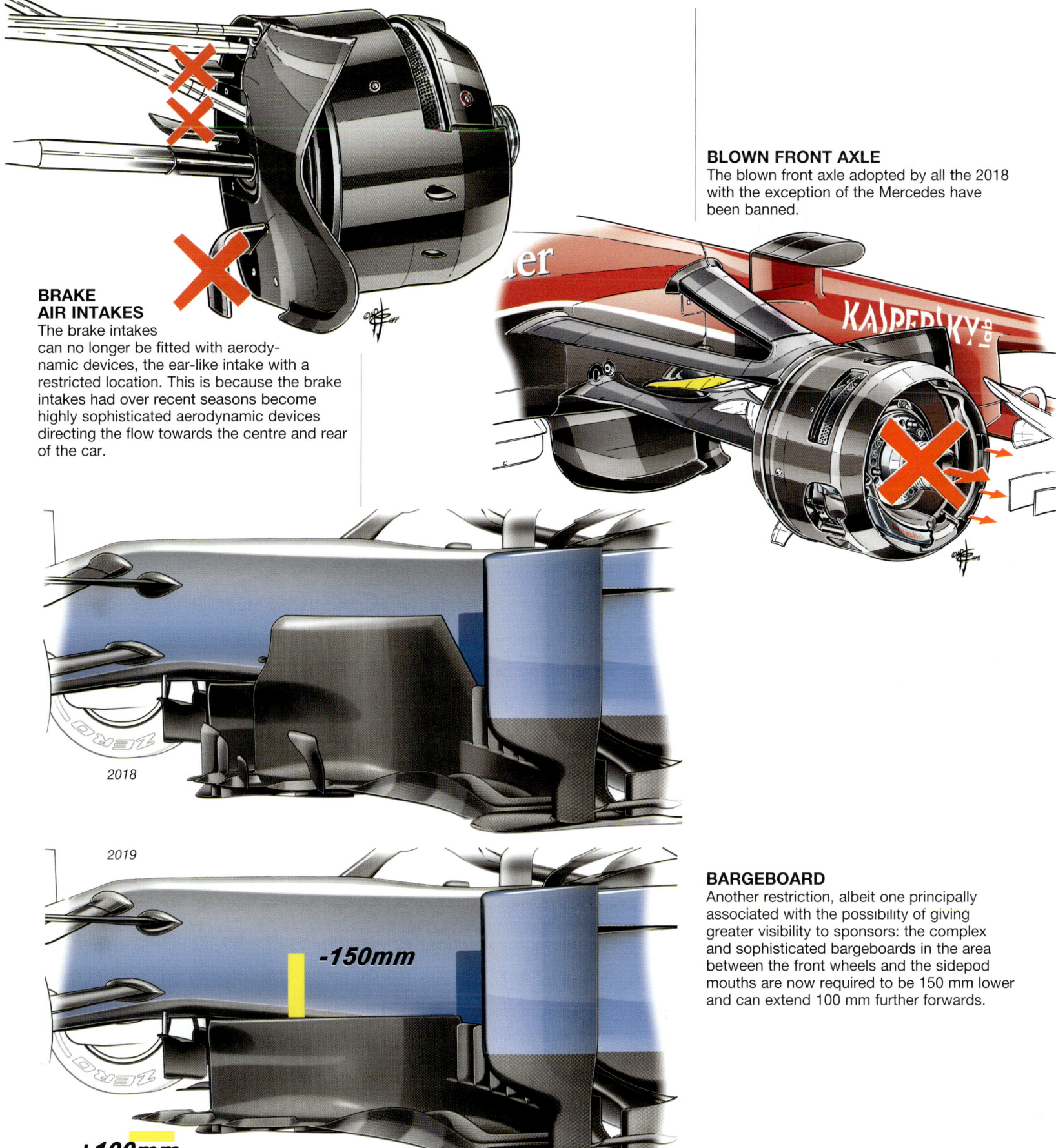

BRAKE AIR INTAKES

The brake intakes can no longer be fitted with aerodynamic devices, the ear-like intake with a restricted location. This is because the brake intakes had over recent seasons become highly sophisticated aerodynamic devices directing the flow towards the centre and rear of the car.

BLOWN FRONT AXLE

The blown front axle adopted by all the 2018 with the exception of the Mercedes have been banned.

2018

2019

-150mm

+100mm

BARGEBOARD

Another restriction, albeit one principally associated with the possibility of giving greater visibility to sponsors: the complex and sophisticated bargeboards in the area between the front wheels and the sidepod mouths are now required to be 150 mm lower and can extend 100 mm further forwards.

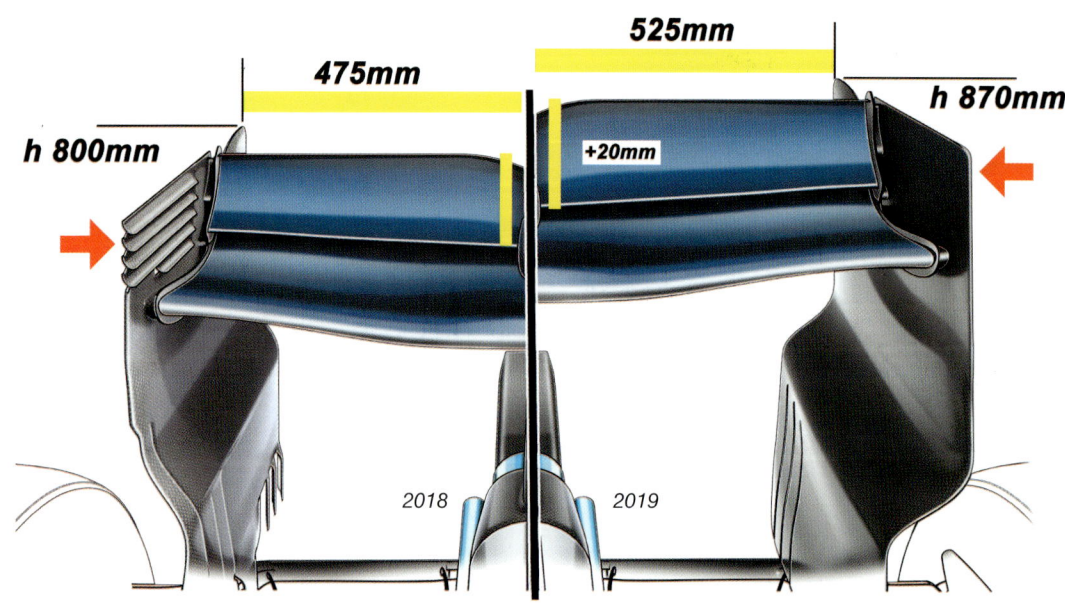

475mm

525mm

h 800mm

h 870mm

+20mm

2018

2019

REAR WING

Extensive changes have been made to the rear wing. The width has been extended from 950 to 1050 mm, the height from 800 to 870 mm from the reference plane and the DRS opening from 65 to 85 mm, thus guaranteeing a notable increase in speed with the device open. Again to improve visibility for sponsors, horizontal slots in the upper leading edge of the end plates have been outlawed. The greater height of the rear wing improves visibility for the drivers thanks also to a revised rear-view mirror location.

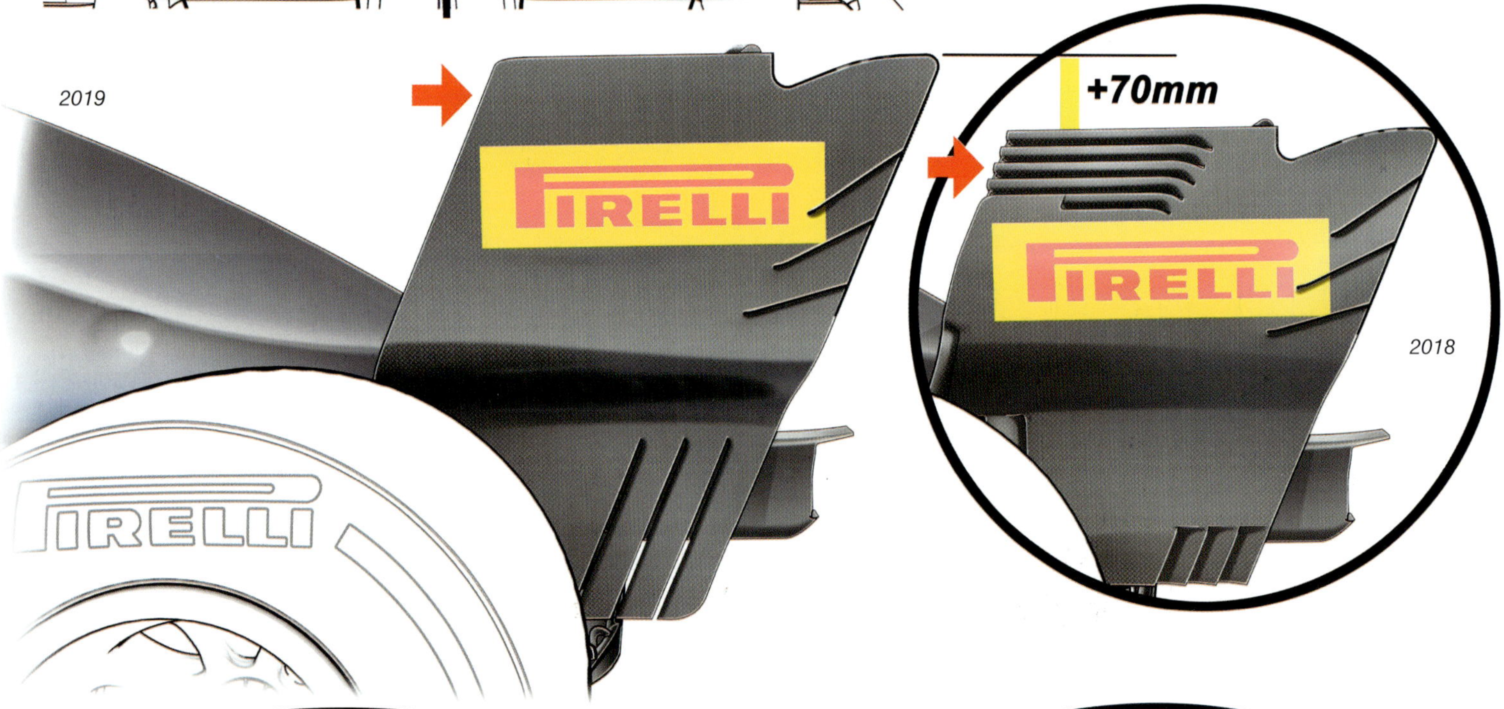

2019

PIRELLI

+70mm

PIRELLI

2018

PIRELLI

2018

65mm

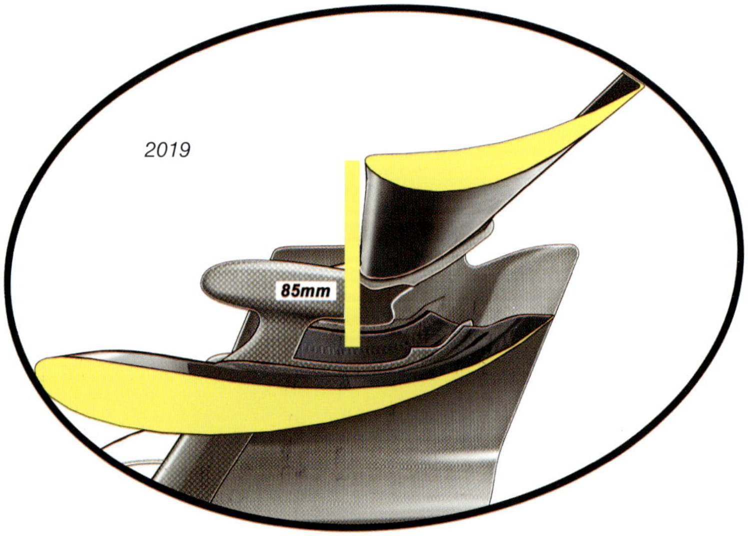

2019

85mm

GIORGIO PIOLAS

REAR VIEW MIRRORS

With the lowering and the greater width of the rear wing for the 2017 season, the rear view mirrors were found to be poorly located to guarantee the drivers good rear visibility, as shown in the comparison drawing. For 2019, not only has the position of the rear wing been modified, but also that of the mirrors.

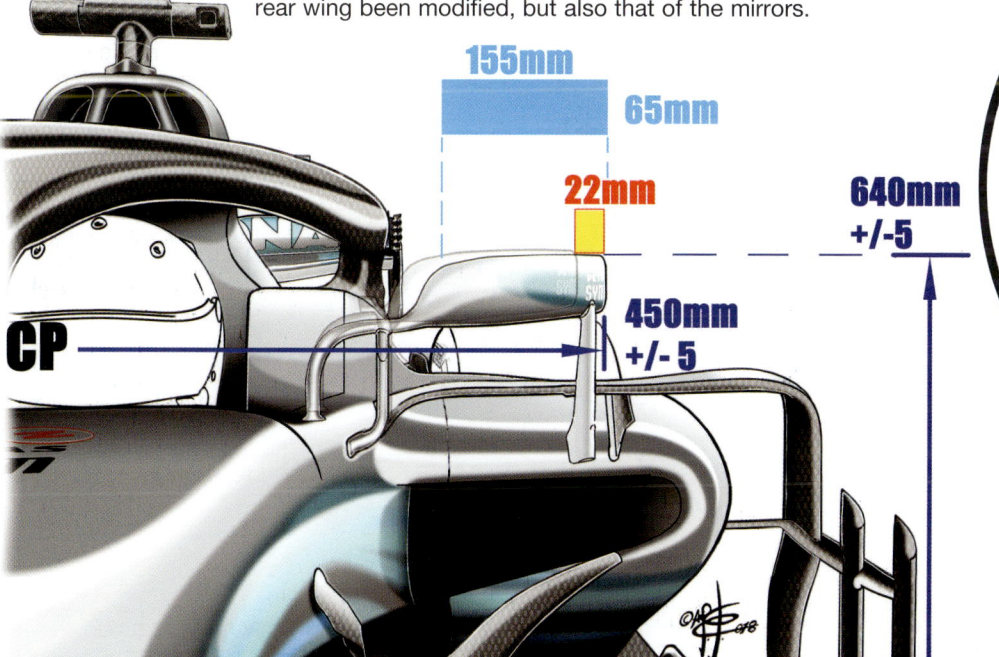

155mm
65mm
22mm
640mm +/-5
450mm +/- 5
CP
RP

REAR LIGHTS

LEDs are now buried in the upper rear part of the end plate and light up together with the rear lights in the case of rain.

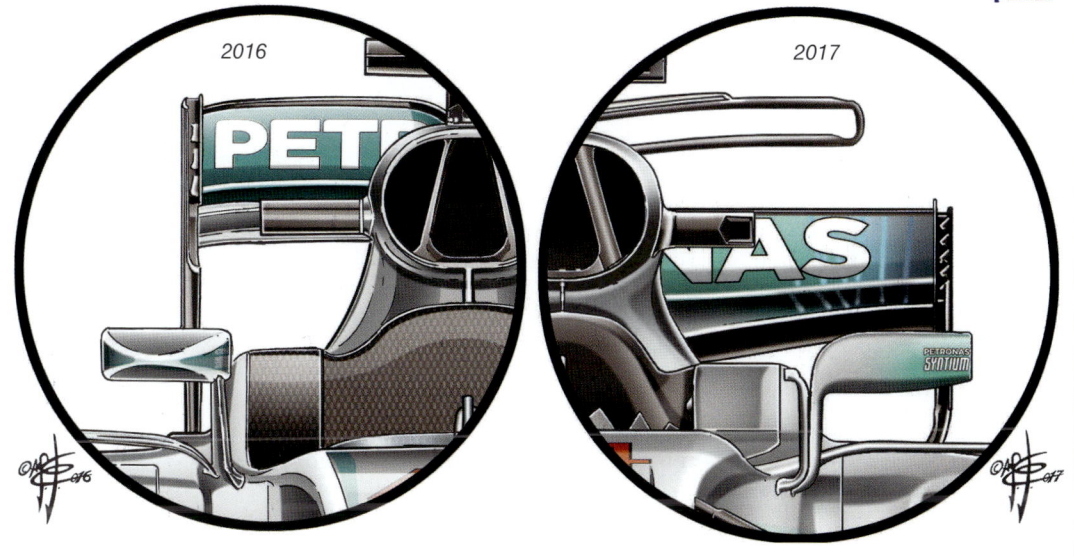

2016

2017

MINIMUM WEIGHT

The overall weight of the car without fuel has been increased by 3 kg against the 5 requested by the teams, reaching a minimum of 743 kg. In the weight of the car, the driver is considered to be equal to 80 kg, with the possibility therefore of adding ballast in the case of a driver weighing less.

The capacity of the fuel tank has been increased from 105 kg to 110 kg, with a hypothetical greater length of the monocoque, which can be adjusted via the length of the spacer between engine and gearbox (in grey).

743 kg

110KG

The **CARS** that made history

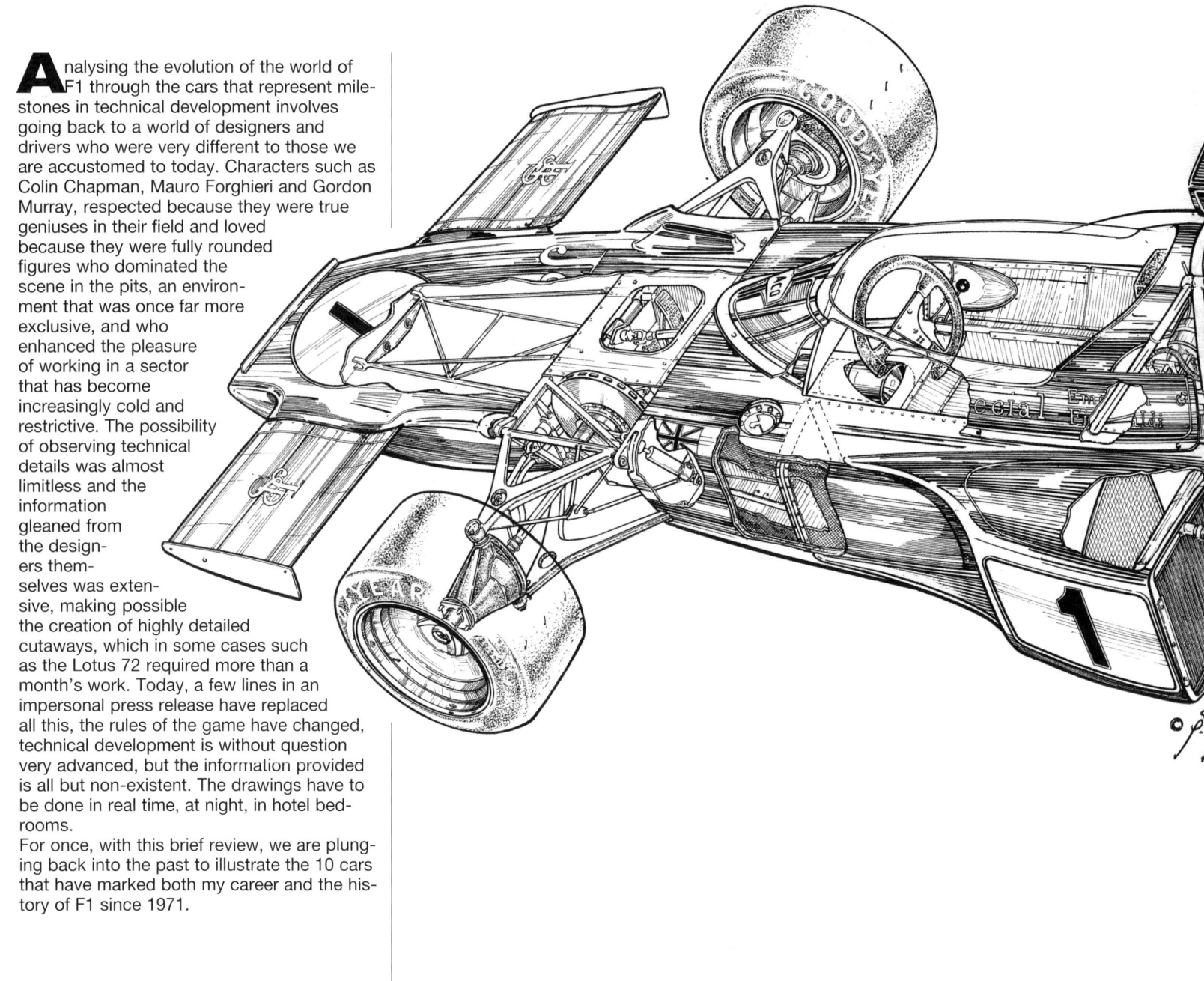

Analysing the evolution of the world of F1 through the cars that represent milestones in technical development involves going back to a world of designers and drivers who were very different to those we are accustomed to today. Characters such as Colin Chapman, Mauro Forghieri and Gordon Murray, respected because they were true geniuses in their field and loved because they were fully rounded figures who dominated the scene in the pits, an environment that was once far more exclusive, and who enhanced the pleasure of working in a sector that has become increasingly cold and restrictive. The possibility of observing technical details was almost limitless and the information gleaned from the designers themselves was extensive, making possible the creation of highly detailed cutaways, which in some cases such as the Lotus 72 required more than a month's work. Today, a few lines in an impersonal press release have replaced all this, the rules of the game have changed, technical development is without question very advanced, but the information provided is all but non-existent. The drawings have to be done in real time, at night, in hotel bedrooms.

For once, with this brief review, we are plunging back into the past to illustrate the 10 cars that have marked both my career and the history of F1 since 1971.

LOTUS 72 (1972-1973)

Enzo Ferrari often repeated that a car was only beautiful when it was winning. While it is true that an F1 car is never judged solely on its beauty, it is just as true that frequently clean, simple and uncomplicated lines achieve an aesthetic level that would make them worthy of a place in a museum of contemporary art. This was certainly the case with the Lotus 72, a true masterpiece of simplicity, from the tip of the wedge-shaped nose that dictated fashions in both racing and road cars. Those lines were due principally to Chapman's decision to relocate the large radiator from the nose, moving it and splitting it into two either side of the car. Even though its aerodynamic sophistication was the most conspicuous aspect of this car, in

other areas to it was highly innovative and featured for example torsion bar suspension, a configuration that while later abandoned on F1 cars did make a reappearance in 1989 on Barnard's Ferrari 640. Another new feature were the inboard brakes front and rear, a choice that sadly cost Jochen Rindt his life in practice for the 1970 Italian GP. The drawing illustrates the "D" version of the 72 that won the 1972 World Championship with Fittipaldi: it differed with respect to Rindt's car for the presence of an unusual T-shaped engine air intake and the fairing of the oil coolers cantilevered below the rear wing, which itself was mounted further back with respect to the rear axle than on other cars of the era. This drawing, almost two metres in length, required no less than 45 days' work.

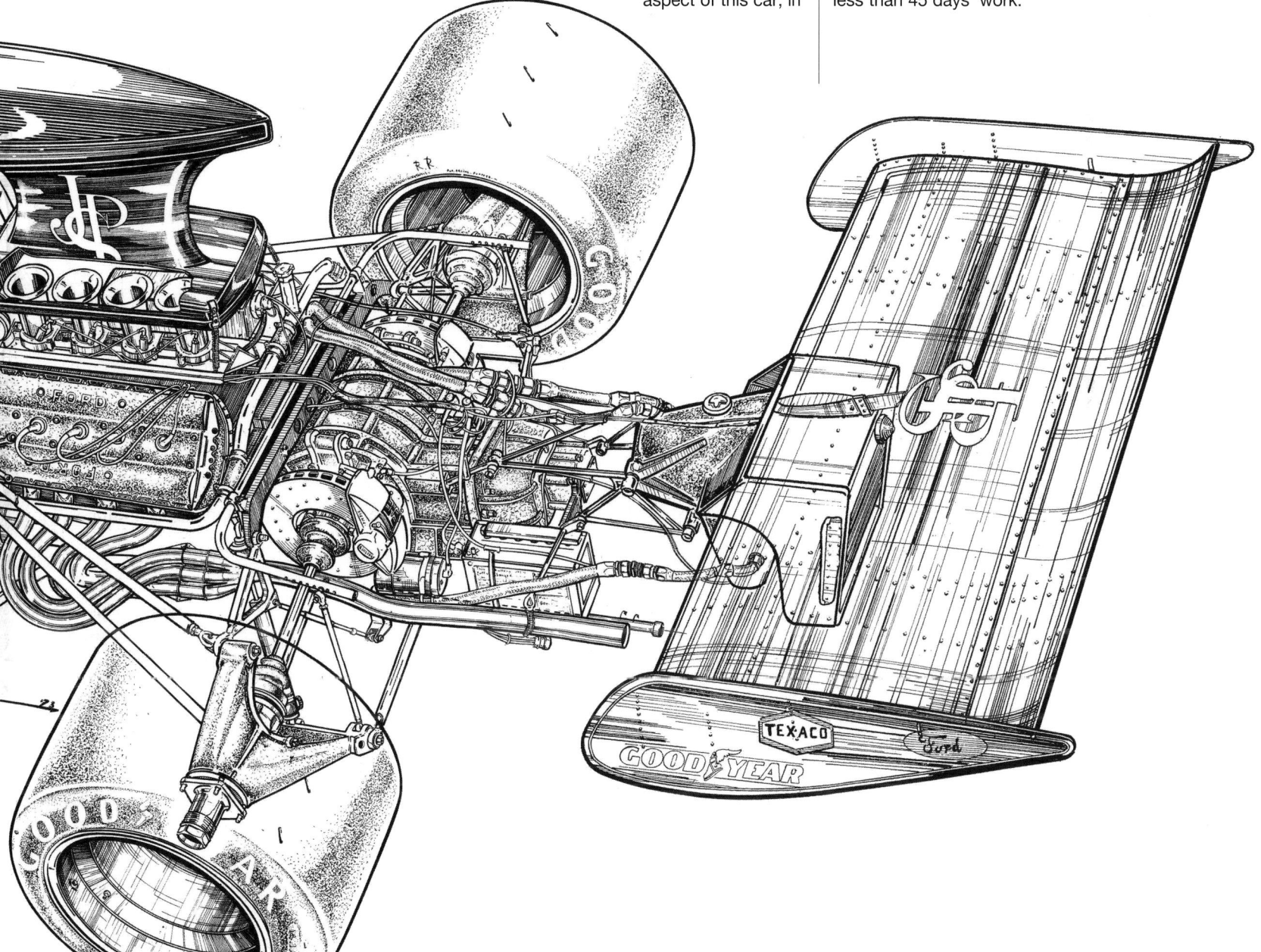

FERRARI 312 B3 (1973-1974)

In order to gain a technical overview of Niki Lauda's conquest of the 1975 World Championship with the Ferrari 312 T, we have to go back to the B3 that appeared in the August of 1973, when Ingegner Mauro Forghieri took over technical direction at Maranello and thoroughly revised the project of the Colombo-Thompson duo. The revolutionised B3 debuted at the Austrian GP in the hands of Arturo Merzario and already boasted many of the technical features of the definitive version from 1974, including the long, oblique radiators in the sidepods that had become very wide and rounded (in the view from above), equipped with long, narrow vents for the hot air in the upper part and the large V-shaped front wing. Before reaching the definitive form of the '74 edition (bottom drawing), Maranello's B3 was subjected to a long series of modifications that affected almost all areas of the car. The final version shows the large double arrow front wing and the rear wing with the arrow-shaped trailing edge. The clean aerodynamics of the rear wing area were very important, an area which on the majority of the British cars was occupied by the oil coolers that in the case of the B3 were instead placed to the sides of the car, ahead of the rear wheels. Another characteristic of this car was the large engine air intake, narrow and vertical. Ferrari just missed out on the World Championship with the B3, in part because the team was focused on Lauda when Regazzoni was actually the more effective of the pair.

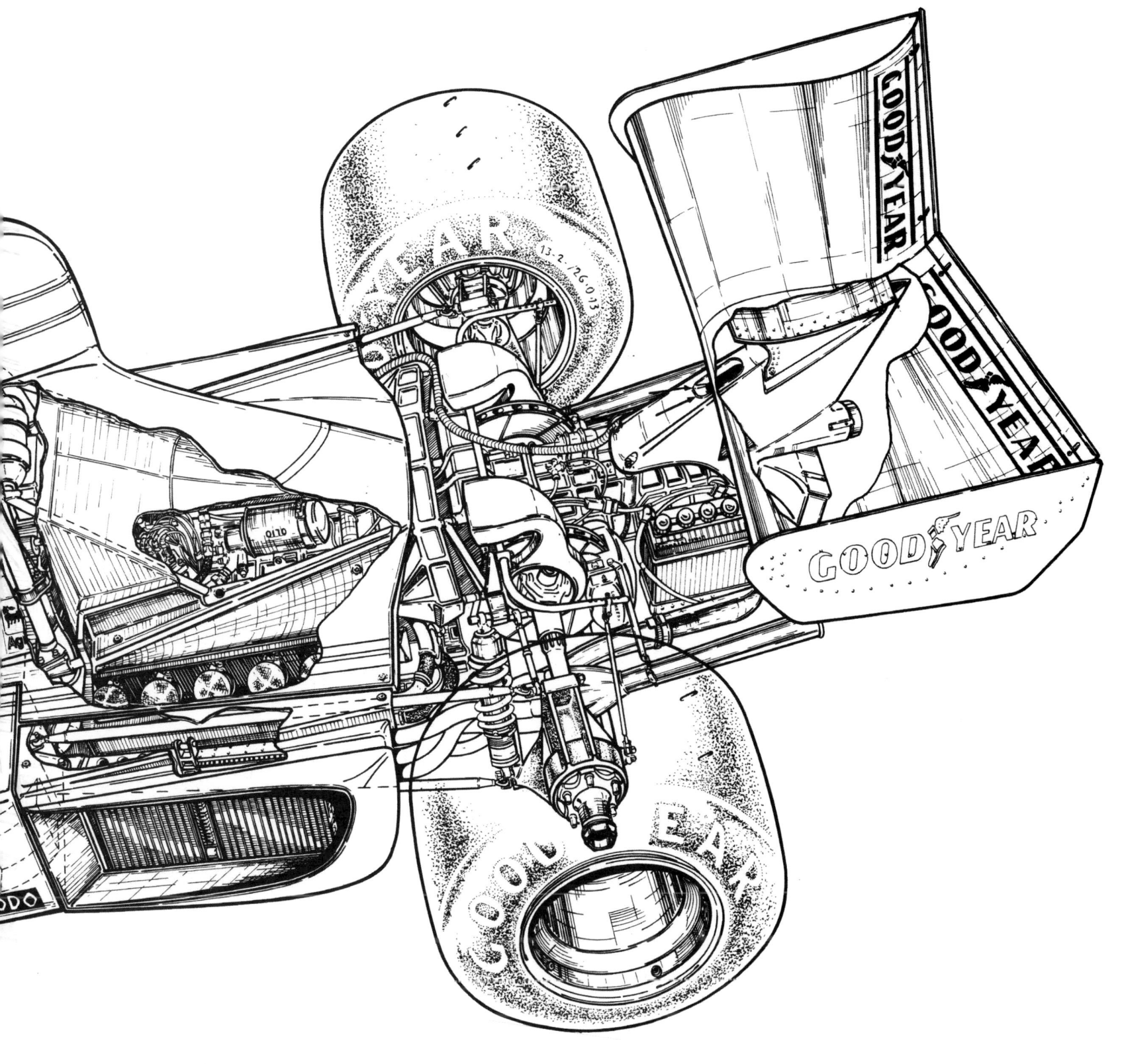

FERRARI 312 T (1975)

The 312 T clearly descended from the B3 while featuring the adoption of a splendid transverse gearbox that was retained on all the Ferraris of the 1970s. A true jewel in cast magnesium, with the front suspension mounts also being cast, the configuration featuring long oblique links and springs located in the central area ahead of the front bulkhead in the aforementioned casting.

Externally, the T retained the general layout of the B3, recognisable thanks to the large air box. Another feature was the simplicity and the attention paid to the installation of the radiators, thanks also to the use of a specific chassis that saw a return to a mixed construction: a tubular spaceframe with riveted aluminium panelling and the fuel tank divided into three separate cells. The sidepods were even wider, concealing the radiators (those for the water behind of the front wheels, the oil coolers instead ahead of the rear wheels) and no longer had the hot air vents in the upper part. Lauda would surely have been able to repeat his World Championship triumph in 1976 too had it not been for the crash and fire at the Nürburgring in a car, the T2, that was a photocopy of the previous season's.

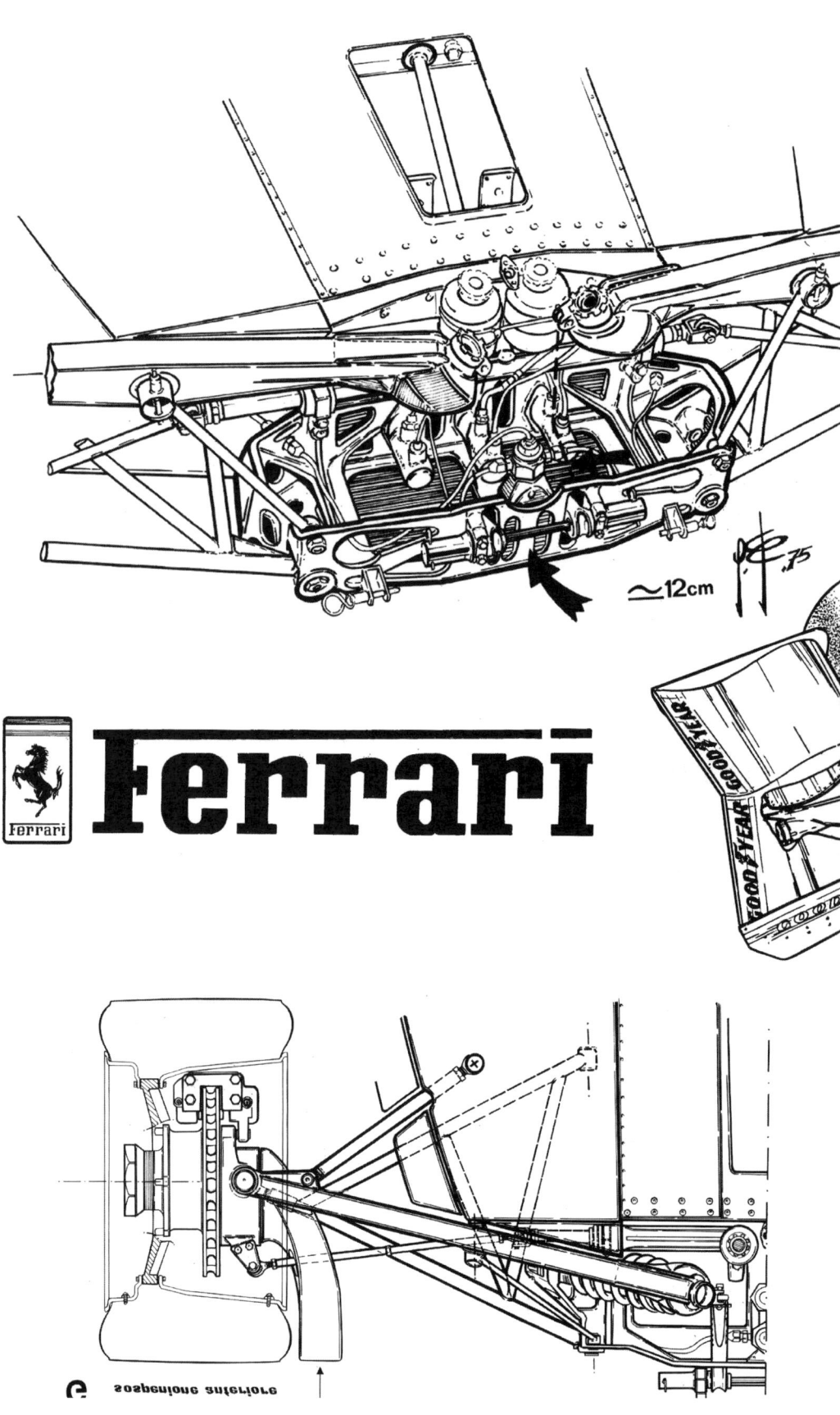

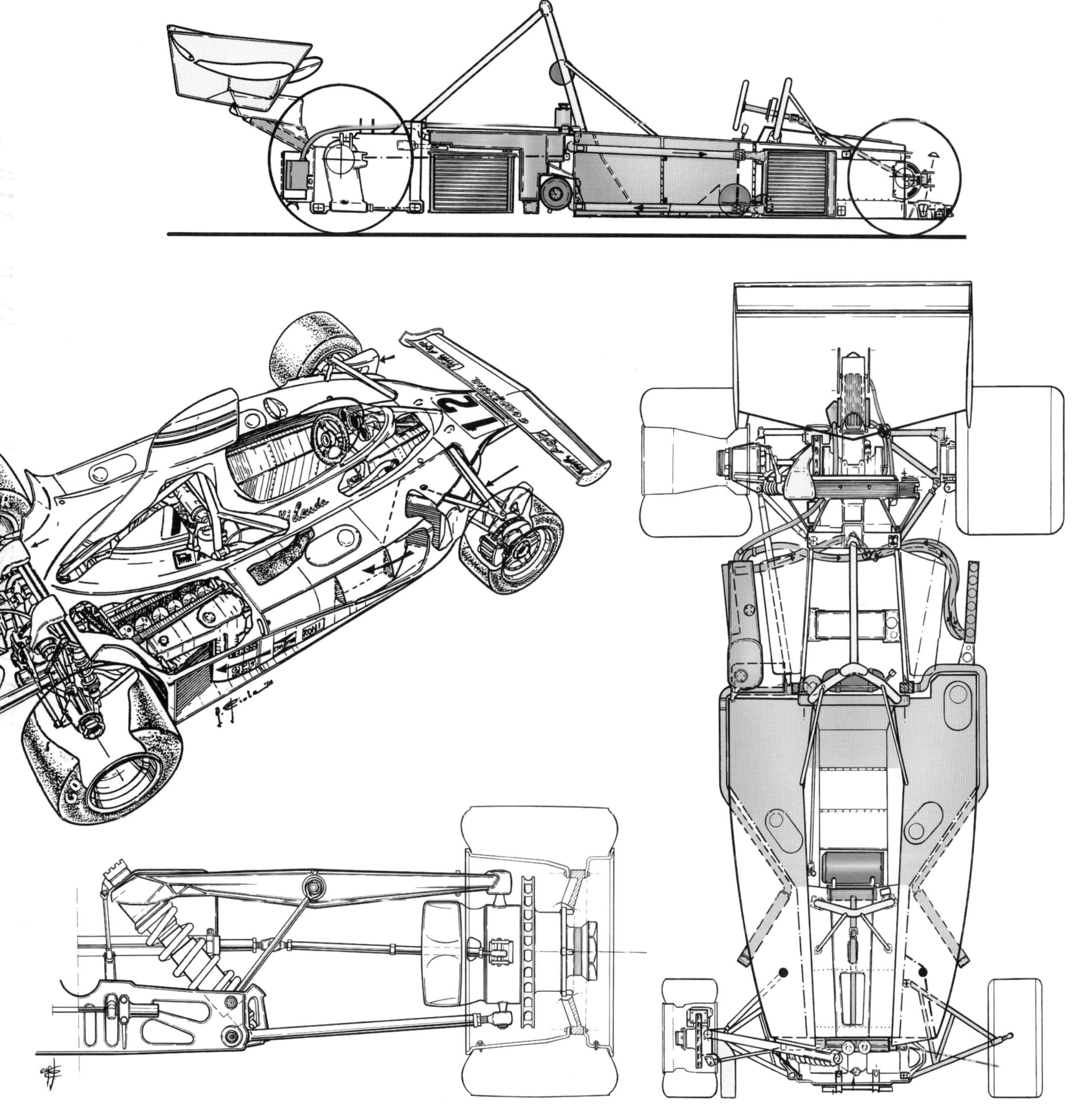

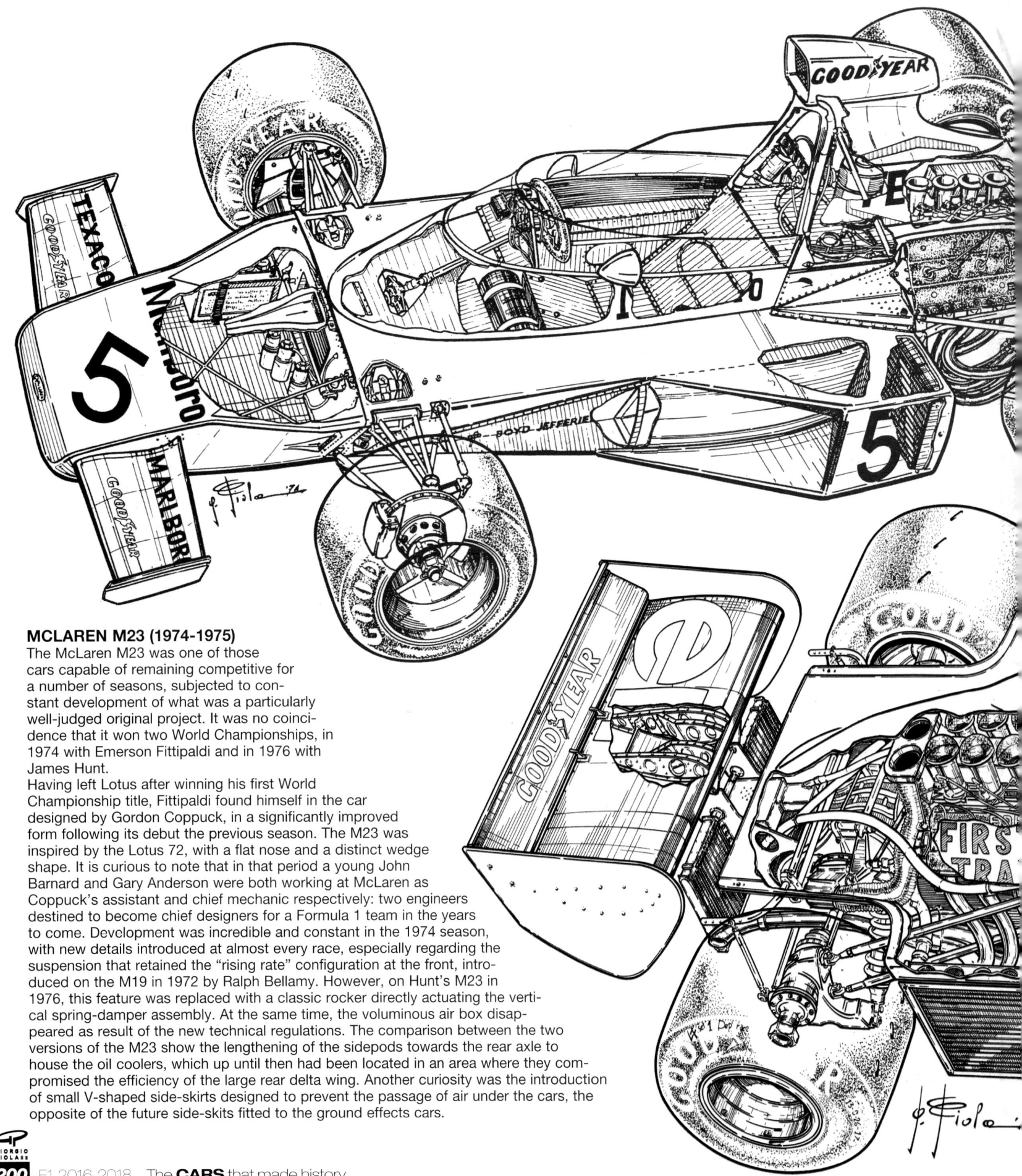

MCLAREN M23 (1974-1975)

The McLaren M23 was one of those
cars capable of remaining competitive for
a number of seasons, subjected to constant development of what was a particularly
well-judged original project. It was no coincidence that it won two World Championships, in
1974 with Emerson Fittipaldi and in 1976 with
James Hunt.
Having left Lotus after winning his first World
Championship title, Fittipaldi found himself in the car
designed by Gordon Coppuck, in a significantly improved
form following its debut the previous season. The M23 was
inspired by the Lotus 72, with a flat nose and a distinct wedge
shape. It is curious to note that in that period a young John
Barnard and Gary Anderson were both working at McLaren as
Coppuck's assistant and chief mechanic respectively: two engineers
destined to become chief designers for a Formula 1 team in the years
to come. Development was incredible and constant in the 1974 season,
with new details introduced at almost every race, especially regarding the
suspension that retained the "rising rate" configuration at the front, introduced on the M19 in 1972 by Ralph Bellamy. However, on Hunt's M23 in
1976, this feature was replaced with a classic rocker directly actuating the vertical spring-damper assembly. At the same time, the voluminous air box disappeared as result of the new technical regulations. The comparison between the two
versions of the M23 show the lengthening of the sidepods towards the rear axle to
house the oil coolers, which up until then had been located in an area where they compromised the efficiency of the large rear delta wing. Another curiosity was the introduction
of small V-shaped side-skirts designed to prevent the passage of air under the cars, the
opposite of the future side-skits fitted to the ground effects cars.

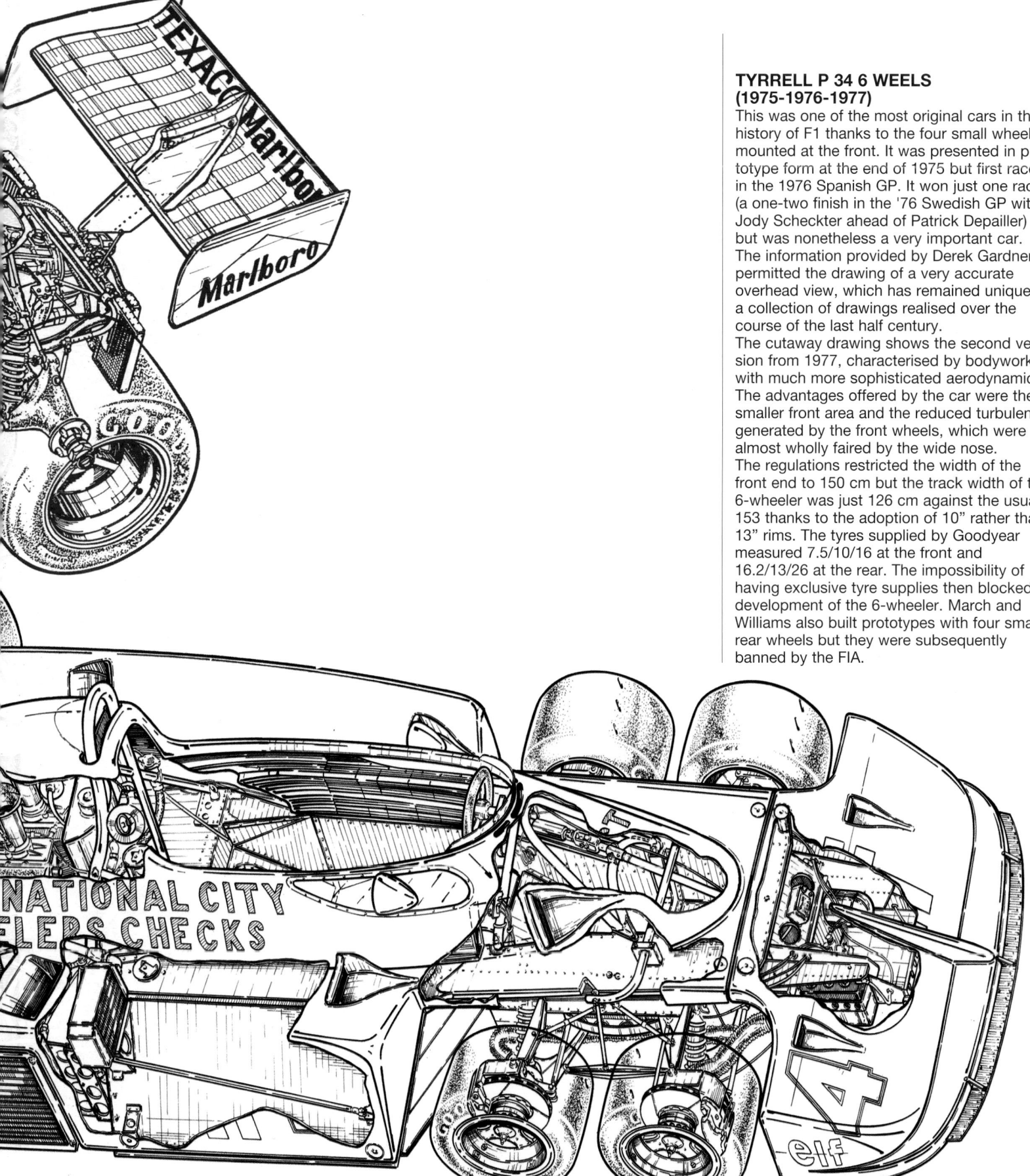

TYRRELL P 34 6 WEELS
(1975-1976-1977)

This was one of the most original cars in the history of F1 thanks to the four small wheels mounted at the front. It was presented in prototype form at the end of 1975 but first raced in the 1976 Spanish GP. It won just one race (a one-two finish in the '76 Swedish GP with Jody Scheckter ahead of Patrick Depailler) but was nonetheless a very important car. The information provided by Derek Gardner permitted the drawing of a very accurate overhead view, which has remained unique in a collection of drawings realised over the course of the last half century.

The cutaway drawing shows the second version from 1977, characterised by bodywork with much more sophisticated aerodynamics. The advantages offered by the car were the smaller front area and the reduced turbulence generated by the front wheels, which were almost wholly faired by the wide nose.

The regulations restricted the width of the front end to 150 cm but the track width of the 6-wheeler was just 126 cm against the usual 153 thanks to the adoption of 10" rather than 13" rims. The tyres supplied by Goodyear measured 7.5/10/16 at the front and 16.2/13/26 at the rear. The impossibility of having exclusive tyre supplies then blocked development of the 6-wheeler. March and Williams also built prototypes with four small rear wheels but they were subsequently banned by the FIA.

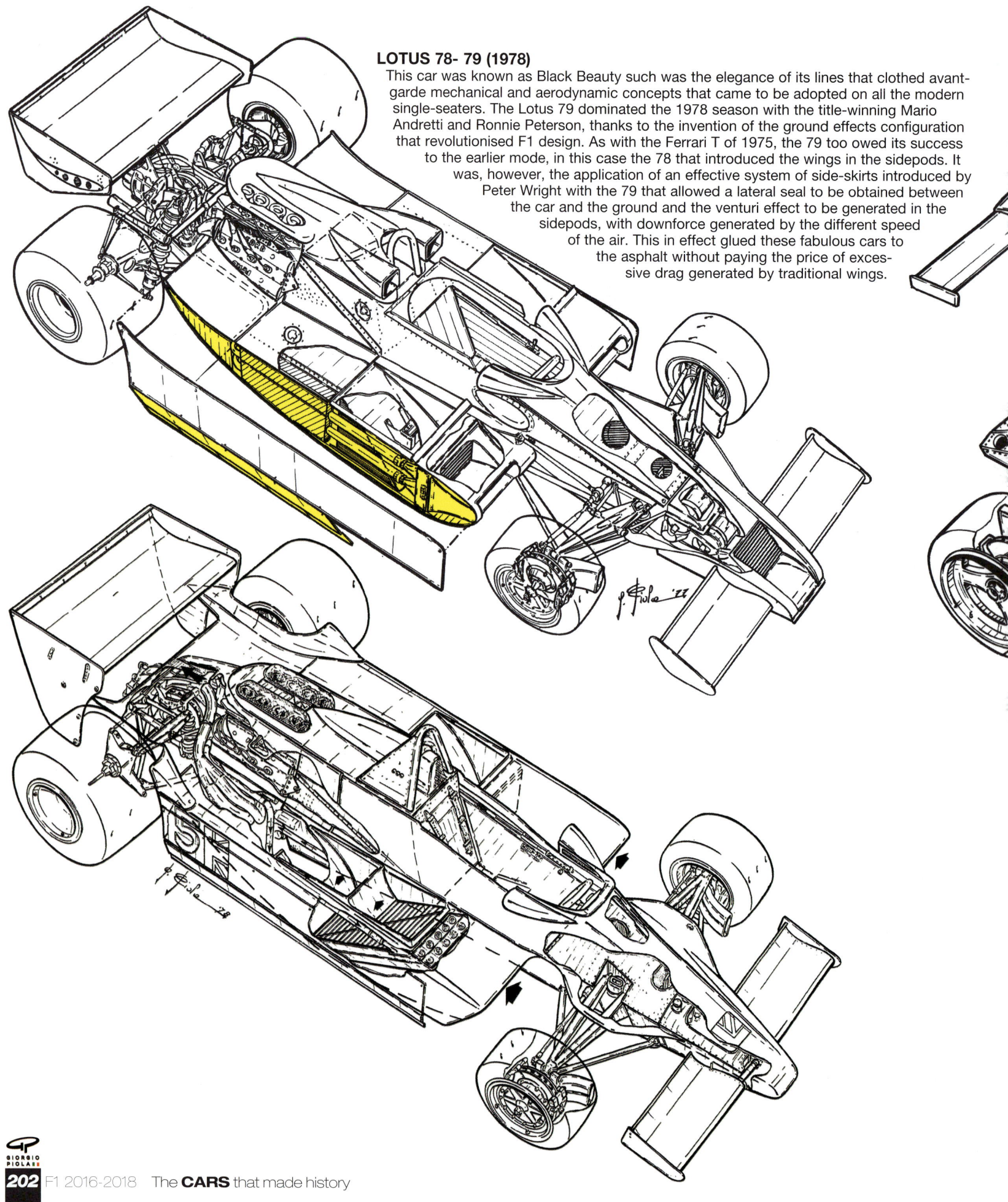

LOTUS 78- 79 (1978)

This car was known as Black Beauty such was the elegance of its lines that clothed avant-garde mechanical and aerodynamic concepts that came to be adopted on all the modern single-seaters. The Lotus 79 dominated the 1978 season with the title-winning Mario Andretti and Ronnie Peterson, thanks to the invention of the ground effects configuration that revolutionised F1 design. As with the Ferrari T of 1975, the 79 too owed its success to the earlier mode, in this case the 78 that introduced the wings in the sidepods. It was, however, the application of an effective system of side-skirts introduced by Peter Wright with the 79 that allowed a lateral seal to be obtained between the car and the ground and the venturi effect to be generated in the sidepods, with downforce generated by the different speed of the air. This in effect glued these fabulous cars to the asphalt without paying the price of excessive drag generated by traditional wings.

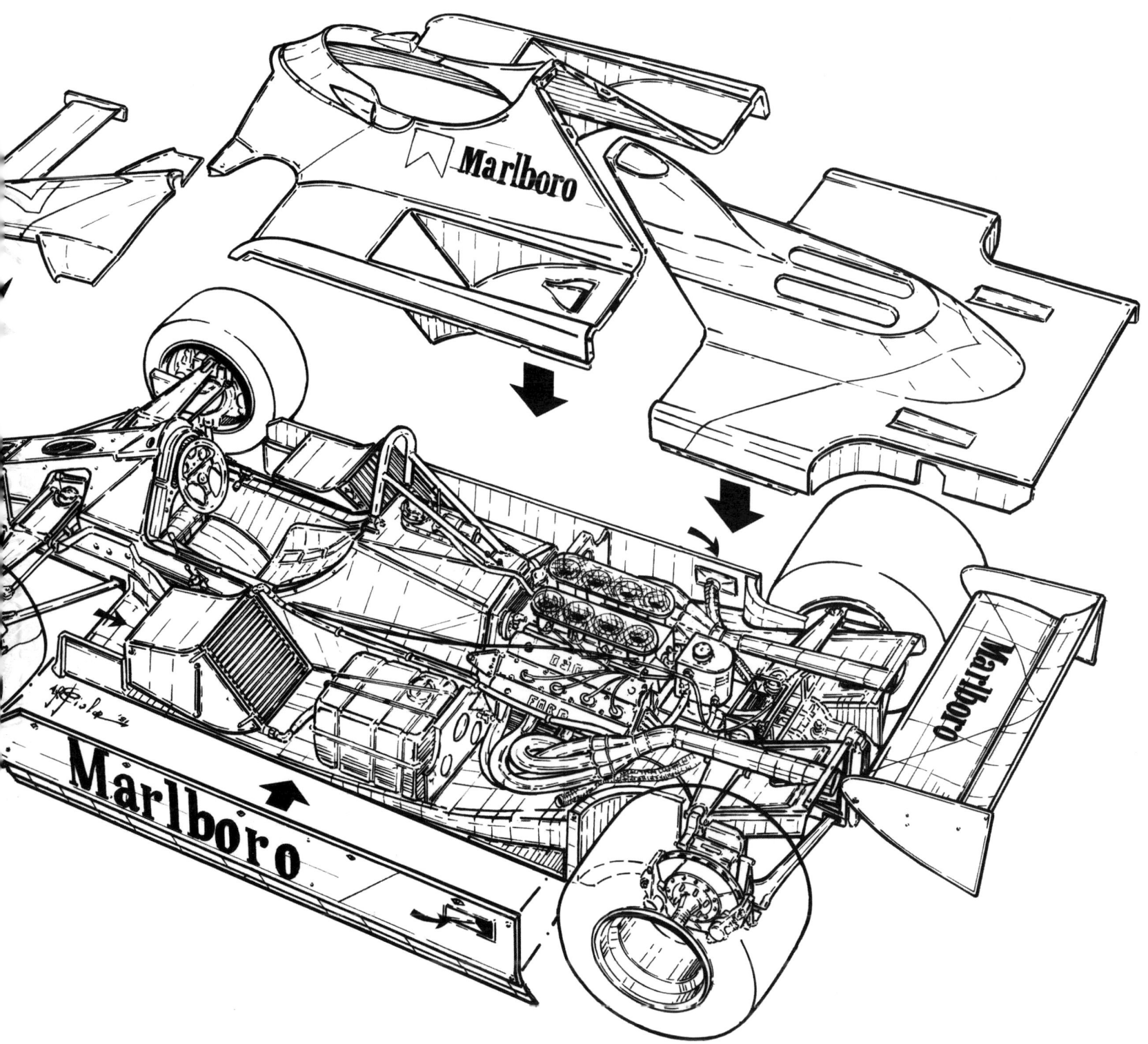

MCLAREN MP4 (1981-1984)

John Barnard is to be credited with having introduced carbonfibre construction technology to F1. In collaboration with Hercules of Salt Lake City, the engineer who arrived with Ron Dennis to replace Gordon Coppuck, his maestro in the 1970s, designed one of the most beautiful cars in the history of F1. The only detail not in carbonfibre was the deformable structure of the nose which was in box-section aluminium and gave rise to protests from other teams as the regulations stated that it should have been in the same material as the chassis.

In practice, the MP4 adopted the same ideas introduced by the Lotus 79 with an attention to detail and overall design that was to become typical of the future designs of this F1 genius who, like Chapman, introduced numerous new features to the world of Formula 1.

The first drawing shows the MP4 from 1981 with wide, square-cut sidepods and the air from the radiators venting from the upper part. 1982 then saw the introduction of the first "Coke bottle" bodywork in the area inside the rear wheels.

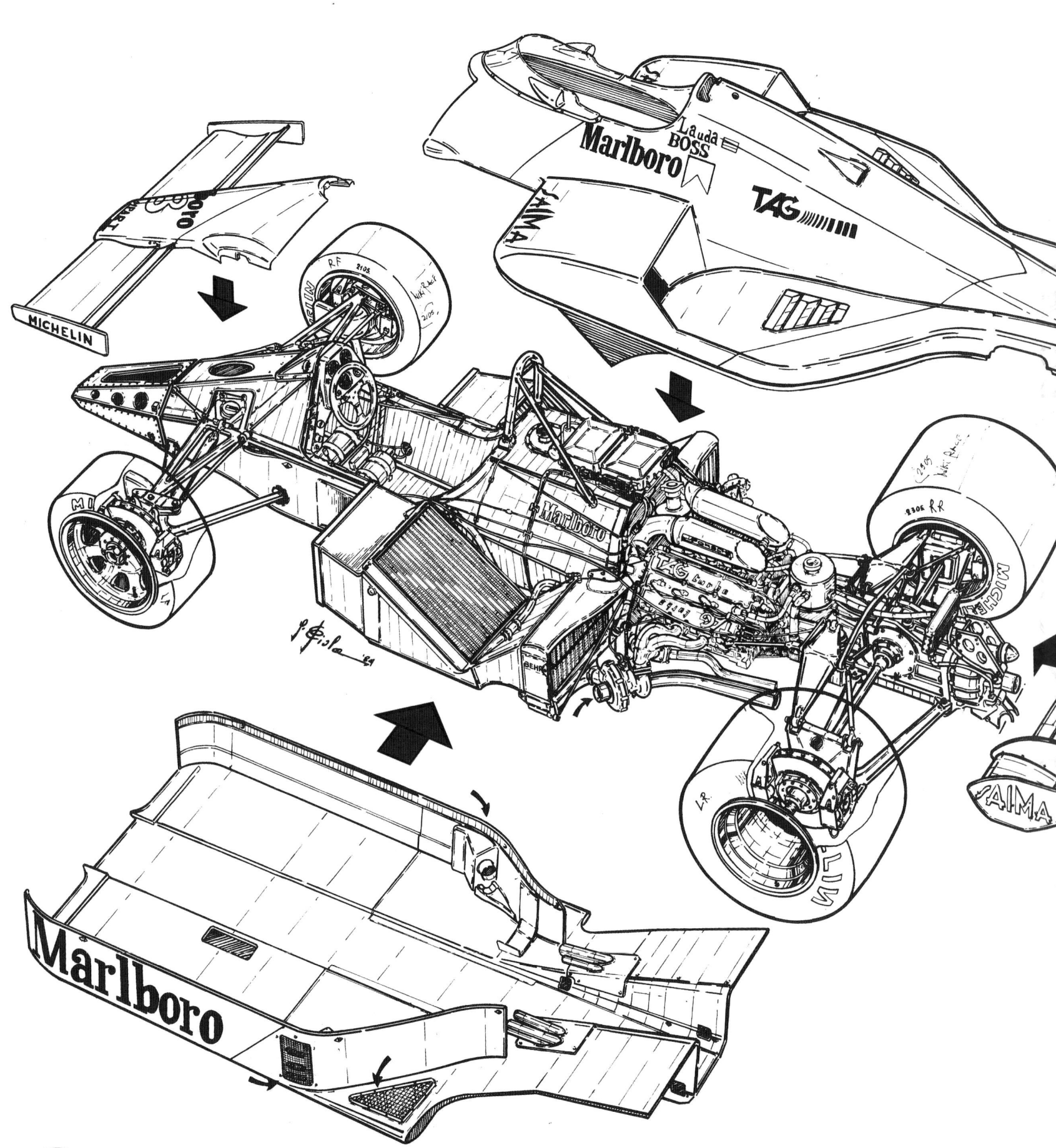

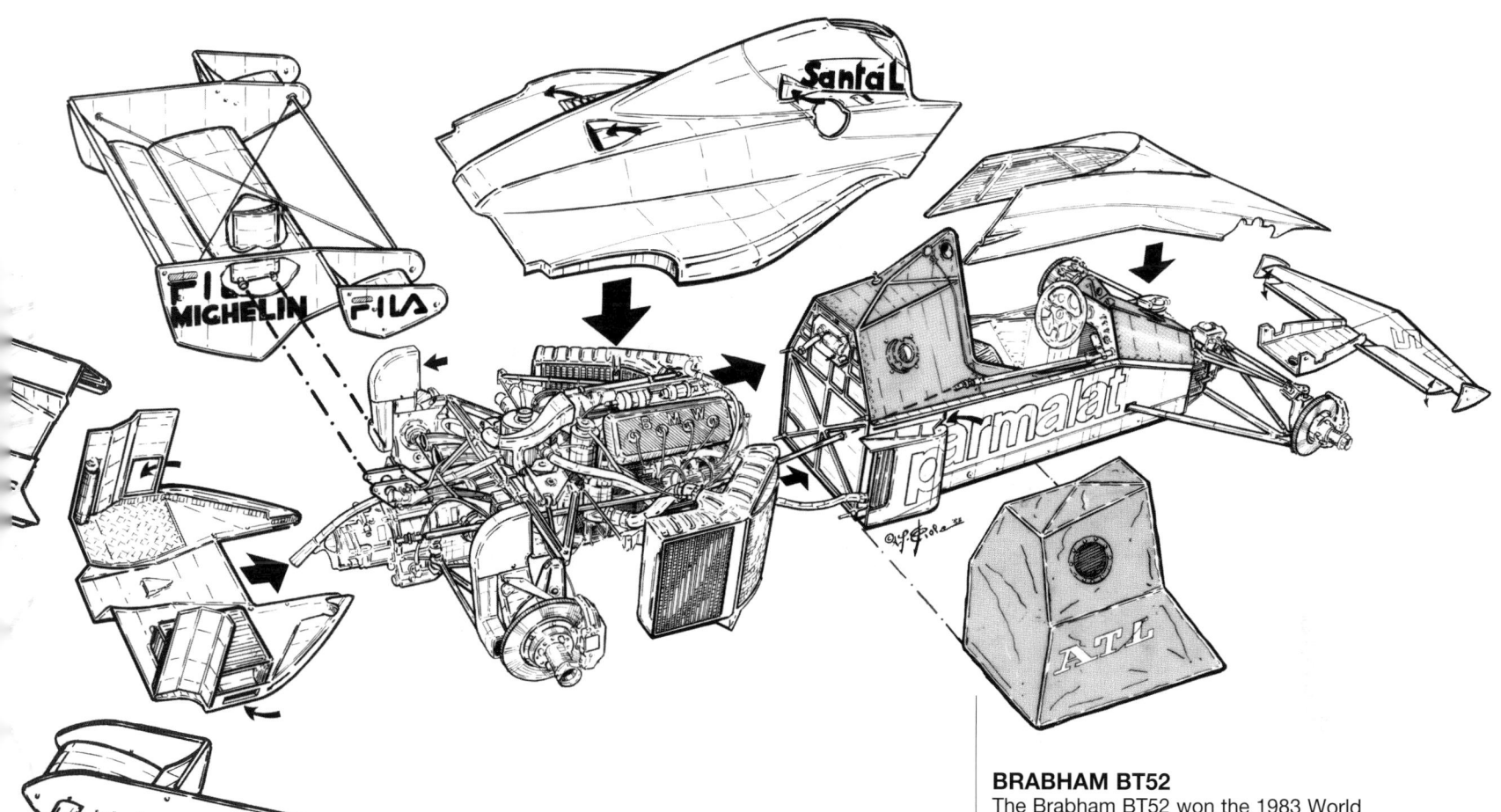

MCLAREN MP4/2

The MP4 was one of the
most long-lived of F1 cars: its
basic configuration endured for
around eight years and was fitted with three different engines:
the Ford Cosworth V8, the six-
cylinder Porsche turbo and the six-
cylinder Honda turbo. This drawing
shows the MP4/2 powered by the
Porsche engine, driven to the World
Championship title by Niki Lauda, with
Alain Prost 2nd, just half a point behind.
A result that on its own is sufficient to make
this car the stuff of legend. The German six-
cylinder was mounted in the McLaren for the
Dutch GP in 1983, an MP4/1 designated with
an E. Note in the drawing the very high sidepods, rounded at the front with slim air vents let
into the upper part. The dual flaps in correspon-
dence with the rear wheels were characteristic of the
1983-1984 season before being banned by the FIA in
order to restrict the speed of the cars. Barnard also
introduced on his cars dual brake calipers manufactured
in-house by McLaren. It was actually a McLaren, Lauda's
MP 4/2, that was the first car to win a race with carbonfibre discs, the South African GP in 1984.

BRABHAM BT52

The Brabham BT52 won the 1983 World
Championship with Nelson Piquet and was
nicknamed "the arrow" thanks to the very
short sidepods with vertical radiators.
A car that swam against the tide in the season that saw the introduction of the flat bottom, eliminating the wing profiles in the sidepods. In practice, the cars could no longer
take advantage of the venturi effect in the
area between the two axles where the underbody was now required to be absolutely flat,
the only area where this was permitted.
The talented Gordon Murray came up with
this very elegant car finished in a simple blue
and white livery at the behest of BMW that
supplied the very powerful four-cylinder tur-
bocharged engine fitted with a single large
Garrett turbo. The BT52 can rightly be considered to be the most interesting car of the
1983 season with its large arrow-head wing,
tapering chassis, weight concentrated on the
rear axle, very short sidepods and cockpit
shifted backwards with the lower body
reduced to a minimum, as can clearly be
seen in the drawing. The V-formation installation of the radiators and heat exchangers
either side of the engine was also a characteristic feature. The strut front suspension
was very interesting, mounted on an external
subframe at front of the chassis.

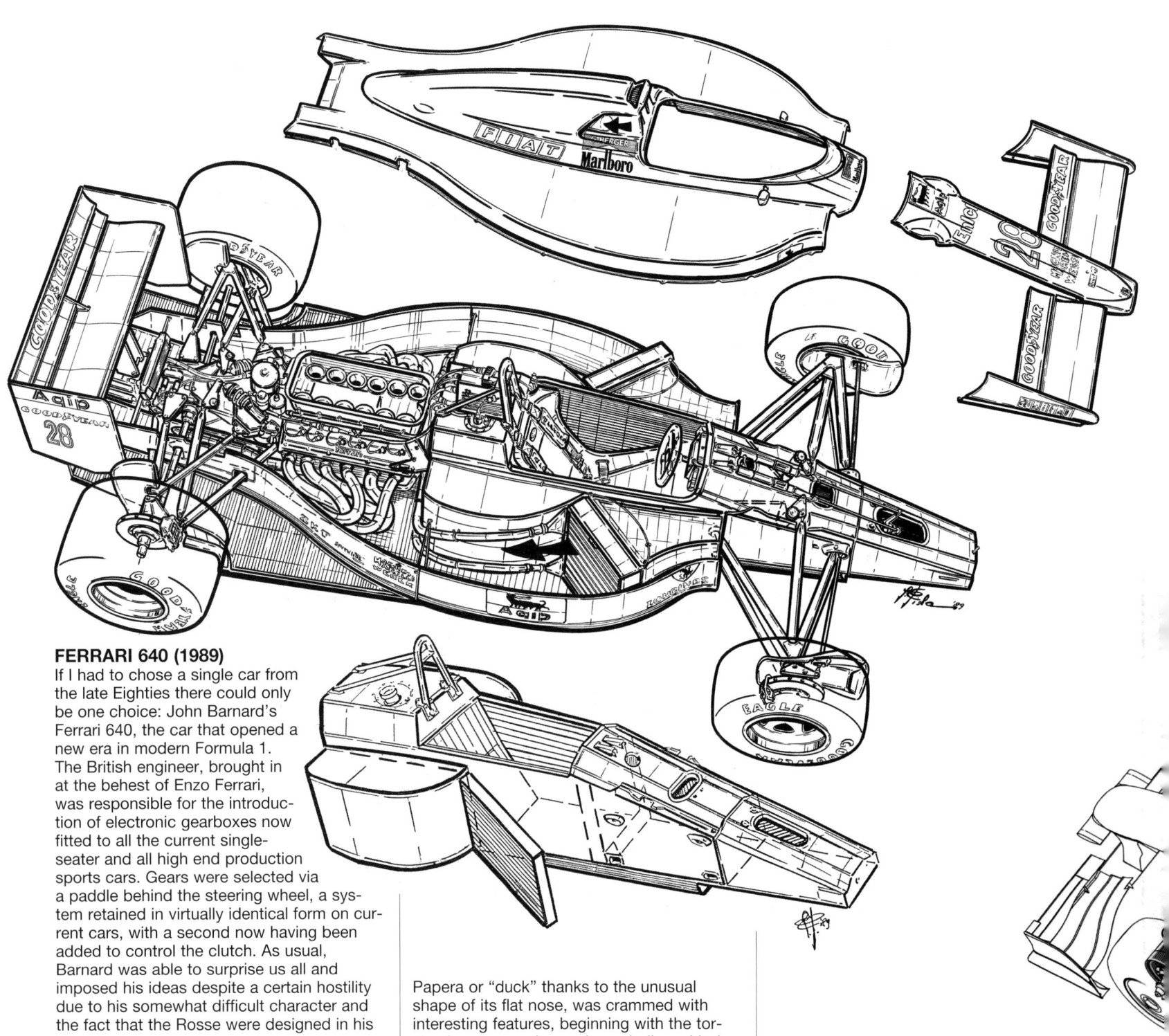

FERRARI 640 (1989)

If I had to chose a single car from the late Eighties there could only be one choice: John Barnard's Ferrari 640, the car that opened a new era in modern Formula 1. The British engineer, brought in at the behest of Enzo Ferrari, was responsible for the introduction of electronic gearboxes now fitted to all the current single-seater and all high end production sports cars. Gears were selected via a paddle behind the steering wheel, a system retained in virtually identical form on current cars, with a second now having been added to control the clutch. As usual, Barnard was able to surprise us all and imposed his ideas despite a certain hostility due to his somewhat difficult character and the fact that the Rosse were designed in his studio in Guildford, England, far from Maranello. The choice of semi-automatic gear selection would not have lasted long on the car had Barnard not also designed the chassis of the 640 in such a way that a classic gear lever would have been impossible to install.

Following Mansell's victory in the debut Grand Prix in Brazil in 1989, there followed an interminable series of retirements that led to the absurd experiment with a manual gear change on the Fiorano track. An avant-garde mechanical design, the 640, nicknamed the

Papera or "duck" thanks to the unusual shape of its flat nose, was crammed with interesting features, beginning with the torsion bar suspension and concluding with the long, sinuous sidepods. It is also worth noting that the suspension layout brought back into fashion by Barnard is now generalised throughout F1. As with all the best examples of cars designed by the Englishman, the basic architecture of the 640 remained in service for a long time, in this case three full seasons. The drawing shows the first version, the car that was victorious with Mansell in Brazil, recognisable thanks to the two small engine air intakes replaced by a more conventional high intake above the driver's head.

RED BULL RB5 (2009)

The Red Bull RB5 was the most innovative and radical car of the last 20 years. The most curious aspect was that three of the most innovative features were actually reprised from old and forgotten F1 designs: the pull-rod rear suspension (4), the V-shape of the chassis in its front section and the low lateral venting of the exhausts (3) in the rear suspension area. We have to go back to the 2001 Minardi to see a pull-rod layout applied to the front suspension, while to find one at the rear we go back as far as 1987 and the Lotus 99T and the successive 100T. Aerodynamics was the factor that tied the whole design together. The pull-rod configuration allowed the whole of the upper part of the strut, torsion bar and damper assembly to be freed up and permitted a diverse differential position with a reduction in height of almost 15 cm.

The reduction in height of the gearbox (5) also allowed the monolithic wing profile wishbone (6) to be aligned with the extractor profile, enhancing its efficiency. The reduction in height of the bodywork in the area at the end of the sidepods

was incredible. There was a new installation for the very long, narrow radiators. (2) The idea of the bargeboards reaching low down to anchor to the diffuser was new. As for the section of the chassis at the front bulkhead, the regulations imposed minimum dimensions: 300 x 350 mm. Newey got around this restriction with a reduction in the centre and raised sides to maintain the same overall section while gaining much in terms of aerodynamic penetration and the installation of the suspension. The lower part of the chassis had a V-shape with what was almost a single lower wishbone mounting point.

The brake calipers (1) were underslung. The RB5 opened the way for what was to become a trend over the following seasons with the exhausts venting towards the diffuser. When it was presented, this feature aroused a degree of perplexity regarding the reliability of the suspension but over the season no problems of this kind occurred.

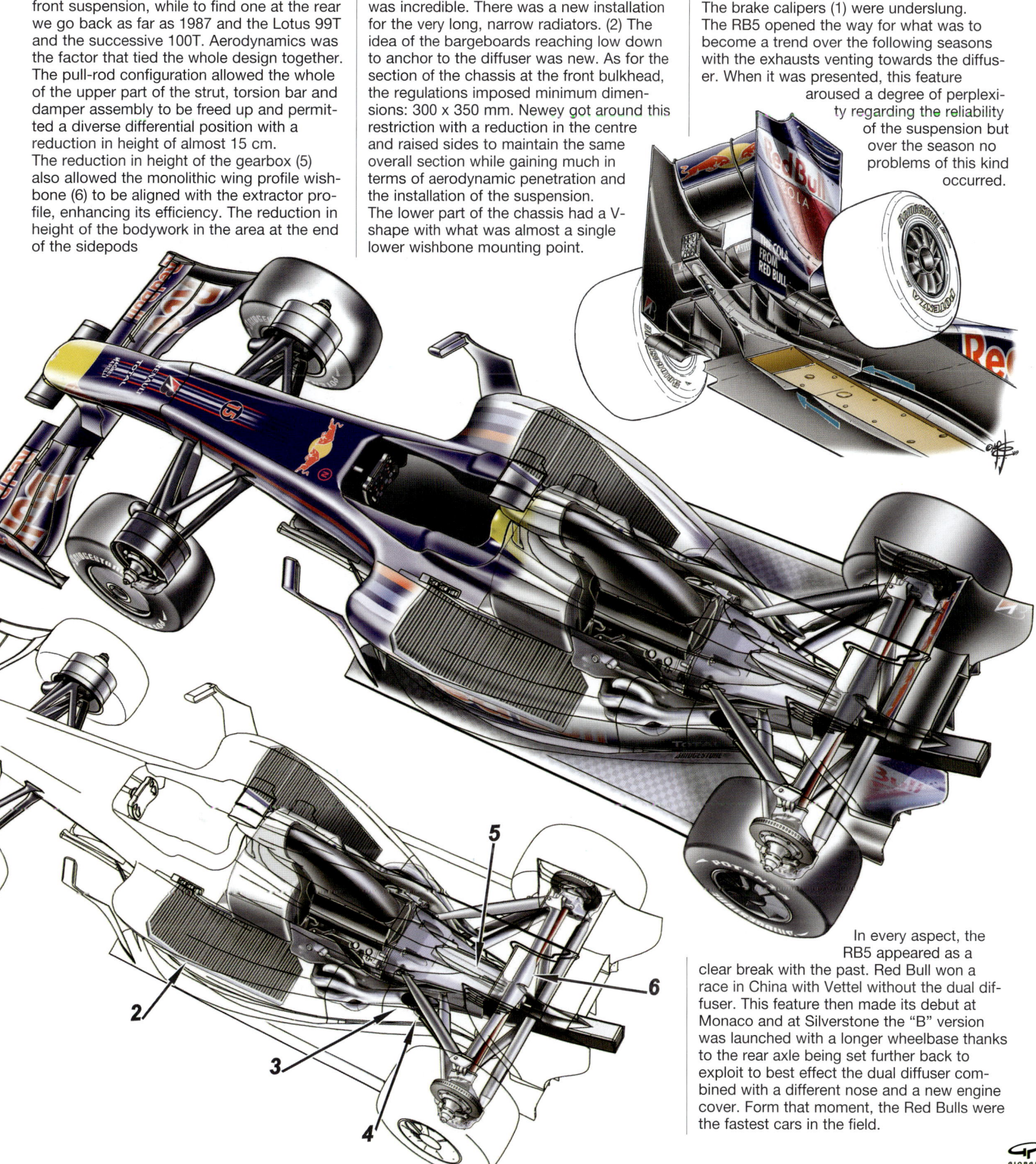

In every aspect, the RB5 appeared as a clear break with the past. Red Bull won a race in China with Vettel without the dual diffuser. This feature then made its debut at Monaco and at Silverstone the "B" version was launched with a longer wheelbase thanks to the rear axle being set further back to exploit to best effect the dual diffuser combined with a different nose and a new engine cover. Form that moment, the Red Bulls were the fastest cars in the field.

Giorgio Nada Editore

Editorial manager
Leonardo Acerbi

Editorial coordination
Giorgio Nada Editore

Graphic design and cover
Aimone Bolliger

Contributors
Franco Nugnes (engines, tyres)

Computer graphic
Alessia Bardino
Giulia Giusto
Camillo Morande
Gisella Nicosia
Paolo Rondelli

3D Animations
Camillo Morande
Annunziata Generoso

Printed in Italy by
D'Auria Printing Spa - Ascoli Piceno
May 2019

© 2019 Giorgio Nada Editore, Vimodrone (Milan, Italy)

Giorgio Nada Editore
Via Claudio Treves,15/17
I - 20090 VIMODRONE MI
Tel. +39 02 27301126
Fax +39 02 27301454
e-mail: info@giorgionadaeditore.it
www.giorgionadaeditore.it

The catalogue of Giorgio Nada Editore publications
is available on request at the above address.

Distribution
Giunti Editore Spa
via Bolognese 165
I - 50139 FIRENZE
www.giunti.it

Formula 1 2016-2018. Technical analysis
ISBN 978-88-7911-684-8

The previous editions of the annual Technical Analysis
are available on www.giorgionadaeditore.it